Techfever
D0212654

Electron Energy Loss Spectroscopy and Surface Vibrations

ELECTRON ENERGY LOSS SPECTROSCOPY AND SURFACE VIBRATIONS

H. Ibach

KFA-Jülich-IGV
West Germany

D. L. Mills

Department of Physics
University of California
Irvine, California

1982

ACADEMIC PRESS

A Subsidiary of Harcourt Brace Jovanovich, Publishers

New York   London   Paris

San Diego   San Francisco   São Paulo   Sydney   Tokyo   Toronto

COPYRIGHT © 1982, BY ACADEMIC PRESS, INC.
ALL RIGHTS RESERVED.
NO PART OF THIS PUBLICATION MAY BE REPRODUCED OR
TRANSMITTED IN ANY FORM OR BY ANY MEANS, ELECTRONIC
OR MECHANICAL, INCLUDING PHOTOCOPY, RECORDING, OR ANY
INFORMATION STORAGE AND RETRIEVAL SYSTEM, WITHOUT
PERMISSION IN WRITING FROM THE PUBLISHER.

ACADEMIC PRESS, INC.
111 Fifth Avenue, New York, New York 10003

United Kingdom Edition published by
ACADEMIC PRESS, INC. (LONDON) LTD.
24/28 Oval Road, London NW1   7DX

Library of Congress Cataloging in Publication Data

Ibach, H.,
    Electron energy loss spectroscopy and surface
vibrations.

    Includes bibliographical references and index.
    1. Electron energy loss spectroscopy. 2. Crystals--
Surfaces. I. Mills, D. L. II. Title.
QC454.E4122   548   81-22938
ISBN 0-12-369350-0   AACR2

PRINTED IN THE UNITED STATES OF AMERICA

82 83 84 85   9 8 7 6 5 4 3 2 1

CONTENTS

v

PREFACE

During the past two decades physicists and chemists have explored both the nature of the outermost atomic layers of solids and the manner in which atoms or molecules bind to the surface. The field presents a formidable challenge. Many common probes of solids (photons, neutrons) penetrate many atomic layers into the material so that information descriptive of the near vicinity of the surface appears as a small feature superimposed on a large signal from the bulk of the material. Particles that sample only the outermost layer or two (electrons, atoms) have limited penetration because they interact strongly with the atoms of the material. Thus sophisticated theories are required before the data may be utilized fully.

This volume is devoted to electron energy loss spectroscopy as a probe of the crystal surface. Electrons with energy in the range of a few electron volts sample only a few atomic layers. As they approach or exit from the crystal, they interact with the vibrational modes of the crystal surface, or possibly with other elementary excitations localized there. The energy spectrum of electrons back-reflected from the surface is thus a rich source of information on its dynamics; as we know from the well-developed fields of vibrational spectroscopy of molecules and solids, the dynamical properties of an entity, along with selection rules, offer insight into its basic structural features. Also, the vibrational modes of molecules adsorbed on the surface provide one with direct information on the nature of the chemical bonds between the molecule and the substrate.

During the past ten years, high resolution electron spectrometers have been developed that are particularly suitable for surface studies. These instruments pass current sufficient to enable vibrations of a fraction of a monolayer to be detected, with resolution comparable to that realized in the optical spectroscopy of surfaces. Paralleling this experimental development, theoretical concepts and selection rules have appeared. These may be employed to analyze data on diverse systems.

At the time of this writing, the number of laboratories engaged in electron energy loss spectroscopy of surfaces is expanding rapidly. The field is in a period of transition from a specialty of a few laboratories, to a technique that will take its place in the array of surface probes available to a modern surface science facility. Thus the time is appropriate for a comprehensive exposition of the subject, including both its experimental aspects and the theoretical concepts used to interpret that data. The purpose of this volume is to put forward a systematic discussion of the method of electron energy loss spectroscopy and the information on surfaces that may be obtained from it. We do not regard it as an extended review article. Instead, we have attempted to develop the material systematically, within a pedagogical framework, illustrating the basic points by selected data that we feel are particularly appropriate. Also, there is both experimental and theoretical material that is original, such as several spectra that appear in the text to illustrate points that are particularly noteworthy. Finally, at various points in the presentation we have tabulated information useful in the intepretation of spectra, and we compare electron energy loss spectroscopy with the optical spectroscopy of surfaces when this is appropriate.

The principal features of this volume are the following: In Chapter 2 we give a detailed analysis of the physics that controls the operation of the monochromator that is the core of the experimental apparatus. The discussion is quantitative, with the key points illustrated by data on operating instruments. In Chapter 3 we analyze the interaction of electrons with vibrational modes of the surface region and with other elementary excitations in the vicinity. While the primary use of electron energy loss spectroscopy to date has been for the study of surface vibrations, it is also a versatile tool for the analysis of other elementary excitations of the surface, such as those associated with the electronic degrees of freedom. Thus, in Chapter 3 we place considerable emphasis on formulating a general description of the interaction of the electron with surface excitations, considering surface vibrations as a special case. Since vibrational spectroscopy is, and will continue to be, a most important application of the method, we devote both Chapters 4 and 5 to an exposition of the lattice dynamics of clean and adsorbate-covered surfaces, with emphasis on those features of particular relevance to surface vibrational spectroscopy. Then, in Chapter 6 we explore a number of selected applications of the method. Here we call on the principles developed earlier in the text for the interpretation of the data and discuss empirical rules that underlie trends in the data.

ACKNOWLEDGMENTS

We wish to express our gratitude to Dr. S. Lehwald and Dr. D. Bruchmann for their productive cooperation. One of us (H. I.) also gratefully acknowledges the hospitality he enjoyed during a sabbatical leave at Irvine. We have benefited greatly from conversations with Dr. J. E. Black and Dr. Talat S. Rahman during the course of this work: both have provided us with theoretical results particularly suitable for inclusion in this volume.

The many figures that appear have been skillfully executed by Miss U. Marx and Ms. Arlene Sanders. Finally, we wish to express our appreciation to Ms. Ersel Williams for typing the manuscript with her usual enthusiasm and good cheer.

CHAPTER 1

INTRODUCTION

1.1 GENERAL REMARKS

During the past two decades, the study of the physical properties of clean and adsorbate covered surfaces, and of chemical phenomena there, has become an active and vigorous area of research. This has led to new and fundamental insights into the physics of condensed matter, and has provided a new domain within which fundamental issues of statistical mechanics may be explored. Also, the elucidation of the structure, geometry, and motions of adsorbates on the surface is required before a microscopic understanding of catalysis and corrosion may be achieved. The field thus bears directly on a most important area of applied science.

Developments in modern high-vacuum technology now allow surfaces to be prepared in a reproducible and well-characterized manner. It is then a challenge to the experimentalist to devise methods that explore the outermost one or two atomic layers of a crystal, and atoms or molecular groups adsorbed there. In recent years, an impressive array of surface spectroscopies have appeared, accompanied by theories which in many cases allow the data to be interpreted in a quantitative fashion. These spectroscopies are diverse in nature, and employ beams of electrons, atoms, ions, and photons.

This volume is devoted to one such spectroscopy: the study of the vibrational motion of atoms and molecules on and near the surface by the analysis of the energy spectrum of low-energy electrons backscattered from it. An electron incident on the crystal with energy E_I may excite a quantized vibrational mode with energy $\hbar\omega_0$ before being backscattered into the vacuum. It thus emerges with energy $E_S = E_I - \hbar\omega_0$, so an analysis of the energy spectrum of the backscattered electrons provides direct information on the vibrational frequencies of the substrate, or on those of atoms or molecules adsorbed on the surface. The method is sensitive to the surface because, with the incident kinetic energy E_I chosen suitably, the electron penetrates, at most, two or three atomic layers into the crystal. The backscattered electrons thus carry information on only the near vicinity of the surface.

Much can be learned of the structure of the surface, and of entities adsorbed on it, through knowledge of the characteristic vibrational frequencies. For many years, vibrational spectroscopy has been of very great importance to the chemist interested in molecular phenomena, and to the solid state physicist as well. Key elements in a molecular or crystal structure may be understood from qualitative features in the vibrational spectrum, while the fine details can reveal subtle features of the geometry. For very similar reasons, vibrational spectroscopy has played an important role in surface science. Furthermore, since the number of laboratories equipped to perform surface vibrational spectroscopy is increasing at the time of this writing, in the near future, the activity in this area will increase substantially.

One of the most important applications of surface vibrational spectroscopy, particularly by the electron energy loss method, is in the analysis of adsorbates on the crystal surface. For example, hydrogen adsorbs readily onto a variety of transition metal surfaces. An issue of fundamental importance is whether the adsorbed entity is an H_2 molecule, or whether the molecule dissociates during the adsorption process to leave two hydrogen atoms, each of which bonds to the surface as a separate entity. Of course, one may envision the possibility of more than one bonding site, and the possibility of both the molecular and atomic form of hydrogen on the surface simultaneously. If hydrogen is adsorbed on the surface in molecular form, then the vibrational spectrum of the surface should show a feature near the gas phase H_2 vibrational frequency of 4560 cm^{-1} (550 meV). This will be the case so long as the hydrogen bond is not dramatically altered by the adsorption process. When the vibrational spectrum of hydrogen on transition metal surfaces is obtained, one obtains features in the range 800–1600 cm^{-1}, well above the substrate vibrational frequencies but far below that of the very high frequency H_2 stretching mode. Thus, from this alone it is clear that the hydrogen adsorbs on the surface in atomic form, rather than as a molecular entity.

Further analysis of the vibrational spectrum allows one to take one more step, and place constraints on the symmetry to the adsorption site. For example, on the tungsten (100) surface, electron energy loss studies show three distinct vibrational frequencies for the adsorbed hydrogen atom. From this, it is clear the hydrogen cannot sit on a site of fourfold symmetry. If this were so, the two vibrational modes parallel to the surface would be degenerate in frequency. The vibrational modes of the adsorbed atom would then consist of one normal to the surface, and two parallel to it, with only two distinct frequencies. The data suggest that the hydrogen sits on a bridge site between two tungsten atoms. This site has twofold symmetry, the vibrational modes parallel to the surface are no longer degenerate, so we have three distinct frequencies as observed.

In contrast to this, when hydrogen is adsorbed on the Pt(111) surface, only two frequencies are observed. The hydrogen cannot be on a bridge site here, but can be only in a site of threefold symmetry, where the frequencies of the two modes polarized parallel to the surface are again necessarily degenerate. Further analysis of the electron energy loss spectrum, through use of a selection rule to be discussed, allows one to select the frequency associated with motion normal to the surface. This shows the frequency of the perpendicular vibration to lie *below* that of the modes parallel to the surface; the hydrogen must lie deep in the hollow site bounded by three platinum atoms on the vertices of an equilateral triangle, since the restoring force for motion normal to the surface is small.

The two examples already discussed show how surface vibrational spectroscopy may be used to infer the nature of the entity adsorbed on the surface from qualitative features in the spectrum, while a more detailed examination of it allows the nature of the adsorption site to be inferred. It is intriguing to see that on the W(100) surface, and the Pt(111) surface, the adsorption sites are qualitatively different. Species such as O_2, CO, and NO can either break up, or remain in molecular form, depending on the substrate. While in a number of cases this had been inferred from earlier work, electron energy loss spectroscopy allows study of these systems under highly controlled conditions.

A particularly important area for application of surface vibrational spectroscopy is the study of surfaces upon which hydrocarbons are adsorbed. Here we begin to come into direct contact with issues of central importance in catalysis. The issues are similar to those already discussed. A complex hydrocarbon may break up into fragments when it is adsorbed on the surface. One wishes to know the identity of these fragments, and any information about the nature of their adsorption geometry that may be extracted from the data. The various bond units, such as the C—H bond, or single, double, or triple C—C bonds have characteristic vibrational frequencies well known from gas-phase vibrational spectroscopy. The bonds which remain intact, along with new bonds that form when a hydrocarbon decomposes on the surface, will produce a characteristic signature in the electron energy loss spectrum. In these cases, the spectra are far more complex than those encountered in the simple examples outlined earlier. Identification of the various features is made easier by a comparison between the vibrational spectrum produced by adsorption of a hydrocarbon of interest and its deuterated analog. Finally, when a complex hydrocarbon is adsorbed on the surface, the composition of the fragments present there depends on the temperature of the substrate. As the temperature is raised, the molecule can decompose into progressively smaller fragments until only carbon remains on the surface. These transformations can be followed in some detail through use of surface vibrational spectroscopy.

In the foregoing discussion a number of the issues that may be addressed, once viable methods of surface vibrational spectroscopy are developed to the point where they may be widely applied, are outlined. In addition to the study of the nature and bonding of adsorbates, the method may also be used to explore vibrational motions of substrate atoms, as we shall see. While this volume is devoted to electron energy loss spectroscopy, there are in fact several techniques that may be used to study vibrational motions of the surface. Before we turn to our principal topic, we review these other methods briefly, with emphasis on how they may be compared with the electron energy loss method.

1.2 A COMPARISON OF SURFACE VIBRATION SPECTROSCOPIES

The experimental study of vibrational motion of atoms in crystal surfaces, and of adsorbates on the surface, is developing rapidly at the time of this writing. While electron energy loss spectroscopy of surfaces has been pursued for a decade, new and exciting experimental developments have appeared in the last two years. In addition, a number of other spectroscopies are at various stages of development. We review these here, with emphasis on comparison of the different experimental techniques.

In an electron energy loss study of surface vibrations, the sample is placed in ultrahigh vacuum. A highly monoenergetic beam of electrons is directed toward the surface, and the energy spectrum and angular distribution of electrons backscattered from the surface are measured. In a typical experiment, the kinetic energy of the incident electrons is in the range of a few electron volts. Under these conditions, the electrons penetrate only the outermost three or four atomic layers of the crystal, so as remarked earlier, the backscattered electrons thus contain information on only the very near vicinity of the surface. As the reader will appreciate shortly, the experimental problem is to achieve high resolution through production of a very mono-energetic and well-collimated electron beam, along with sensitive detectors. This must be done without degrading the signal below the detectable level. At present, the best resolution that may be obtained is 30 cm^{-1} (3.7 meV), although to detect weak signals, it is sometimes necessary to operate with lower resolution. The resolution offered by electron energy loss spectroscopy is thus substantially less than is customarily encountered in optical spectroscopies, though it is adequate for study of both the vibrational modes of adsorbed molecules, and of the substrate.

The electron energy loss spectroscopy of interest here employs an externally generated electron beam to probe the surface. It is also possible to perform vibrational spectroscopy with the conduction electrons of metals. This may be done in tunnel junctions, where two metals (possibly dissimilar) are separated by an insulating oxide barrier. If a dc electrical voltage V_0 is applied across such a structure, then a dc current $I(V_0)$ flows through the oxide barrier. To the conduction electrons in the metal, the oxide is a repulsive barrier of finite width, so the electrons pass from the first metal to the second by quantum mechanical tunneling. If a molecule with a vibrational normal mode of frequency ω_0 resides in the barrier, there is a certain probability that the electron may tunnel through the barrier inelastically, after exciting the vibrational mode of the molecule. It suffers the energy loss $\hbar\omega_0$ in the process. For this to occur, the condition $eV_0 > \hbar\omega_0$ must be satisfied, since otherwise no electrons have kinetic energy sufficient to enter the empty states above the Fermi level of the second metal. Thus, as the voltage V_0 is increased, when V_0 exceeds $\hbar\omega_0/e$, a new tunneling channel through the barrier opens, and the junction conductance $G(V_0) = dI/dV_0$ jumps discontinuously.

In the experiment, it is the derivative $dG(V_0)/dV_0$ that is detected, and this displays a feature at $\hbar\omega_0$ with width limited by the linewidth of the vibrational normal mode, or the amount of "thermal smearing" of the Fermi surfaces, whichever is larger. In practice, it is the "thermal smearing" which limits the resolution of the technique, and cryogenic temperatures clearly are necessary if high resolution is required. The resolution is typically in the range 8–40 cm^{-1} (1–5 meV) depending on the temperature employed. (A collection of theoretical and experimental papers on this topic may be found in Ref. [1].) It is possible to detect 0.1 monolayer of material by this means.

Quite clearly, inelastic electron tunneling spectroscopy probes samples very different in nature than those examined through use of an external electron beam. The early tunneling experiments explored molecules trapped within the oxide barrier of metal–oxide–metal structures. More recent work explores the vibrational modes of molecules adsorbed on supported catalyst particles. One may evaporate small particles made from catalytically active metals, such as rhodium, on the surface of an oxidized aluminum electrode. A top electrode is laid over the particles, and the conductance derivative $dG(V_0)/dV$ shows structure at the vibrational frequencies of molecules adsorbed on the particles. These experiments explore molecules on particles very similar to those employed in real catalytic processes, and further development of the method may bring it into direct contact with important issues in catalysis.

Optical spectroscopy is a powerful means of probing the vibrational motions of atoms and molecules in bulk material. This may be done by infrared absorption, or by the inelastic scattering of light (Raman scattering) from the quantized vibrational motions of matter. The two techniques are frequently complementary, because different selection rules apply to each.

In general, the resolution offered by optical spectroscopy is superior to that encountered in either electron spectroscopy already described. However, when optical methods are applied to the study of adsorbates on surfaces, the signals are very weak and sometimes difficult to detect against the background.

Consider, for example, the study of infrared absorption by molecules adsorbed on a metal surface. The reflectivity of such a sample will show dips when the frequency of the radiation matches that of the adsorbate. The reflectivity of the metal is in fact very close to unity in the infrared, and accurate measurement of the small deviation from unity is difficult. Nevertheless, impressive sensitivities have been achieved by exploiting the fact that only p-polarized light is absorbed by the adsorbed molecules, which allows one to detect 1/1000 of a monolayer in favorable cases. Quite recently, experiments have been reported in which the energy absorbed by the sample is measured directly, rather than the reflectivity. If material is adsorbed on a thin metal film, then when the photon frequency is tuned through an absorption band of the substrate, the temperature of the film will rise. The effect may be studied at low temperatures, where the specific heat of the film is small. By this method, submonolayer coverages of argon on copper films have been detected [2]. Techniques such as this may improve the sensitivity of the infrared absorption method to the point where this tool is used more widely in surface science.

Raman spectroscopy has yet to prove a useful probe of surface vibrations, though the area is in a state of rapid evolution. If one considers a monolayer of adsorbate molecules on the surface, and assigns to the molecules a cross section for Raman scattering typical of that found in the gas or liquid phase, then under typical conditions, one must detect somewhere between one and ten photons per second to observe the Raman signal. This must be done in the presence of a background produced by photons in the wing of the laser line scattered elastically by defects or imperfections in the surface, and also by fluorescence from the substrate. Such Raman signals can be detected with sensitive equipment, but this is a difficult task for the experimentalist.

There are circumstances under which the Raman cross section can be enhanced over the gas or liquid state values by many orders of magnitude. Such giant enhancements of the Raman cross section have been observed for scattering from molecules adsorbed on metals in electrochemical cell environments, and for molecules adsorbed on thin, polycrystalline metal films. For selected molecules on particular metals, most notably silver, the cross section may be larger than the gas or liquid phase value by a factor of 10^4 or possibly 10^6. At this time, the physical origin of these very large enhancements is not well understood. It is clear that in all experiments where the effect is observed, the surfaces are rough, with roughness present on the scale of 50 to 500 Å. In the electrochemical cells, the enhancement is observed only after the cell is subjected to several electrochemical cycles. Degradation of the metal electrode, followed by deposition of new material during the course of the cycle roughens the surface here. In the experiments with molecules adsorbed on thin metal films, one has highly nonuniform films, possibly in the form of island films. Experiments which systematically vary the degree of roughness, or the concentration of steps on the surface, show these features are essential for observation of the enhanced signals.

If enhanced Raman cross sections are found to be observed commonly for molecules adsorbed on surfaces or at interfaces, and the phenomenon is not confined to a small number of molecule–substrate combinations, the light scattering method will emerge as a very important probe of surfaces and interfaces. When compared with methods that employ particle beams (electrons, atoms, ions), one advantage of the light scattering method is that it can allow exploration of a surface covered with gas at high pressure, or liquid. This is so if the gas or liquid is transparent to the incident and scattered Raman photons. Thus, with this method, one can in principle examine surfaces under conditions where they are in contact with a working environment.

On smooth surfaces, where the intrinsic Raman cross section is small and the signal difficult to detect, schemes exist that enhance the signal from one to two orders of magnitude. One may incorporate the substrate of interest as the outer layer of a multilayer structure designed to enhance the electric field of the incident and scattered photon near the outer surface, or carry out the experiment through a prism placed near the surface. In the latter case, for suitable separation between the prism and substrate, there are particular angles where the incident and scattered photon couple strongly to surface electromagnetic waves (surface polaritons) on the metal, to enhance the magnitude of the surface electric field associated with both the incident and scattered photon. Both of these methods have been employed to enhance Raman signals from surface and interfaces.

A final method for study of surface vibrations potentially capable of very high resolution is inelastic scattering of low-energy atom beams from the surface. In the experiment, a highly monoenergetic beam of atoms is directed toward the surface, and the energy spectrum of atoms scattered off the surface is detected. If an atom incident on the surface has kinetic energy E_1, and scatters inelastically from the surface after creation (or absorption) of a vibrational quantum with energy $\hbar\omega_0$, then it emerges with kinetic energy $E_I - \hbar\omega_0(E_I + \hbar\omega_0)$. Thus, analysis of the energy spectrum of atoms scattered inelastically from the surface provides access to the frequency spectrum of surface vibrations.

The inelastic atom scattering method is conceptually very similar to electron energy loss spectroscopy. However, the basic parameters that characterize it are very different. In atom or ion-beam spectroscopy, time-of-flight spectroscopy is the principal means of obtaining this information, and sophisticated detection schemes are required.

It is difficult to produce highly monochromatic beams of neutral atoms. A number of laboratories have created beams with energy spread $\Delta E/E_I \simeq 0.2$, and with such a beam, surface vibrational frequencies may be measured. Quite recently, highly collimated He beams have been prepared with incident kinetic energy of 20 meV, and with an energy spread $\Delta E/E_I = 0.016$ full width at half-maximum [3]. The energy spread in the incident beam is then only 0.32 meV, which is more than one order of magnitude smaller than can be obtained from the highest quality electron beams that can be prepared presently. The existence of such highly monoenergetic atomic beams provide for this method the potential of obtaining surface vibrational spectra comparable to the best offered by the optical spectroscopies presently under development.

While the inelastic scattering of atoms from surfaces is a potentially powerful probe of surface vibrations, the field is in an early stage of development at this time. The prospects are exciting, but full interpretation of the spectra may require an improved understanding of the nature of the atom–surface interaction. It is encouraging to see that in recent years, the very complex energy variations of measured cross sections for elastic atom–surface scattering can be brought into quantitative contact with the theory, with the consequence that the atom–surface interaction potential may be accurately deduced from the data. An important issue in the interpretation of the inelastic scattering data will be the extent to which multiphonon processes influence the energy spectrum of the scattered atoms. If the spectra are dominated by one-phonon scatterings, then one may expect that much can be learned without the need for detailed theories of atom–surface scattering, while the interpretation may be much less simple if this is not the case.

From this section, it is evident that we have in hand or under development a variety of methods that probe vibrations of atoms and molecules on the crystal surface. All of these methods present a formidable challenge to the experimentalist. Some, such as infrared spectroscopy and electron energy loss spectroscopy have been pursued actively for several years, while the others are much newer and at a much less mature stage. The coming decade will see these emerge, so we shall have available a number of approaches, each of which will have its particular virtues. We now turn to our principal topic, the study of surface vibrations by the electron energy loss method.

1.3   THE ELECTRON ENERGY LOSS METHOD OF SURFACE VIBRATIONAL SPECTROSCOPY: INTRODUCTORY REMARKS

A schematic drawing of a typical electron energy loss experiment is given in Fig. 1.1. To form a monoenergetic beam of electrons, an electrostatic deflecting system in combination with entrance and exit slits is used to select electrons of well-defined energy from those emitted by an appropriate source. The electrons which pass through the monochromator strike the sample, and a second electrostatic system is used as an analyzer of the energy spectrum of the scattered electrons. The monochromator and analyzer in some systems are fixed in orientation, and the sample may be rotated, as indicated in the figure. By this means, the distribution of scattered electrons in angle, as well as energy may be sampled. The experiment is very simple in concept, but the construction of the spectrometer involves a number of subtle considerations, if these devices are to achieve energy resolution sufficient to probe vibrational losses, and pass sufficient current at the same time. A detailed discussion of the apparatus will be given in Chapter 2, with emphasis on the physical principles that control the performance of the device.

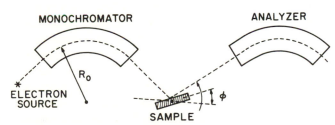

Fig. 1.1   Schematic diagram of an electron energy loss experiment. Electrons from a cathode pass through a monochromator, strike the sample, and the energy spectrum of the scattered electrons is probed by a second monochromator.

The discussion of the inelastic scattering process requires one to address the nature of the interaction between the incoming electron and the vibrating atoms in the solid. This is a complex problem in general, since the electron interacts strongly with the substrate, and any attempt to describe this coupling in simple terms necessarily leads to a schematic and limited picture.

Nonetheless, it proves useful to distinguish between three limiting cases, though one recognizes that there are no sharp and precisely defined boundaries between them: these are dipole scattering, inelastic scattering via an intermediate negative ion resonance, and impact scattering. As an atom on a surface or in a molecule vibrates, it modulates the electric dipole moment of its environment in a time-dependent fashion. An electron in the vacuum above the crystal senses a long-ranged electric field of dipolar character, and this produces small-angle scatterings typically substantially more intense than the scatterings observed at large deflection angles. One observes a "lobe" of inelastics sharply peaked about the specular direction, when dipole scattering is present. The principal features of this mechanism will be outlined in simple terms later in this chapter, and in detail in Chapter 3. Negative ion resonances are observed frequently in elastic collisions between electrons and molecules in the gas phase [4]. On a surface, one may expect the lifetime of such resonances to be rather short when a chemisorbed molecule is coupled to the substrate electron states. In fact, up to now, negative ion resonances for molecules adsorbed on surfaces have not been observed except in a few experiments (see Chapter 7).

We shall see that for small deflection angles, with dipole scattering dominant, it is possible to obtain a simple and useful form for the cross section without resort to a microscopic description of the electron–substrate interaction. At large deflection angles, it is necessary to turn to a fully microscopic description to describe the cross section theoretically. This large deflection angle regime, outside the "dipolar lobe," is often referred to as the impact scattering regime. Theoretical methods distinctly directed toward this regime have appeared in recent years; of course, in an experiment as one scans the angular distribution of electrons scattered inelastically from the surface, one moves *continuously* from the dipole dominated small-angle scatterings out into the "wing" where a fully microscopic analysis is required. Thus, it is clearly artificial to use different terminology for each region. There are, however, two limits to the problem from the point of view of the theorist, and in much of the data the "dipole lobe" appears as a clear and distinct feature. As a consequence, in this volume, we draw a distinction between the two limits by use of the terminology just described, and the reader should recognize the limitations of this procedure.

It is important to realize that not all vibrational modes of a molecule will be equally strongly excited. For negative ion resonances, for example, the vibration modes excited strongly are determined by the symmetry of the compound state that is formed with the electron. Likewise, a certain selection rule may be formulated for impact scattering, as we see from the next paragraph.

The theory of inelastic electron–phonon scattering is similar to x-ray or

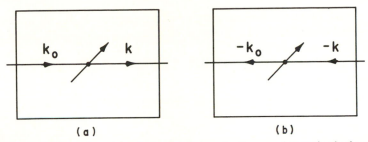

Fig. 1.2   An illustration of the electron trajectory for the case of scattering in the specular direction. We assume the crystal is viewed from above, and we show (a) a given trajectory and (b) its time-reversed partner.

neutron scattering with the complication that multiple elastic scattering must be fully taken into account for electrons. The cross section for inelastic scattering of neutrons or x rays contains the scalar product of $\mathbf{k}^{(S)} - \mathbf{k}^{(I)}$ and the polarization of the normal mode of the vibration, where $\mathbf{k}^{(S)}$ and $\mathbf{k}^{(I)}$ denote the wave vectors of the scattered and incident beam. For the case of specular reflection, $\mathbf{k}^{(S)} - \mathbf{k}^{(I)}$ is perpendicular to the surface, which makes the cross section zero for modes polarized parallel to the surface. This argument can be extended to a general vibration. In Fig. 1.2a we depict a top view of a surface, the electron trajectories of a specularly reflected electron, and some molecular species. The electron is assumed to be sufficiently fast so that its trajectory is not changed by the interaction with the vibrational modes ($k^{(S)} \sim k^{(I)}$). The probability for the excitation of molecular vibration is then simply described by the matrix element $\langle \psi_0 | V | \psi_1 \rangle$, with ψ_0 and ψ_1 the vibrational ground state and the first excited state of the molecule. Then V is the electron–oscillator interaction, treated here as a time-dependent potential generated by a point electron on a classical trajectory. By going to the time-reversed situation (Fig. 1.2b), it is easily seen that V is even with respect to inversion. Since the ground state is always even to all symmetry elements of the point group of the adsorbed molecule, the excited state ψ_1 must also be even. Therefore the cross section for excitation should vanish for all vibrations which belong to the same irreducible representation as a two-component vector parallel to the surface. For the simple case of an atom adsorbed in a site of sufficiently high symmetry, this means that the cross sections for the vibrations parallel to the surface vanishes. While this selection rule is strictly valid only in the case $k^{(S)} = k^{(I)}$, it was shown recently by Tong *et al.* [5] that the cross section for parallel motions drops several orders of magnitude within a few degrees around the specular direction even for electron energies of 30 eV. It is therefore likely that a significant drop in intensity for parallel modes is found for even smaller energies as long as the energy E_I is large compared to $\hbar\omega$.

As a result of the short-range portion of the electron–molecule interaction, the angular distribution of the electrons scattered inelastically from a molecular vibration is relatively broad. Typical high-resolution spectrometers accept a very small solid angle only. Therefore typical count rates of inelastically scattered electrons from impact scattering are small. In addition to the broad angular distribution of inelastic electrons having experienced impact scattering, inelastically scattered electrons very sharply peaked about the specular direction are observed as a consequence of the dipole scattering. The integrated intensity of the near specular peak of inelastically scattered electrons is typically smaller than that of the specular elastic beam by roughly three orders of magnitude, for scattering from the vibrational motions of a monolayer of adsorbed molecules.

The dipole scattering which leads to the near specular peak may be described as follows. Consider a unit cell on a perfectly smooth, clean, crystal surface. There is a static electric dipole moment $\mathbf{P}_0$ of the charge distribution in this unit cell, since there is no inversion symmetry at a surface site. If a molecule is adsorbed onto this unit cell, the static dipole moment becomes $\mathbf{P} \neq \mathbf{P}_0$. Now if a vibrational normal mode of the adsorbed molecule with frequency ω_0 is excited, the dipole moment $\mathbf{P}$ is modulated to become $\mathbf{P} + \mathbf{p} \exp(-i\omega_0 t)$.

The oscillating component of the electric dipole moment sets up electric fields in the vacuum above the crystal, and these oscillating electric fields scatter the incoming electron inelastically. The dipole field is long ranged, and as a consequence the scattering cross section is peaked strongly around the forward direction, which in this case is the specular direction, since the electron is reflected from the crystal surface before it strikes the detector. A simple argument to be presented shortly can be used to isolate the key parameter that controls the angular distribution of the electrons scattered inelastically by the dipole mechanism.

For the case of impact scattering, we formulated a selection rule using time-reversal symmetry. Although no specific assumption was made about the nature of the interaction, the quasiclassical argument may become questionable when the matrix element becomes a sharply peaked function of angle around the specular direction. Nevertheless, the same selection rule can be deduced for dipole scattering by a different argument. This may be appreciated by examining a special case. Consider a simple diatomic molecule adsorbed on a metal surface, and which stands vertically. The CO molecule adsorbed on the surface of a transition metal provides an example of this geometry. If the molecule stands vertically, the CO stretching vibration leads to an oscillating dipole moment $\mathbf{p}$ perpendicular to the surface. This dipole produces an image dipole in the substrate parallel to itself; an electron which approaches the surface thus sees an oscillating dipole poten-

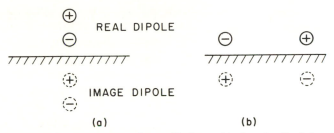

Fig. 1.3 A dipole placed near a surface and its image (a) when the dipole is perpendicular to the surface and (b) when the dipole lies parallel to it.

tial from a dipole with total strength **2p**. This is illustrated in Fig. 1.3a. If the molecule stands vertically, as illustrated in Fig. 1.3a, and if the adsorbate site has symmetry sufficiently high that two mutually perpendicular reflection planes normal to the surface exist, then all the normal modes of the molecule (six in number, with four distinct frequencies) must describe vibrational motions either strictly perpendicular to the surface or strictly parallel to it. The two modes that involve motions normal to the surface (the CO stretch, and the C-surface mode) will produce the near specular dipole peak, but the modes parallel to the surface will not. Excitation of those CO vibrations would now produce an oscillating dipole moment parallel to the surface. The image dipole is now antiparallel to the oscillating dipole moment of the molecule, so that the total oscillating dipole moment of the molecule–substrate combination vanishes (Fig. 1.3b). Thus the absence of these modes in a spectrum show that the molecule stands upright, while additional modes in the spectrum would indicate that the CO axis is tilted away from the surface normal.

From the foregoing discussion, it is clear that from electron energy loss spectroscopy, one may obtain information not only on the vibrational frequencies of adsorbate and substrate atoms, but on the surface geometry as well. We shall discuss the various selection rules in a more complete and satisfactory fashion in Chapters 3 and 4.

The key features of the dipole scattering mechanism, namely, the narrow angular distribution, may be extracted from an elementary argument. For a dipole placed just above the metal surface, the potential seen by an electron in the vacuum above the crystal is,

$$V(\mathbf{r}) = 2(pz/r^3) \exp(-i\omega_0 t) + \text{c.c.,} \tag{1.1}$$

where the factor of 2 has its origin in the image dipole induced in the metal substrate, and the z axis is normal to the surface.

If $\mathbf{r}_{||}$ is the projection of $\mathbf{r}$ onto the plane parallel to the surface, and $\mathbf{Q}_{||}$

is a two-dimensional wave vector that also lies in the plane parallel to the surface, then we may rewrite Eq. (1.1) as

$$V(\mathbf{r}) = p e^{-i\omega_0 t} \int \frac{d^2 Q_{||}}{\pi} e^{i\mathbf{Q}_{||} \cdot \mathbf{r}_{||}} e^{-Q_{||} z} + \text{c.c.} \tag{1.2}$$

In Eq. (1.2), the dipole potential is synthesized from a linear combination of two-dimensional waves, each of which is localized near the surface. The component of wave vector $\mathbf{Q}_{||}$ has a field which extends into the vacuum the distance $l(Q_{||}) = Q_{||}^{-1}$. Thus, the long-wavelength contributions to Eq. (1.2) have fields which extend very far into the vacuum. This is no accident, but rather is a consequence of the requirement that the electrostatic potential in the vacuum must obey Laplace's equation. Thus, if a contribution to the total potential varies like $\exp(i\mathbf{Q}_{||} \cdot \mathbf{r}_{||})$ for fixed z, Laplace's equation requires its z dependence to be $\exp(-Q_{||} z)$.

Suppose an electron approaches the crystal surface with energy E_I, and wave vector $\mathbf{k}^{(I)}$. If it is scattered by the potential to a final state of energy $E_S = E_I \pm \hbar\omega_0$ and wave vector $\mathbf{k}^{(S)}$, then in the Born approximation this is accomplished by the contribution to Eq. (1.2) with $\mathbf{Q}_{||} = (\mathbf{k}_{||}^{(S)} - \mathbf{k}_{||}^{(I)})$, where $\mathbf{k}_{||}^{(S)}$ and $\mathbf{k}_{||}^{(I)}$ are the projections of the scattered and incident electron onto the plane parallel to the surface.

Suppose the incident electron approaches the crystal along the normal, and that it emerges after inelastic scattering with wave vector that makes an angle θ_S with the surface normal. Then for small-angle deflections, $Q_{||} \cong k^{(I)}\theta_S$. Since the potential extends into the vacuum above the crystal the distance $l(Q_{||}) = Q_{||}^{-1}$, the time spent by the electron in the dipole potential, is $\Delta t = 2l(Q_{||})/v_0$, where v_0 is its velocity. The factor of 2 enters because the electron experiences the potential both as it approaches and as it exits from the crystal. The expression for Δt may be arranged to read $\Delta t = \hbar/E_I\theta_S$, where E_I is the incoming kinetic energy. Now, if the angular deflection θ_S is so large that $\Delta t\,\omega_0 \ll 1$, the electron passes through the dipole potential so rapidly that the probability of exciting the vibrational mode is very small. On the other hand, if $\Delta t\,\omega_0 \gg 1$, we are in the adiabatic limit and the electron exits without exciting the mode. The excitation cross section is thus a maximum when $\Delta t\,\omega_0 \approx 1$, and this gives $\hbar\omega_0/E_I$ as the most probable deflection angle. A detailed theory shows that it is $\hbar\omega_0/2E_I$ that enters. Thus, under the usual experimental conditions, $\hbar\omega_0 \ll E_I$, the electric dipole scattering mechanism produces a contribution to the energy loss cross section peaked sharply around the specular direction.

It is interesting to insert some numbers into the preceding expressions. A typical energy loss is the range of 1000 cm^{-1} (124 meV), and impact energies of 5 eV are frequently used. Then $\hbar\omega_0/2E_I \cong 0.7°$. The electron excites the vibrational mode when it is roughly 60 Å above the surface, on the average. At this distance from the adsorbed molecule, a multipole expansion of the oscillatory component of the potential is fully justified, the dipolar contribution is the first nonvanishing term unless it is forbidden by symmetry considerations, and the dipolar contribution is larger than the next term by more than one order of magnitude. It is for this reason that near specular inelastic scatterings are well described by the simple picture just outlined.

A full discussion of the angular distribution of the inelastically scattered electrons, including those which suffer large-angle deflections, requires a fully microscopic theory which takes account of the details of the potential encountered by the electron. Such calculations have been carried out, and we shall turn to a discussion of these later in the present volume. It is clear from elementary physical considerations that at large deflection angles (large values of $Q_{||}$), the dipole picture breaks down, and so does the selection rule which limits scattering only to modes with the oscillating dipole moment normal to the surface. Thus, the energy loss spectrum of electrons which scatter through large angles contains features from all possible vibrational modes of an adsorbed molecule; a comparison between the energy loss spectrum associated with large-angle scatterings and the near specular loss spectrum allows one to identify which modes have symmetry that produce an oscillating dipole moment normal to the surface. This places strong constraints on the nature of the adsorption site, and the molecular orientation in many cases.

In the present discussion, we have confined our attention to energy losses produced by the inelastic scattering from vibrational motions of adsorbed molecules. The method can also be used to study vibrational motions of the substrate atoms, and furthermore to study electronic properties of both clean and adsorbate covered surfaces, as we shall see.

REFERENCES

1. T. Wolfram, ed., *"Inelastic Electron Tunneling Spectroscopy"* Springer-Verlag, Berlin and New York, 1978.
2. R. B. Bailey and P. L. Richards, *Surface Sci.* **100**, 8626 (1980).
3. G. Brusdeylins, R. B. Doak, and J. P. Toennis, *Phys. Rev. Lett.* **44**, 1417 (1980)
4. G. J. Schulz, *Rev. Mod. Phys.* **45**, 423 (1973).
5. S. Y. Tong, C. H. Li, and D. L. Mills, *Phys. Rev. B* **24**, 806 (1981).

CHAPTER 2

INSTRUMENTATION

Vibration spectroscopy of surfaces and adsorbed molecules has become feasible through improved electron optical devices which allow the measurement of electron energy losses of a few millielectron volts only. The electron optics involved in the construction of electron spectrometers of high-energy resolution is very different from the electron optics of other instruments such as microscopes, microprobes, or photoelectron spectrometers. Many aspects important in conventional electron spectroscopy have little or no bearing on high-resolution spectrometers. In fact, one encounters new physical and technical problems in high-resolution spectroscopy. This chapter on instrumentation deals specifically with those aspects of electron optics that are unique to high-resolution spectroscopy.

Electron energy loss spectroscopy uses electrons as a means of excitation, as well as the entities that carry information back from the surface. Therefore, not only must the analyzer be capable of high-energy resolution, but also the incident beam must be highly monochromatic, i.e. it must contain electrons within an energy window not broader than a few millielectron volts. No physical source of electron emission is known with such a narrow energy distribution. Therefore, spectrometers must use a thermionic or field emitter followed by an electron optical device which acts as an energy filter with small energy window. This filter will be called the monochromator. The monochromator is typically followed by a lens system that allows the energy of the electrons at the target to be independently chosen from the monochromator pass energy. In Fig. 2.1, a block diagram of the complete experimental system is given. As one sees from this diagram, a similar lens system

Fig. 2.1   Block diagram of an electron energy loss spectrometer.

plus an energy selective analyzer is then used to detect the electrons inelastically scattered from the target.

2.1 PHYSICAL REQUIREMENTS OF SPECTROMETERS

The natural linewidth of a vibrational level of an adsorbed molecule is significantly larger than typical of the gas phase. This is due to the coupling between the vibrational motions of the adsorbate and the vibrational and electronic excitations of the substrate. Typical vibrational linewidths for adsorbed molecules are 0.5–1 meV. This value sets the goal for the desired *energy resolution* of an electron spectrometer. In present generation experimental systems, the typical energy width that may be realized is between 3 and 8 meV, so there is substantial room for further improvement.

As we shall see in Section 2.2, the need for high resolution requires the monochromating devices to operate at low-pass energies, below 1 eV. The need for this low-pass energy leads to the most significant problems specific to high-resolution spectroscopy. On an energy scale of 1 eV or less, a real metal surface can no longer be considered an equipotential surface; the surface potential (work function) of a given material may vary as much as 1 eV, for surfaces parallel to different crystallographic planes. Patchy surface potentials can also result from contamination that is spatially nonuniform. In fact, even in an ultrahigh vacuum, a metal surface is rarely perfectly clean, but the surface is covered by a carbonaceous layer of unknown composition and character. This layer may be semiconducting, or even nonconducting. Therefore, a "metal surface" subject to electron bombardment may charge up to several tenths of a volt.

Uncontrolled surface potentials are therefore notorious in high-resolution spectroscopy, and even technical measures such as covering all surfaces with graphite do not solve the problem completely. The consequences of uncontrolled surface potentials are twofold. First, as the pass energy of a monochromator is lowered in order to achieve high resolution, one finds in practice that the energy width ΔE typically levels off at some point, or even degrades. To the extent that one is then not limited by the electronic noise of power supplies, and rf pickup in the feed cables, one then has achieved the optimum energy width ΔE allowed by the patchy surface potentials in a given experimental situation and pretreatment history. The resolution of the spectrometer can be improved further only by dealing with the physical problem of fluctuations in the work functions that differ from point to point in the apparatus.

Another consequence of the fluctuations in the surface potential is that elaborate calculations to design electron lens systems are of little value. There may even be a virtue in using relatively simple and short lens systems, since the patchy work functions will reduce the transmission of a lens system, the longer and more complicated it is. Therefore, the whole field of electron optics which deals with image formation has a limited bearing on the construction of high resolution spectrometers. As a consequence, we do not treat this topic in the present volume.

As in any other spectroscopy, one has a trade-off between energy resolution and the available signal. However, in high-resolution electron spectroscopy, the decrease in signal intensity with increasing resolution is more severe than is typical, since the monochromatic current at the sample varies approximately with the square of the energy width ΔE. Therefore, in spectrometer design, a matter of particular concern is improving the available current for a given resolution.

One can see why the monochromatic current is limited from the following argument. Suppose a spectrometer is operated with a monochromatic current lower than the maximum value. This may be achieved by using a low current in heating the cathode. If, then, the monochromatic current (at the sample location) is observed as a function of the cathode heating current, one first finds the monochromatic current increases as expected. But then it levels off, or possibly passes through a maximum. This behavior results from one of two different reasons. Both are associated with the space charge which builds up within the dense electron beams, and the additional forces the electrons experience as a result of this space charge. One reason is that the emission system which feeds the monochromator is no longer matched to the monochromator geometry: the image size may become larger than the entrance aperture of the monochromator, or more typically, the angular spread of the input beam becomes too large to be handled by the monochromator. We shall call this situation "cathode limited." The other reason is that as a result of the increasing space charge, the imaging properties of the monochromator deteriorate. We call a current limited for this reason "monochromator limited."

Obviously, in any particular case, it is important to understand which of these is limiting the spectrometer, since otherwise one might try to improve the wrong part of the system. Even without entering the theory of monochromating devices deeply, it is easy to see that an important parameter in determining which of the two limits controls the maximum spectrometer output is the angular spread α_i of the feed beam of the monochromator. The current provided by the cathode system will increase with α_i. A monochromator always has a limited acceptance angle. Since the space charge inside the monochromator will cause the beam to diverge and thus add to the angular spread already present, a monochromator can pass more current the smaller the initial angular spread of the electron beam (Fig. 2.2). Thus,

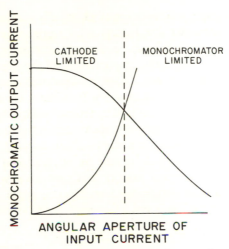

Fig. 2.2   Qualitative sketch of the output current of a monochromator fed by a cathode system as a function of the angular aperture. The monochromatic current is either limited by the input current as supplied by the cathode (cathode limited) or by the maximum current that can be properly handled by the monochromator.

there is an optimum value of α_i for a given spectrometer design, but from spectrometer to spectrometer the two limiting curves in Fig. 2.2 will differ. To find the optimum usable resolution, one must understand how the design parameters of the spectrometer control the current limit. Much of this chapter will be devoted to this question.

Another matter of concern in spectrometer performance is the *background*, i.e., unwanted electrons counted by the detector. Here it is convenient to distinguish between two different experimental situations. Let us first consider a situation where the spectrometer is set up in such a way that the monochromated beam is directly focused into the analyzer without a sample in between ("direct beam focusing"). If one then scans a "loss spectrum", typically by offsetting the analyzer and some lens potentials, one may observe electron counts in the loss regime where there should not be any. These electrons result from scattering of the beam inside the analyzers. Sometimes "ghost peaks" are observed. The means of reducing this spurious background will be discussed in Section 2.7. The background intensity must be smaller than 10^{-5}–10^{-6} of the intensity of the direct beam. This limit is imposed because in an actual loss experiment, metal surfaces used as a target will produce a continuous background of this order of magnitude, for reasons that will be discussed later.

While the situation described so far applies to an experiment where the loss spectrum is measured for a scattering angle near the specularly reflected beam (where the sample produces many elastically scattered electrons) it is also important to consider the background signal when the analyzer is positioned in such a way that the direct beam cannot enter the analyzer. This is the background relevant to loss experiments off the specular beam, where loss intensities can be very low. This background should therefore not exceed the dark count rate of the detector (less than one count per second).

The intensity of a loss spectrum is also determined by the *acceptance angle* of the spectrometer, and it therefore appears advantageous to operate with relatively low angular resolution in order to collect as much signal as possible. Unfortunately, here again one is limited by physical and technical considerations. As already mentioned, energy analyzers have a finite acceptance angle α_i. This, then, also limits the convergence angle ϑ_c at the sample position because of Abbe's sine law [1], which reads for this case

$$\sqrt{E_a}\, y_a \sin \alpha_i = \sqrt{E_s}\, y_s \sin \vartheta_c, \tag{2.1}$$

where E_a and E_s are the kinetic energies of the electron at the analyzer and sample positions, respectively, and y_a and y_s are characteristic linear dimensions of the analyzer entrance aperture and the imaged area on the sample, respectively. The left-hand side of Eq. (2.1) is more or less fixed by resolution requirements imposed on the analyzer (see Section 2.2), and the potential at the sample is typically higher than at the analyzer. Thus, the only way to increase ϑ_c would be to have the image size y_s small. This, however, would require lenses of short focal length, with the sample close to the focus. This would then put severe constraints on the size of the sample and the precision of the sample positioning. Furthermore, consideration of the spherical aberration of lens systems shows that for $E_s/E_a \sim 10$, the ratio y_s/y_a cannot be made much smaller than one. Therefore, in typical electron spectrometers constructed so far, the acceptance angle ϑ_c is smaller or, at the most, equal to the acceptance angle of the analyzer.

A final and important aspect of the spectrometer performance is the ease and reliability in *handling* and *stability* with time and exposure of the surfaces to reactive gases. We shall give consideration to this part of the problem in Section 2.8.

2.2 ENERGY DISPERSIVE SYSTEMS

Any focusing system for electrons imaging an object with chromatic aberration might in principle be used for monochromating an electron beam. Low-energy electron spectrometers generally employ energy dispersive systems of the electrostatic deflector type. In high-energy electron transmission spectroscopy, a Wien filter (crossed electric and magnetic fields) was also used successfully, and in fact the best resolution reported so far ($\Delta E = 1.8$ meV) was obtained with a Wien filter spectrometer [2]. In this section, we shall discuss the various types of energy analyzers only briefly. For a more detailed discussion, the reader might consult a recent review by Roy and Carette [3], and further references contained in this article.

The simplest device for the energy analysis of electrons is the parallel plate condensor, or plane mirror analyzer; see Fig. 2.3a. In the uniform deflecting field, the trajectories are simply parabolas. Obviously, focusing can be obtained in one direction only. In order to calculate conditions under which the entrance slit is imaged onto the exit slit, trajectories are expanded in a power series in α, the angle between the central path (dotted line in Fig. 2.3a) and the trajectory at the entrance slit. The condition for focusing is then that the deviation of the trajectory from the center path vanishes in first order

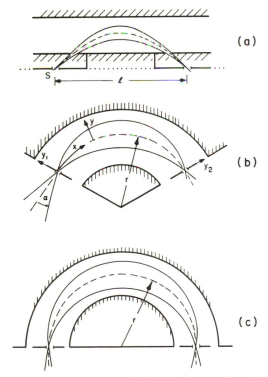

Fig. 2.3 (a) Parallel plate mirror, or cylindrical mirror analyzer when the device has rotational symmetry around the dotted axis; (b) 127° cylindrical deflector; (c) 180° spherical deflector with two-dimensional focusing.

in α. Unfortunately, higher terms in the power series in α remain. This causes an imperfect image of the entrance slit, and a reduction in transmission and resolution. Ultimately, this is one reason for the limited current that can be passed through analyzers.

The energy resolution of electrostatic deflectors as depicted in Fig. 2.3 may be expressed in a rather general way. The base resolution ΔE_B, i.e., the energy difference between the maximum and the minimum energy transmitted through the analyzer is given by [3].

$$\Delta E_B/E_0 = As + B\alpha^n + C\beta^n, \tag{2.2}$$

with E_0 the pass energy, s the slit width, and α and β the (semi) angular apertures in the plane, and perpendicular to the plane of deflection, respectively. The coefficients A, B, and C for the different systems are summarized in Table 2.1.

TABLE 2.1

Coefficients in Eq. (2.2) for Commonly Used
Electrostatic Deflector Analyzers

Analyzer	A	B	C	n
Plane mirror 30°	$3/l$	9.2	1	3
Cylindrical mirror 42°	$2.2/l$	5.55	0	3
Cylindrical deflector 127°	$2/r$	4/3	1	2
Spherical deflector 180°	$1/r$	1	0	2

Based on these coefficients, one may assess the suitability of different deflectors for various applications. Such an assessment has been made in Ref. [3]. The result of this assessment is that the figures of merit of deflectors in various applications might differ as much as 2 or 3. The main factor which determines the order of magnitude of the sensitivity and signal-to-noise ratio in energy loss spectroscopy, however, is, as already discussed, the monochromator current that is available. It is therefore this question that shall be our primary concern in the following. This problem will be discussed in detail for the 127° cylindrical deflector spectrometer. For this system, the influence of space charge on the electron trajectories constitutes a one-dimensional problem, which simplifies the analysis. Also, the cylindrical deflector is the system with which the authors have the most practical experience. This will allow us to compare the results of theoretical models to experimental data at each stage of development. The results are then easily adapted to the spherical deflector and the mirror analyzers (Section 2.6).

2.3 THE ANALYSIS OF THE CYLINDRICAL DEFLECTOR

2.3.1 Trajectories and Resolution

An electron with mass m and velocity v in the deflecting field $\mathscr{E}_r$ of the deflecting device travels along the center path if the centrifugal force and the force from the deflecting field balance. When this condition is achieved, we have

$$mv^2/r = 2E_0/r = e\mathscr{E}_r, \tag{2.3}$$

where r is the radius of the center path, and E_0 the energy of the electron. The radial electric field $\mathscr{E}_r$ is determined by the potential difference ΔV between the outer and the inner electrode with radii of curvature r_o and r_i, respectively. One has

$$\mathscr{E}_r = \frac{\Delta V}{\ln(r_o/r_i)}\frac{1}{r}. \tag{2.4}$$

The pass energy E_0 is therefore a linear function of the deflecting potential ΔV,

$$E_0 = \tfrac{1}{2}e\,\Delta V/\ln(r_o/r_i). \tag{2.5}$$

To describe trajectories of orbiting electrons, it is convenient to introduce geometrical parameters that measure deviations from the center path. Let y_1 be the radial deviation at a certain selected point, and α the angle between the trajectory and the central path at the same point. Then we shall let δE be the difference between the energy of an electron from the pass energy E_0. We let x be the distance from the selected point to the point on the central path and y the radial deviation of the electron trajectory measured by dropping a perpendicular down to the central path. If terms up to second order in α are retained, the equation of the electron trajectory is [4]

$$y = y_1 \cos\left(\sqrt{2}\,\frac{x}{r}\right) + r\,\frac{\delta E}{2E_0}\left(1 - \cos\left(\sqrt{2}\,\frac{x}{r}\right)\right) + \frac{r\alpha}{\sqrt{2}}\sin\left(\sqrt{2}\,\frac{x}{r}\right)$$

$$+ r\alpha^2\left(\frac{2}{3}\cos\left(\sqrt{2}\,\frac{x}{r}\right) - \frac{7}{24}\cos\left(2\sqrt{2}\,\frac{x}{r}\right) - \frac{3}{8}\right). \tag{2.6}$$

First-order focusing is achieved when the term linear in α vanishes. This requires the length x_f of the central path to be

$$x_f = r\pi/\sqrt{2}, \tag{2.7}$$

which is equivalent to a deflecting angle $\phi = \pi/\sqrt{2} \cong 127°$. For this choice of deflecting angle, Eq. (2.6) reduces to, when $x = x_f$,

$$y_2 = -y_1 + r(\delta E/E) - \tfrac{4}{3}r\alpha^2, \tag{2.8}$$

where y_1 is the deviation from the central path at the selected point, and y_2 is the deviation at a distance x_f downstream.

Thus, when δE and α are both zero, a given point y_1 is imaged onto y_2 with a magnification of -1. When a slit is placed so it limits the maximum values of y_1, the image of this slit will be of the same width. The *position* of the image, however, depends in general on the energy deviation δE, which makes the system select electrons in a particular energy window, i.e., the system is energy resolving. Since the magnification is -1, the exit slit should have the same width s as the entrance slit for optimum transmission.

The maximum positive energy deviation δE_+ from the pass energy E_0 through the device is then, from Eq. (2.8),

$$\delta E_+/E_0 = (s/r) + \tfrac{4}{3}\alpha_i^2 \tag{2.9}$$

with α_i the angular aperture of the input beam. The maximum negative energy deviation that may be tolerated is

$$\delta E_-/E_0 = -s/r \tag{2.10}$$

so the base width of the transmitted energy distribution is (compare with Table 2.1)

$$\Delta E_B/E_0 = 2s/r + \tfrac{4}{3}\alpha_i^2. \tag{2.11}$$

In Eq. (2.11), the effect of angular aberration on the energy resolution is displayed clearly.

So far, only trajectories that lie within the plane of deflection have been considered. However, Eq. (2.6) also holds for an electron which has a velocity component perpendicular to the plane of deflection (the z direction). This component of velocity remains constant, since the field is radial. The velocity of an electron with a trajectory out of the plane of deflection by the angle β is roughly

$$v = v_r(1 + \tfrac{1}{2}\beta^2), \tag{2.12}$$

which corresponds to the kinetic energy, again for β small,

$$E = E_0(1 + \beta^2). \tag{2.13}$$

When trajectories tipped out of the plane by angles which range from 0 to β are allowed to pass through the deflector, an additional term β^2 appears in Eq. (2.11). If one introduces the slit height h, we have

$$\beta^2 = h^2/2\pi^2 r^2. \tag{2.14}$$

It is easy to see that this term can be made very small by choosing the slit height appropriately. In practical spectrometers, one may also design the lens system between monochromator and analyzer (Fig. 2.1) such as to limit the imaged value of β, or the slit heights. Therefore, we shall neglect the influence of the β^2 term in our subsequent considerations.

While the base width of the energy distribution is easy to calculate, as we see from the foregoing example, it is not a precisely defined quantity in experimentally observed energy distributions, since it is not possible to feed an analyzer with an input beam which has a sharp cutoff at a particular angle α_i. For comparisons between experimentally observed energy distributions, the full width at half-maximum $\Delta E_{1/2}$ is more useful. A rough estimate of $\Delta E_{1/2}$ is obtained by dividing the width ΔE_B calculated from the foregoing considerations by a factor of 2. A better procedure, however, would be to model the transmitted energy distribution in a computer experiment. In Fig. 2.4, the transmission probability is plotted as a function of energy, assuming a uniformly illuminated entrance slit (curve labeled $\hat{\alpha} = 0$), and an angular distribution uniform up to the angle $\hat{\alpha}$ indicated. For small angular apertures $\hat{\alpha}$, the distribution is simply the convolution of the entrance slit with the exit slit, i.e. a triangular shaped function. The transmission is 100% for electrons which have the pass energy E_0, because of the magnification factor of -1 in Eq. (2.8). For larger angular apertures, however, the transmission for electrons with precisely the pass energy becomes smaller, since the image is no longer perfect. The energy of optimum transmission is

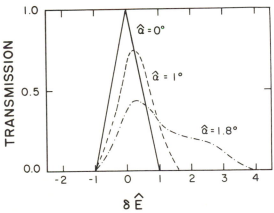

Fig. 2.4 Energy distribution of electrons transmitted through a cylindrical deflector assuming a feed beam with homogeneous energy and angular distribution up to $\hat{\alpha} = \alpha_i(4r/3s)^{1/2}$. The energy scale is also normalized ($\delta\hat{E} = \delta E r/s E_0)_\alpha$ which makes the energy distributions independent of pass energy E_0, radius r, and slit width s. The distribution $\hat{\alpha} = 1$ marks the limit in the angular aperture of the feed beam up to which the device should be used.

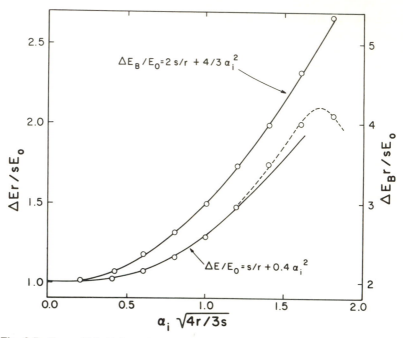

Fig. 2.5 Base width (right scale) and FWHM (left scale) of the transmitted energy distribution assuming homogeneous energy and angle distribution in the feed beam.

also shifted to an energy higher than the nominal pass energy E_0. Furthermore, the energy distribution comes to resemble a Gaussian shape rather than the triangular form found for very small α_i. Finally, asymmetry sets in, with a tail that extends to high energies for larger values of α_i.

By utilizing sets of such computer-generated distribution curves, the full width at half-maximum $\Delta E_{1/2}$ and the base width ΔE_B may be calculated, as a function of the (reduced) angular aperture (Fig. 2.5). As one can see from Fig. 2.5, the base width is well represented by Eq. (2.11) while the half-width $\Delta E_{1/2}$ is approximately

$$\Delta E_{1/2}/E_0 = (s/r) + 0.4\alpha_i^2. \qquad (2.15)$$

The verification of this relation by experimental data is unfortunately not quite as straightforward as one might think. First, α_i is seldom known independently. Second, another analyzer of substantially better resolution would be required in order to measure the energy distribution from the device under investigation. One might try, by using an experimental arrangement as sketched in Fig. 2.7, to improve the resolution of the analyzer with respect to the monochromator by lowering the pass energy of the analyzer.

Then, however, by virtue of Abbe's sine law, the input angular aperture of the analyzer increases, thus reducing the resolution. In fact, by combining Eqs. (2.1) and (2.15), it is easy to show that in the limit where the α^2 term dominates, the resolutions stay equal when either the monochromator or the analyzer pass energy is lowered alone. The simplest and most reliable procedure for measuring the resolution of a spectrometer built with identical monochromator and analyzer is therefore to measure the overall resolution with identical pass energy in the monochromator and analyzer, and a lens system with lens potentials arranged symmetrically around the sample. Then time-reversal symmetry of the electrostatic force laws ensures that the exit slit of the monochromator is imaged onto the entrance slit of the analyzer with the same angular aperture. If one varies the analyzer potential by a small fraction of the pass energy in order to scan the energy distribution, the image will not change significantly. Then the energy resolution of either the monochromator or analyzer can be obtained by dividing the half-width of the overall distribution by $\sqrt{2}$. This factor arises because the half-width of the convolution of two Gaussian-shaped distributions add geometrically,

$$\Delta E_{\text{tot}} = (\Delta E_1^2 + \Delta E_2^2)^{1/2}. \tag{2.16}$$

2.3.2 Transmission

The question we address in this section is what fraction of the trajectories which pass through the entrance slit also pass through the exit slit of the monochromator.

First, let us consider trajectories with the angular deviations α and β, introduced in the previous subsections, very close to zero. Then the cylindrical condenser obviously forms a perfect image of the entrance slit at the focal point calculated from Eq. (2.7). As the energy of the electrons varies, the image of the entrance slit moves across the exit slit in the radial direction. This gives rise to the triangular-shaped energy distribution, as illustrated in Fig. 2.4. The transmission probability for electrons with the pass energy E_0 is unity, as discussed in the previous subsection. For a trajectory with angular deviation $\alpha \neq 0$, the image of the slit is shifted to negative values of y_2 due to the second-order term in α in Eqs. (2.6) and (2.8). Therefore, only a fraction of the electrons that have the pass energy E_0 will be transmitted, as we see from the curves in Fig. 2.4 with $\hat{\alpha} \neq 0$. We shall call the function which describes this transmission probability $T(\alpha)$. For a bundle of trajectories with angles in the range $-\alpha_i \leq \alpha \leq +\alpha_i$, with α_i the input aperture angle, the transmission function $T(\alpha)$ is averaged over a range of angular deviations to yield an average transmission factor

$$T_{\alpha_i} = (2\alpha_i)^{-1} \int_{-\alpha_i}^{+\alpha_i} T(\alpha)\, d\alpha. \tag{2.17}$$

This may be used to calculate the fraction of electrons with energy E_0 which pass through the exit slit. As we see from Fig. 2.4, the α^2 term in Eq. (2.8) causes a shift in the energy distribution to energies higher than E_0. As a consequence, the transmission at the maximum of the energy distribution is slightly higher than that at the nominal pass energy E_0, as illustrated in Fig. 2.4 by the curve for $\hat{\alpha} = 1.8°$. For the sake of simplifying the analytic treatment that follows, we shall neglect this minor difference.

By using Eq. (2.8), the transmission function $T(\alpha)$ is easily calculated, since $T(\alpha)$ is simply the fraction of the entrance slit that is imaged on the exit slit. Since the image of the slit is shifted by the amount $\frac{4}{3}r\alpha^2$ for a fixed value of α, we have

$$T(\alpha) = (s - \tfrac{4}{3}r\alpha^2)/s = 1 - (4r\alpha^2/3s). \tag{2.18}$$

This relation introduces a critical angle α_m, defined by

$$\alpha_m = (3s/4r)^{1/2}. \tag{2.19}$$

For trajectories characterized by the angular deviation α_m, the transmission of electrons with the nominal pass energy E_0 is zero, while electrons of still higher energy may pass. From Eq. (2.8), one may easily calculate the transmission function for any electron energy. Since, however, we are interested principally in electrons with energy near the nominal pass energy E_0, we use the analytical form for $T(\alpha)$ in Eq. (2.18) for all electrons. Then the average transmission probability T_{α_i} introduced earlier is

$$T_{\alpha_i} = \begin{cases} 1 - \frac{4}{9}(r/s)\alpha_i^2, & \alpha_i < \alpha_m, \\ \frac{2}{3}\alpha_i^{-1}(3s/4r)^{1/2}, & \alpha_i > \alpha_m. \end{cases} \tag{2.20}$$

In Fig. 2.6, the transmission factor T_{α_i} for electrons with energy E_0 is compared to the transmission at the maximum of the energy distributions shown in Fig. 2.4. The difference between the two curves becomes significant only for large values of α_i, i.e., where the energy distributions become too broad to be useful in practical applications. The analytical form of T_{α_i} in Eq. (2.20) is therefore a reasonable approximation, in spite of the simplicity of the expression.

So far, in this subsection, we have been concerned only with trajectories in the plane of deflection. The transmission of the cylindrical deflector will be reduced further by a factor T_z which takes account of the reduced transmissions from nonplanar trajectories. In Fig. 2.7, we show two different situations for focusing in the z coordinate, perpendicular to the plane of deflection. Figure 2.7a applies when ideal slot lenses are used between cathode and the first entrance slit, with the effect of possible acceleration or deceleration of the electrons neglected. Then no focusing in the z direction occurs,

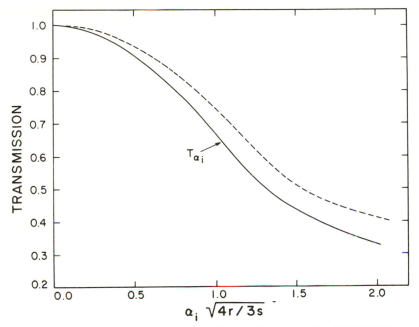

Fig. 2.6 Transmission T_{α_i} at pass energy E_0 as calculated by Eq. (2.20) (full line) and transmission at the maximum of the energy distribution curves (Fig. 2.4) as calculated from computer simulations.

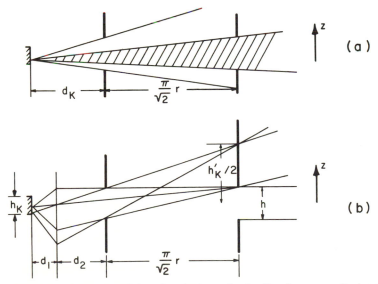

Fig. 2.7 Two focusing conditions for the beam in the direction perpendicular to the deflecting plane (z direction).

and the factor T_z is

$$T_z = d_K/[d_K + (\pi/\sqrt{2})r],\tag{2.21}$$

where d_K is the distance from the cathode to the entrance slit, as shown in Fig. 2.7.

Another situation one may consider to be more favorable is illustrated in Fig. 2.7b, where it is assumed that the cathode is imaged onto the exit slit of the monochromator. The cathode in Fig. 2.7b is also imaged *in the deflection plane* onto the exit slit by virtue of a first image at the entrance slit, and a second provided by the monochromator. This situation then represents image formation in two dimensions at the exit slit. The transmission factor T_z in this case is the ratio of the slit height h to the height of the image h'_K of the cathode,

$$T_z = \frac{h}{h'_K} = \frac{h}{h_K}\frac{d_1}{d_2 + (\pi/\sqrt{2})r},\tag{2.22}$$

with d_1 and d_2 defined in Fig. 2.7b. For small extensions h_K of the cathode in the vertical direction, this transmission factor can be very close to unity. One must, however, consider that at low energy, electrons will be drawn from a rather extended space charge cloud near the cathode. Likewise, the distance d_1 from the cardinal plane cannot be increased arbitrarily, since the position of the cardinal plane also determines the maximum input current in a given solid angle. From a practical viewpoint, the transmission factors in the two cases do not differ very substantially, and we shall use Eq. (2.21) further on in the discussion of the monochromatic current. We shall also see that the space-charge limitations on the transmission of the monochromator are such that any gain in T_z is partly compensated, since the space-charge effects then become more effective in degrading the imaging properties of the spectrometer.

2.3.3   Possible Methods of Improving Transmission in the Monochromator

Transmission through the cylindrical deflector and through any other dispersive device is limited by angular aberrations. The analytical forms for the critical angle α_m encountered in Section 2.3.2 and the average transmission T_{α_i} are controlled by the term second order in α in the expression for the electron trajectory. This describes the broadening of the image of the entrance slit at the first-order focusing point.

One is led to inquire how one might improve the transmission of the monochromator and the shape of the energy distribution, to achieve performance superior to that described in the preceding sections. Inspection of

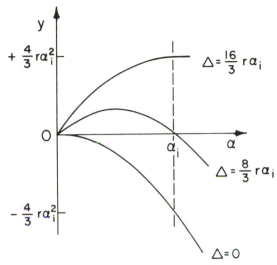

Fig. 2.8   Deviation from the center path y for electrons with pass energy E_0 which originate from the center of the entrance slit, as a function of α at various positions $x = x_f - \Delta$.

the trajectory equation shows that while the minimum trace width of the trajectories is at the first-order focal point when positive and negative values of α are supposed equally supplied by the incoming beam, the minimum trace width shifts to smaller deflecting angles when only positive values of α are allowed. In a series of papers, Roy and Carette [4] have proposed exploiting this effect, to achieve better transmission and energy distribution curves. They have done this through use of computer calculated energy distributions. However, the effect can also be studied analytically, and the analytic treatment proves illuminating.

Let us consider electrons with the pass energy E_0 embarking on the trajectory at the center of the entrance slit [$y_1 = 0$ in Eq. (2.6)]. Then at the focal point, only the α^2 term in Eq. (2.6) remains. To calculate the deviation y at the exit slit as a function of α for deflecting angles other than $\pi/\sqrt{2} \cong 127°$, we expand Eq. (2.6) around the focal point, by letting $x = x_f - \Delta$, where $x_f = \pi r/\sqrt{2}$. This gives

$$y = \alpha\Delta - \tfrac{4}{3}r\alpha^2 + O(\Delta^2). \tag{2.23}$$

This function is plotted for $\Delta > 0$, $\alpha > 0$ in Fig. 2.8. Obviously, the difference between the maximum and the minimum value of y, i.e., the trace width, is different for different values of Δ. The difference is

$$y_{max} - y_{min} = \frac{3}{16}\frac{\Delta^2}{r} + \max\begin{cases}-\alpha_i\Delta + \tfrac{4}{3}r\alpha_i^2, \\ 0.\end{cases} \tag{2.24}$$

Within the brace, one uses the largest of the two quantities indicated. The minimum trace width for trajectory bundle with positive α only is therefore found when the condition

$$-\alpha_i \Delta + \tfrac{4}{3} r \alpha_i^2 = 0 \tag{2.25}$$

is fulfilled. This defines the position of the minimum trace width for any given angular aperture,

$$\Delta = \tfrac{4}{3} \alpha_i^2 r. \tag{2.26}$$

The trace width is then, when this condition is realized,

$$\Delta y = \tfrac{3}{16}(\Delta^2/r) = \tfrac{1}{3} r \alpha_i^2. \tag{2.27}$$

This must be compared with the trace width one obtains if one allows for both positive and negative values of α, and uses the normal first-order focal point for the exit slit. There the trace width is simply

$$\Delta y = \tfrac{4}{3} r \alpha_i^2. \tag{2.28}$$

When the two results are compared, one must realize that α_i has to be made twice as large when only positive values of α are used, in order for the total angular width of the beam to be the same in both cases. Therefore, asymmetric feeding of the cylindrical deflector as proposed by Roy and Carette brings no advantage as long as the aperture of the input beam is controlled by external constraints.

This conclusion differs from that of Roy and Carette. The reason is that in their computer models, the deflecting plates of the analyzer were used to define the angular aperture. The aperture is then energy dependent, since incoming electrons with large α *and* high energy will be cut off earlier than those with large α, but with low energy. This symmetrizes the energy distribution, since the angular aberrations have the tendency to make the energy distribution asymmetric toward higher energies, as discussed earlier and displayed in Fig. 2.4. Since the use of deflecting plates as the control on the angular aperture, or one of the deflecting plates at least, is a must in this asymmetric use of the cylindrical deflector, the asymmetric mode is advisable only in those cases where the generation of secondary electrons is not a matter of concern. In high-resolution spectrometers, where the pass energy is of the same order of magnitude of typical energy losses, the generation of scattered electrons and secondaries must be kept as low as possible. Otherwise artificial loss structures ("ghost peaks") may appear in the loss spectrum. The asymmetric use of the cylindrical deflector, and also other electrostatic deflectors, is therefore not recommended, except for premonochromators followed by a second monochromator.

2.3.4 Fringing Field Effects

The electron trajectories calculated from Eq. (2.6) are for an idealized field in the cylindrical deflector. Entrance and exit apertures were assumed not to disturb the field. In practical designs, it is more convenient to make the apertures equipotential surfaces. The radial field must then fall to zero on the aperture elements, and one must consider the effect of this on the imaging properties of the monochromator. A detailed discussion of this question has been given by Herzog [5] and by Wollnik and Ewald [6]. A solution to the problem of practical importance is obtained when the distance l of the apertures from the end of the deflecting plates (see Fig. 2.9) is such that the angle ϕ subtended by the deflecting plates is equal to the effective angle of an equivalent ideal cylindrical field. According to Herzog, for this case to be realized, the distance l must assume the value

$$l = 0.265(r_o - r_i), \tag{2.29}$$

provided the slit width is small compared to $(r_o - r_i)$.

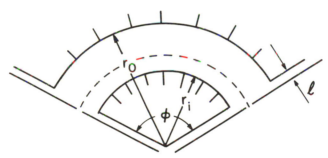

Fig. 2.9 A cylindrical deflector terminated by metal apertures and the potential of the center pass.

The fact that the two slits are no longer within the cylindrical field, but assume a position at the same distance l from the boundary of the ideal field, requires a reduction in the deflecting angle, in order to ensure that the entrance slit is properly imaged onto the exit slit. Wollnik and Ewald [6] have calculated a set of transfer matrices for arbitrary deflecting fields, and these are most helpful in the analysis. If one uses their solution for symmetric image distances for the entrance and the exit slit, the reduction $\Delta\phi$ in the required deflection angle from the ideal value $\pi/\sqrt{2}$ is

$$\Delta\phi = -2(l/r) + O((l/r)^2). \tag{2.30}$$

From this result, we see that the angle of deflection, as measured between the entrance and exit slit, along with the total path length between the entrance and exit slit remain the same as in the ideal deflector, and only the angle subtended by the deflector plates is reduced. Corrections from the fringing field on the energy resolution and transmission characteristics are small, and in fact can be demonstrated to be second order in l/r.

In deriving these results, the potentials of the entrance and the exit apertures were assumed to be equal to that of the center path, midway between the deflecting plates. In terms of the deflection voltage ΔV, the potential of the aperture in this case is

$$V(r) = V(r_i) - \frac{\Delta V}{\ln(r_o/r_i)} \ln[\tfrac{1}{2}(r_o + r_i)/r_i]. \qquad (2.31)$$

In practice, it is frequently more convenient to make the potential on the aperture the average of that between the potential of the deflecting plates

$$V(r) = \tfrac{1}{2}[V(r_o) + V(r_i)], \qquad (2.32)$$

from which it follows that the radius of the center path must be the geometrical average of the radius of the outer and inner deflection plates,

$$r = (r_o r_i)^{1/2}. \qquad (2.33)$$

2.4 SPACE-CHARGE EFFECTS

2.4.1 The Ribbon-Shaped Beam

Under the influence of its own space charge, a ribbon-shaped beam formed from initially parallel trajectories spreads out, as one moves down the beam (Fig. 2.10). This spreading is easily calculated for a homogeneous, monochromatic beam of electrons. If one assumes the perturbation of the trajectories by the space charge is small, one may ignore the field components along the direction of the path. From Gauss's law, where $\mathscr{E}$ is the electric field generated by the space charge, we have

$$\nabla \cdot \mathscr{E} = 4\pi\rho, \qquad (2.34)$$

and if the charge density ρ is a function only of y, the distance from the center of the beam (Fig. 2.10), the force on an electron in the y direction is

$$F = e\mathscr{E}_y = 4\pi e \int_0^y dy' \, \rho(y'). \qquad (2.35)$$

The equation of motion of one particular electron is then

$$\ddot{y} = \frac{4\pi e}{m} \int_0^y dy' \, \rho(y'). \qquad (2.36)$$

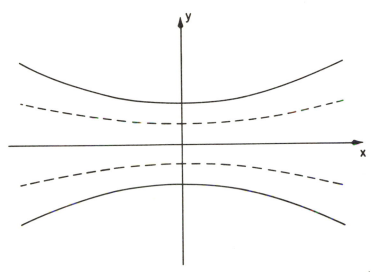

Fig. 2.10 Parabolic shape of a laminar beam under the influence of its own space charge.

If y is chosen at the edge of the beam, then necessarily the integral on the right-hand side of Eq. (2.36) remains constant in time. This is because the current j carried by the beam is, for a ribbon of unit width,

$$j = 2h\dot{x} \int_0^y dy'\, \rho(y') \tag{2.37}$$

with $\dot{x}$ the velocity of the electrons as they move along the central path. The factor of 2 enters because the width of the beam is $2y$.

Thus, the radial acceleration on an electron at the edge of the beam is constant in time, and is given by

$$\ddot{y} = 2\pi e j/mh\dot{x}, \tag{2.38}$$

and the trajectory of the electron on the outer edge of the beam is then

$$y = (\pi e j/mh)(x^2/\dot{x}^3) + y_0, \tag{2.39}$$

where y_0 is the half-width of the ribbon at the entrance slit. The expression in Eq. (2.39) is an equation for the shape of the ribbon of electrons, as it propagates through the monochromator. If, furthermore, the charge density $\rho(y)$ is independent of y, Eq. (2.39) describes the trajectory at any point within the beam, provided the total current j is replaced by the current $j(y) = 2h\dot{x}\rho y$ bounded by the trajectory of concern. If E is the energy of the electrons, then Eq. (2.39) may be arranged to read

$$y = (j/4kE^{3/2})(x^2/h) + y_0, \tag{2.40}$$

where the constant $k = (8\pi^2 e^2 m)^{-1/2} = 1.04 \times 10^{-5}$ A/(eV)$^{3/2}$.

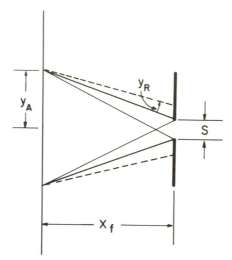

Fig. 2.11   Model for the increased image size at the entrance aperture of a monochromating device due to space charge.

2.4.2   Space-Charge Effects and the Feed Beam

The corrections to the electron trajectories from the space-charge effects just discussed already allow us to make rough estimates of the maximum current that can be fed into a cylindrical deflector. In Fig. 2.11, we show a model of image formation on the entrance slit. Image formation within the plane of deflection is depicted.

The aperture of the lens and the slit width roughly define the boundary of the electron beam; we shall assume the potential to be constant along the path. Unlike the condition just considered, the bundle is not laminar. Nevertheless, we can use the expression derived in the preceding section to estimate roughly the effect of the space-charge fields on the beam, as long as the convergence angle is small. The additional broadening of the bundle at the entrance slit then adds roughly

$$y_R \approx x_f^2/4hkE^{3/2} \qquad (2.41)$$

to the width of the beam.

A consequence of this broadening is that the image of the cathode at the slit position will become larger, and only a fraction of the current can now pass the entrance slit. If j_o is the current which passes through before the space-charge effect is included, and j_i afterward, then if the entrance beam would completely fill the entrance slit in the absence of the space-charge effect, we have

$$j_i = j_o[s/(s + 2y_R)]. \qquad (2.42)$$

In the limit $y_R \gg s$, where space-charge effects dominate the ability to feed current into the monochromator, we have

$$j_i = 2hkE^{3/2}(s/x_f^2) \qquad (2.43)$$

or

$$j_i = 2hkE^{3/2}(\alpha_i^2 s/y_A^2) \qquad (2.44)$$

for a finite lens aperture, with y_A illustrated in Fig. 2.11. This result is an estimate of an upper limit to the input current which is independent of the brightness of the cathode. When $2y_R \gg s$, an increase in the current density j_o simply leads to an linear increase in y_R, so a smaller fraction of the electrons get through the entrance slit as the cathode brightness is increased. If one considers the case of laminar flow between two parallel plates, where the electrons are accelerated from zero kinetic energy to a fixed energy E, then a result similar to Eq. (2.44) is obtained, but with a prefactor smaller by a factor of 2/9.

The pass current estimated from Eq. (2.44) is, at low kinetic energies, much smaller than the theoretical limit expected from the maximum emissity of the cathode. If one assumes the cathode is imaged onto the image slit with lenses of circular shape, the current density in the slit, j_i/hs, is related to the current density i_o emitted by the cathode, in the absence of space charge effects,

$$j_i/hs = i_o/M^2, \qquad (2.45)$$

where M is the linear magnification factor in the image formation. With use of Abbe's sine law, this becomes Langmuir's equation

$$j_i = i_o hs[(E_0 + k_B T)/k_B T] \sin^2 \alpha. \qquad (2.46)$$

With typical emission current densities of roughly 1 A/cm^2, and energies E_0 the order of 1 eV, the current j_i calculated from Eq. (2.46) is several orders of magnitude larger than that allowed by the space-charge limit established in Eq. (2.44). Cathodes used for feeding high-resolution spectrometers are therefore operated far below their maximum brightness, simply because space-charge effects limit the current that can be passed through the entrance slit, and there is no gain achieved by operating the cathode at high brightness levels.

2.4.3 Anomalous Energy Broadening (Boersch Effect)

In electron beams of high density, the energy distribution is broader than the initial Maxwellian distribution emitted by the cathode. This effect was first described by Boersch [7] and is therefore frequently referred to as the

Boersch effect. In previous publications, it has been suggested that the Boersch effect is responsible for the limited current one can obtain from monochromating devices [8].

The physical basis for the Boersch effect is the following: Electrons emitted from a thermal source at temperature T have the Maxwellian distribution

$$f(E)\,dE \sim E^{1/2} \exp[-E/k_B T]\,dE, \tag{2.47}$$

where $f(E)\,dE$ is the number of electrons emitted by the cathode with energy between E and $E + dE$.

In an accelerating potential V_1, all electrons are accelerated at the same rate, and increase their energy by the amount $E_1 = eV_1$. The distribution remains unchanged in shape, save for a uniform shift in energy. For $E > E_1$, we then have

$$f(E)\,dE \sim (E - E_1)^{1/2} \exp[-(E - E_1)/k_B T]\,dE, \qquad E > E_1. \tag{2.48}$$

If $eV_1 \gg k_B T$, the velocity distribution becomes narrowly peaked about the value $v_1 = (2eV_1/m)^{1/2}$. We have, if $v_\parallel$ is the velocity parallel to the direction of acceleration,

$$f(v_\parallel)\,dv_\parallel \sim \frac{1}{(2m)^{1/2} v_1} (v_\parallel^2 - v_1^2)^{1/2} \exp\left(-\frac{m}{2k_B T}(v_\parallel^2 - v_1^2)\right),$$

$$v_\parallel > v_\perp. \tag{2.49}$$

With $v_\parallel - v_1 = \Delta v \ll v_1$, this becomes

$$f(v_\parallel)\,dv_\parallel \sim \left(\frac{\Delta v}{mv_1}\right)^{1/2} \exp\left(-\frac{mv_1}{k_B T}\Delta v\right). \tag{2.50}$$

The width of this distribution is proportional to v_1^{-1}. On the other hand, the width of the velocity distribution perpendicular to the direction of acceleration remains unchanged. Any electron–electron scattering that occurs within the beam will tend to equalize the width of the velocity distributions parallel and perpendicular to the beam axis, since perpendicular components of momentum may be partially converted to parallel components (note, however, that electron–electron collisions cannot change the average drift velocity, since they conserve each Cartesian component of *total* momentum). A broadening of the velocity distribution parallel to the beam corresponds to a broadening of the energy distribution, which was unchanged by the acceleration process itself.

These considerations show that the magnitude of the effect depends necessarily on the density of the beam, its distance of travel, and the magnitude of the acceleration it has experienced. A detailed discussion of the effect for various beam geometries was recently given by Knauer [9]. For a cylindrical beam, the energy broadening per unit path length l was calculated to be

$$\frac{d(\overline{\Delta E^2})}{dl} = 2\pi^{3/2}e^3\left(\frac{m}{k_B T}\right)^{1/2} j_o \ln\left[\frac{k_B T}{e^2}\left(\frac{2e^2 E_1}{mj_o^2}\right)^{1/6}\right], \qquad (2.51)$$

or

$$\frac{d(\overline{\Delta E^2})}{dl} = \frac{3.74 j_o}{T^{1/2}} \ln\left[0.126T\left(\frac{E_1^{1/2}}{j_o}\right)^{1/3}\right], \qquad (2.52)$$

where in Eq. (2.52), the temperature is in °K, j_o is in A/cm^2, E is in eV, and $d(\overline{\Delta E^2})/dl$ is then given in eV/cm.

It is easily shown that the additional broadening provided by electron–electron collisions amounts to only a few percent of the thermal energy spread of the cathode, for input current densities used at the low pass energies necessary in high-resolution spectroscopy. Therefore, in contrast to the viewpoint sometimes offered, the deterioration of the energy distribution in the input beam via the Boersch effect is not the reason for the limitation on monochromator current.

The Boersch effect should be considered, however, for the problem of transport of a monochromatic beam over longer distances. The additional broadening provided by it may be calculated from Eq. (2.52). The kinetic energy $k_B T$ is to be replaced by the kinetic energy perpendicular to the direction of acceleration, in the case where the electron energy distribution in this direction is nonthermal,

$$k_B T \to \tfrac{1}{2}mv_\perp^2 = \tfrac{1}{2}mv_\parallel^2 \alpha^2, \qquad (2.53)$$

or

$$T \cong (\alpha_i^2/2k_B)E_0.$$

When the monochromating device is used with an angular aperture α_m [Eq. (2.19)], then $\alpha_i^2 E_0 \cong \Delta E$ [Eqs. (2.15) and (2.19)]. The additional broadening ΔE_z caused by transporting the monochromatic beam is then

$$\overline{\Delta E_z^2} = 4.85 \times 10^{-2} \frac{j_o l}{(\Delta E)^{1/2}} \ln\left[732\,\Delta E\left(\frac{E_1^{1/2}}{j_o}\right)^{1/3}\right], \qquad (2.54)$$

where j_o is in A/cm^2, l in cm, ΔE and E_1 are in eV, and ΔE_z^2 is then given in (electron volts)^2.

The broadening is nearly independent of E_1 and l, when considered a function of the combination $j_o l$, the product of the current density and the length of the beam (Fig. 2.12). While the Boersch effect is small for typical current densities that lead to 5-meV resolution, it may become considerable when small apertures or extended path lengths are employed.

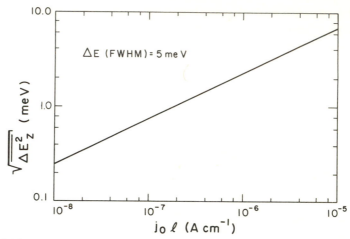

Fig. 2.12   Additional broadening of an accelerated beam with an initial width $\Delta E = 5$ meV as a function of the current density times pass length. The curve is nearly independent of the acceleration energy E_1 as long as $E_1 \gg E_0$.

2.5   SPACE-CHARGE EFFECTS ON THE CYLINDRICAL DEFLECTOR

2.5.1   Outline of the Problem

If, in the discussion of space-charge effects in Section 2.4.1, the charge density $\rho(y)$ is uniform, then, as remarked at the end of this section, all electron trajectories are parabolic. If we consider one such trajectory, then we refer to its image through the central path as the conjugate trajectory. In the limit of laminar flow, the space charge contained between a pair of conjugate trajectories is constant, and it was this consideration that led us to the parabolic form for the electron paths. In our earlier discussion, which estimated image distortions produced by space-charge effects, we considered an idealized monochromatic beam. In energy-dispersive systems, however, the broadening of the beam produced by energy dispersion becomes significant, and can no longer be neglected. In fact, we may expect the influence of space charge to be most significant in the region immediately after the entrance slit. For electrons with energy near the pass energy E_0, the space-charge density decays rapidly as more and more electrons stray from the the initial bundle as a result of energy dispersion. This allows the energy dispersive system to handle more current than a simple lens. On the other hand, we shall find that the presence of space charge reduces the transmission. The maximum monochromatic current is ultimately determined as an interplay between the two effects.

Previous treatments of the maximum current carried by monochromators have ignored the role of energy dispersion in limiting the maximum input current. Such calculations notoriously overestimate the maximum achievable monochromatic current and, what is worse, fail to provide insight into the relevant parameters that in the end control the available current in real devices. In this chapter, we now carry the problem one step further to take into account the influence of space-charge effects on the transmission. We do this by a method that treats the space-charge effect as a first-order perturbation; by this term, we mean that for a given pair of conjugate trajectories of electrons with the pass energy E_0, we calculate the effective space charge through use of the trajectory equations without space charge. In a second step, we study the influence of that space charge on the trajectory equations, and this finally allows us to calculate the transmission as a function of the input current. We shall see that the result compares favorably with the experimental data. The calculation will be carried out for the cylindrical condenser, but the basic scheme is easily carried through for other electrostatic deflector configurations.

2.5.2   The Decay of Space Charge Enclosed by Conjugate Trajectories

The extension of the space-charge zone will be found to be only a fraction of the total pass length in the cylindrical condenser. We may therefore use an expansion of the trajectory equations [Eq. (2.6)] for small pass length,

$$y = y_1 + (\delta E/2E_0 r)x^2 + \alpha x, \qquad (2.55)$$

so all trajectories are parabolic at this level of approximation. The intersection of such a parabola for a trajectory with energy deviation δE with a beam of pass energy E_0 (Fig. 2.13a) depends on the starting points y_1 and the angle α. For very small extensions of the space-charge zone, the dependence on α may be neglected, while for large extensions the starting position y_1 becomes irrelevant. We shall find favorable operating conditions of the spectrometer to be such that the second case is the appropriate model. We therefore may consider a beam with an infinitesimally small slit width, and angular aperture α_i (Fig. 2.13b). Trajectories for electrons with the pass energy are then straight lines, while electrons with an energy deviation δE are again parabolas intersecting any given ray of electrons with pass energy E_0 and an angle α at some distance x_n from the slit. We assume that the energy distribution of incoming electrons is symmetric around E_0, and as a result the electron charge distribution is necessarily symmetric around the initial path. Just as in the case of the laminar beam, the relevant space charge

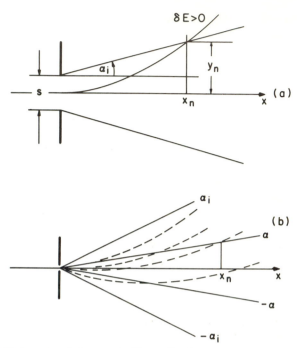

Fig. 2.13 Trajectories of electrons inside a cylindrical deflector with the pass energy E_0 and the energy $E = E_0 + \delta E$. For any given pair of trajectories with $\pm\alpha$ the effective space charge is decaying with the path length as a result of the energy dispersion. For small slits s, shown in (b), the extension of the space-charge zone is independent of s.

for electrons with energy $E = E_0$ and an angle α is that contained within the two conjugate beams at $+\alpha$, and $-\alpha$, while the forces from space charge outside this central section vanish. The difference between the present case, and our earlier laminar flow example, is that electrons with energy deviation δE are leaving or moving into the central sector, depending on their energy deviation δE and their initial angle α.

We now assume a uniform distribution of initial angles α_n, so that for any given energy deviation δE, a set of trajectories $n = 0, 1, 2, \ldots, N$, is characterized by the angle α_n, where

$$\alpha_n = -\alpha_i + 2\alpha_i(n/N). \tag{2.56}$$

These trajectories intersect the trajectory with $\delta E = 0$ and the angle $+\alpha$, and thus leave the central section at a pass length x_n given by

$$\alpha x_n = (\delta E/2E_0 r)x_n^2 - \alpha_i x_n + 2\alpha_i(n/N)x_n. \tag{2.57}$$

From this, we calculate dn/dx_n, the number of trajectories which leave the central sector per unit path length,

$$dn/dx_n = -(\delta E/4E_0 r)(N/\alpha_i). \tag{2.58}$$

The same number per unit path length is also entering the central sector, however, until the trajectory with $\alpha = -\alpha_i$ ($n = 0$) has entered. Therefore, the fraction of the total current contained in the central sector remains constant (this equals α/α_i) up to the path length

$$x = 2[(\alpha_i - \alpha)/\delta E]E_0 r, \tag{2.59}$$

and the fraction of the current contained in the central sector decays linearly from that point on, until the last trajectory has left the central sector at

$$x = 2[(\alpha_i + \alpha)/\delta E]E_0 r. \tag{2.60}$$

Because of the linear decay of the space charge between these two path lengths, we may replace the values by their average without great error, and assume the space charge vanishes at path lengths greater than

$$x = 2(\alpha_i E_0 r/\delta E). \tag{2.61}$$

Now we must fold in the influence of the energy distribution of the incoming electrons. For simplicity, we fold in a rectangular distribution between $-(\Delta E_K/2) \le \delta E \le +(\Delta E_K/2)$, where ΔE_K is identified as the full width at half-maximum of a real distribution entering the monochromator. The fraction of electrons that is still within the central sector at a path length x remains equal to the initial value for x smaller than x_0, where

$$x_0 = 4\alpha_i E_0 r/\Delta E_K, \tag{2.62}$$

and this fraction decays roughly as x_0/x for $x > x_0$.

Furthermore, we have to take account of the divergence of the beam in the direction perpendicular to the deflection plane. Through use of focusing conditions such as those given in Fig. 2.7a, we can finally write the fraction of the input current contained within the central sector at path length x as

$$f(x) = f_0 \frac{d_K}{d_K + x} \begin{cases} 1, & x < x_0, \\ x_0/x, & x > x_0, \end{cases} \tag{2.63}$$

where $f_0 = \alpha/\alpha_i$.

This result allows us to calculate the additional force due to the space charge, and therefore a first-order correction to the trajectory equation. In order to make the calculation more convenient, we replace the space charge expressed in Eq. (2.63) by one that remains constant at the initial value up to a path length x_s defined by

$$x_s = \int_0^\infty \frac{f(x)}{f_0}\, dx \tag{2.64}$$

and by zero for $x > x_s$. This simplified scheme still contains all relevant parameters, but on the other hand makes the correction to the trajectories very easy to calculate. From Eq. (2.63), we have

$$x_s = \int_0^{x_0} \frac{d_K}{d_K + x}\, dx + \int_{x_0}^\infty \frac{d_x x_0}{x(d_K + x)}\, dx \tag{2.65}$$

or

$$x_s = d_K \ln\left(1 + \frac{4E_0 r \alpha_i}{d_K \Delta E_K}\right) + \frac{4E_0 r \alpha_i}{\Delta E_K} \ln\left(1 + \frac{d_K \Delta E_K}{4E_0 r \alpha_i}\right). \tag{2.66}$$

At this point, it proves most instructive to study the extension of the space-charge zone as a function of the parameters d_K and α_i. As seen from Fig. 2.14, the space-charge zone increases in length, as both d_K and α_i increase. Small-angle apertures are therefore preferable, in order to keep the distortion of the trajectories as small as possible. One must keep in mind, however, that at the beginning of this section we have assumed α_i to be

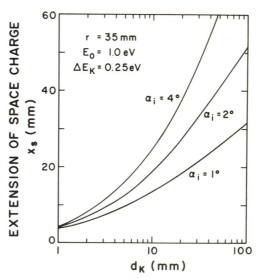

Fig. 2.14 Extension of the space-charge zone x_s as defined by Eq. (2.66) as a function of the distance of the cathode from the entrance slit d_K and the aperture angle α_i for $r = 35$ mm, $E_0 = 1$ eV and $\Delta E_K = 0.25$ eV.

larger than $s/2x_s$. For $\alpha_i \leq s/2x_s$, the space-charge zone is determined by the slit width s, and does no longer decrease with decreasing α_i. Too small a value for the slit width s will, in turn, lead us into the region where the cathode space-charge limit is realized [Eq. (2.44)]. The dependence of x_s on the characteristic distance d_K of the cathode from the entrance slit is also quite interesting. With Eq. (2.21), we have found the transmission to *increase* with d_K, and here we find the length of the space-charge zone also increases with d_K. As a result, the net effect of changes in d_K on the maximum monochromator current will be small. For this reason, it is not important which focusing is used for the cathode in the z direction (Fig. 2.7).

For small values of α_i, in the range $\alpha_i \leq s/2x_s$, where the preceding formulas do not apply, the extension of the space-charge zone is readily estimated. If, for simplicity, we consider only the boundary trajectories, the space charge for these trajectories decays as (note $f_0 = 1$ here)

$$f(x) = 1 - (\Delta E_K/4E_0 rs)x^2 \tag{2.67}$$

which after integration as in Eq. (2.64) leads to

$$x_s = d_K\left[\left(\frac{d_K}{x_1} - \frac{1}{2}\right) + \left(1 - \frac{d_K^2}{x_1^2}\right)\ln\left(1 + \frac{x_1}{d_K}\right)\right], \tag{2.68}$$

where, in Eq. (2.68),

$$x_1^2 = 8E_0 rs/\Delta E_K. \tag{2.69}$$

2.5.3   Trajectories in the Presence of Space Charge

The model just described, in which a constant space-charge density exists between the pair of conjugate trajectories with angles $\pm\alpha$, allows one to calculate the correction to the trajectories which results from the presence of the space charge. In the first approximation, one simply supplements the expression for the electron trajectory in the absence of space charge by the quadratic term in x (for $x \leq x_s$) which applies to the parabolic form for the trajectories discussed previously. Thus, for an electron with kinetic energy equal to the pass energy, we have

$$y \cong \alpha x + \frac{j_0}{4khE_0^{3/2}}\frac{\alpha}{\alpha_i}x^2, \qquad x \leq x_s, \tag{2.70}$$

and the trajectories are straight lines for $x > x_s$. Here the current contained between a given trajectory and its conjugate at $-\alpha$ is $j_0(\alpha/\alpha_i)$ [see Fig. 2.15].

Note from Fig. 2.15 that the space-charge-corrected trajectories have a different apparent angle α' and appear to be originating from a focal point shifted by the amount Δ, when $x \geq x_s$. The angle α' and the focal point

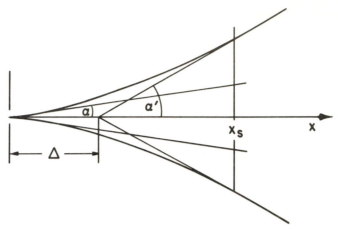

Fig. 2.15   The focal shift Δ and the effective angle α' for a trajectory resulting from space charges.

shift are given by

$$\alpha' = \alpha(1 + 2cx_s), \tag{2.71}$$

$$\Delta = cx_s^2/(1 + 2cx_s), \tag{2.72}$$

where

$$c = j_o/(4hkE_0^{3/2}\alpha_i). \tag{2.73}$$

As a consequence of the shift Δ in the focal point, the image of the entrance slit (the point of first-order focusing) is no longer on the exit slit, but shifted by the same amount Δ farther along the path length. This obviously must have a substantial influence on the transmission as soon as

$$\Delta \alpha_i' = cx_s^2 \geq \tfrac{1}{2}s. \tag{2.74}$$

Correction of the spectrometer for the space-charge-induced focal shift by making the total angle of deflection larger is, while possible, not recommendable since the necessary correction is current dependent. Therefore, a spectrometer "optimized" by increasing the pass length would show lower transmission, and what is worse, lower resolution when operated with low input current. Instead, we shall find another mode of operation more favorable: the deviation of the trajectories from the center path at the exit slit position is

$$y = \alpha cx_s^2 - \tfrac{4}{3}\alpha^2 r(1 + 2cx_s)^2. \tag{2.75}$$

For $\alpha > 0$, this function has a maximum at an angle

$$\alpha_M = \frac{3}{4} \frac{x_s}{r} \frac{c x_s}{(1 + 2 c x_s)^2},$$ (2.76)

and at this angle, the value of y is

$$y_M = \frac{3}{16} \frac{x_s^2}{r} \frac{c^2 x_s^2}{(1 + 2 c x_s)^2}.$$ (2.77)

Therefore, if one shifts the angular distribution of the incoming beam to be centered around α_M, instead of around $\alpha = 0$, first-order focusing ($dy/d\alpha = 0$) is retained. Such a shift is easily accomplished when the electron optics of the cathode system has one or two divided lenses. The particular virtue of this scheme is that the proper angle for any given current is automatically found by optimizing the monochromatic current for a fixed cathode temperature.

The shift in α is accompanied by a shift in y [Eq. (2.77)], which is equivalent to a somewhat lower, current-dependent pass energy $E_0(j)$:

$$E_0(j) = E_0 \left(1 - \frac{3}{16} \frac{x_s^2}{r^2} \frac{c^2 x_s^2}{(1 + 2 c x_s)^2} \right).$$ (2.78)

This reduction in E_0 is rather small. For high currents, and a typical value of x_s^2/r^2 of 0.2, the shift is only about 1% of E_0, and therefore has no influence on the transmission characteristics of the monochromator. The shift is, nonetheless, quite noticeable, and in fact is observed always when loss spectra are recorded with high- and low-input currents. The observed shift may therefore be used to estimate the extension x_s of the space-charge zone directly from experiments.

2.5.4   The Influence of Space Charge on Transmission

While the asymmetric operation of the cylindrical deflector under space-charge conditions saves first-order focusing for any input current, the second-order aberrations also increase with the input current. The second-order term in α [Eq. (2.75)] is of the same form as without space-charge effects, so the influence of the input current appears as an additional factor that controls this term. We can therefore directly employ the expressions for the transmission as derived in Section 2.3.2 and correct for the space-charge effects. Thus, we have

$$T(\alpha) = 1 - \frac{4}{3} \frac{r}{s} \alpha^2 (1 + 2 c x_s)^2$$ (2.79)

and

$$
T_{\alpha_i} = \begin{cases}
1 - \dfrac{4}{9}\dfrac{r}{s}(1 + 2cx_s)^2\alpha_i^2, & \alpha_i < \dfrac{1}{1 + 2cx_s}\left(\dfrac{3s}{4r}\right)^{1/2}, \\[3mm]
\dfrac{2}{3\alpha_i}\left(\dfrac{3s}{4r}\right)\dfrac{1}{1 + 2cx_s}, & \alpha_i > \dfrac{1}{1 + 2cx_s}\left(\dfrac{3s}{4r}\right)^{1/2}.
\end{cases} \tag{2.80}
$$

Thus, the second-order term in α limits the angle accepted by the spectrometer more and more with increasing current. Another way of saying this is that for any given input current, the transmission may be enhanced by reducing the intended input angle α_i. The limit to this procedure is the available input current when α_i is small. With Eq. (2.80), we now have in hand quantitatively what was discussed in Section 2.1 from the conceptual view point (also see Fig. 2.2).

At this point, a comparison to experimental data would be useful. Measurements have been performed on a spectrometer characterized by the parameters $r = 35$ mm, $s = 0.15$ mm, $h = 3$ mm, and $d_k = 13$ mm. The measured half-width of the input energy distribution was $\Delta E_K = 0.45$ eV. The angular aperture α_i cannot be measured directly. An estimate of $\alpha_i \approx 2°$

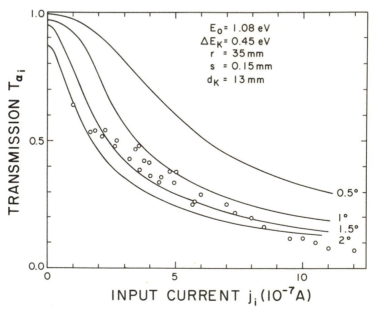

Fig. 2.16 Calculated transmission of a cylindrical deflector as a function of the input current j_i and the aperture angle α_i according to Eq. (2.80). The experimental data approximately fit the curve with $\alpha_i = 1.5°$.

was made from the measured resolution as a function of the pass energy E_0, using Eq. (2.15). In Fig. 2.16, we show the experimentally determined transmission T_{α_i} as a function of the input current. The theoretical curves [Eq. (2.80), with Eqs. (2.66) and (2.73)] are also plotted with α_i as a parameter. In view of the fact that no parameter is fitted, the agreement is quite reasonable. The model for the transmission under space-charge-limited conditions (in the monochromator) that we have developed in this chapter therefore provides a reliable scheme for the engineering of cylindrical deflectors. With some adjustments to be discussed briefly later, the model can be used for other deflector geometries.

2.5.5   The Monochromatic Current

As discussed in the introduction to this chapter, the available current at a given resolution is the most severe limitation on electron energy loss spectroscopy, and therefore the design parameters of cylindrical deflectors should be chosen to provide the maximum current. With the help of the considerations of the previous sections, we are now in a position to obtain an expression for the monochromatic current and its dependence on the radius r, the slit width s, the slit height h, the parameter d_K that describes the distance of the cathode from the entrance slit, and the energy width ΔE_K of the current supplied by the cathode. With Eq. (2.80) we obtain

$$j_0 = (\Delta E/\Delta E_K)j_i T_\alpha T_z, \qquad (2.81)$$

or to be more explicit,

$$j_0 = \frac{\Delta E}{\Delta E_K}\frac{d_K}{(d_K + (\pi/\sqrt{2})r)}\,j_i \begin{cases} 1 - \dfrac{4}{9}\dfrac{r}{s}a^2\alpha_i^2, & \alpha_i < \dfrac{1}{a}\left(\dfrac{3s}{4r}\right)^{1/2}, \\[2ex] \dfrac{1}{\alpha_i}\dfrac{2}{3}\left(\dfrac{3s}{4r}\right)\dfrac{1}{a}, & \alpha_i > \dfrac{1}{a}\left(\dfrac{3s}{4r}\right)^{1/2}, \end{cases} \qquad (2.82)$$

with $a = 1 + j_i x_s/(2hkE_0^{3/2}\alpha_i)$.

In discussing this result, we must keep in mind that the resolution is also affected by the space-charge effect. If we use the basic equation, Eq. (2.15), that describes the energy resolution of the device, we find upon replacing the input angular aperture α_i by the broadened angle $a\alpha_i$ due to the space charge,

$$\Delta E/E = (s/r) + 0.4a^2\alpha_i^2. \qquad (2.83)$$

This set of equations defines a maximum current j_0 for any pair α_i, ΔE. The maximum current may be found by numerical analysis of the equations. One finds, however, that the theoretical maximum for a reasonable set of parameters lies in the regime of large input current (large a) where the transmission T_α is low. This means that to operate the deflector with input current near the theoretical maximum, one also operates in a regime where the second term in the equation for the resolution [Eq. (2.83)] is large. This is not a desirable mode of operation for several reasons. First, to obtain the highest possible resolution in view of the problem of patchy work functions, one would like to keep the second term in Eq. (2.83) small. For the same reason an extra high-current load of the monochromator, which may result in charging, should be avoided. Furthermore, the angular spread of the incoming beam may deteriorate at higher input currents. In any case, the use of high input current and relatively low transmission brings one closer to the cathode limit as discussed in Section 2.4.2 without a substantial gain in the monochromatic current. Finally, the output current is a very broad function of the input current near the mathematical optimum. Therefore, by reducing the input current substantially, one looses only a small and technically unimportant amount of the monochromatic current.

Thus, in discussing the influence of various parameters on the monochromatic current, we use as a side condition the fact that the second term in Eq. (2.83) should be small and have the value

$$0.4a^2\alpha_i^2 = 0.3s/r \qquad (2.84)$$

which makes

$$a\alpha_i = (3s/4r)^{1/2}, \qquad (2.85)$$

i.e., equivalent to the critical angle above which beams cease to contribute any more to the transmitted image of the entrance slit (see Section 2.3.2). Obviously, in order to allow for any current, this means that α_i must be smaller than

$$\alpha_i < (3s/4r)^{1/2}. \qquad (2.86a)$$

With this condition and Eq. (2.80), the transmission T_α is equal to $\frac{2}{3}$ and the monochromatic current becomes

$$j_\sigma = 0.90 \frac{\Delta E^{5/2}}{\Delta E_K} \frac{d_K}{d_K + (\pi/\sqrt{2})r} \frac{hrk}{sx_s} \left[1 - \alpha_i \left(\frac{4r}{3s} \right)^{1/2} \right], \qquad (2.86b)$$

where the extension of the space zone x_s [Eq. (2.66)] may also be expressed as a function of ΔE using Eqs. (2.83) and (2.84). The constant k is that given after Eq. (2.40).

In Fig. 2.17, the typical behavior of the monochromatic current as a function of resolution ΔE is plotted for a set of parameters together with experimental values. The monochromatic current is approximately pro-

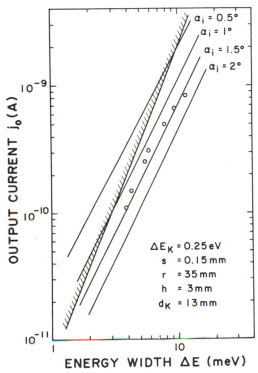

Fig. 2.17 Monochromatic current of a cylindrical deflector as a function of the energy width (FWHM) ΔE for various aperture angles of the feed beam. The current is approximately proportional to the second power of ΔE. The hatched line indicates the limit where the feed beam of the monochromator becomes space-charge saturated. The circles are experimental data.

portional to $(\Delta E)^2$. The hatched line indicates the estimated cathode limit using Eq. (2.44) and assuming $x_f \sim d_K/2$. We remember, however, that Eq. (2.44) was only an estimate, not quite as well founded as the analysis of the cylindrical deflector.

2.5.6   The Optimum Cylindrical Deflector

By using the expression for the monochromatic current, we now calculate the current as the function of the basic parameters, and discuss the optimum practical design. We first note that the monochromatic current is left invariant by an overall scaling of the dimensions. We therefore scrutinize the effect of the individual parameters by starting from the following reference frame: $\Delta E = 0.005$ eV, $r = 35$ mm, $\Delta E_K = 0.25$ eV, $s = 0.15$ mm, $h = 3$ mm, and $d_K = 13$ mm.

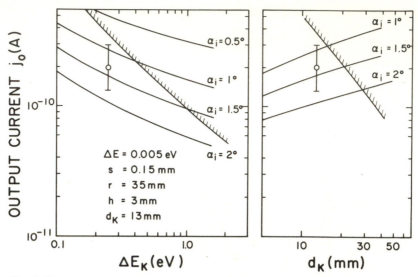

Fig. 2.18 Monochromatic current as a function of the energy width of the feed beam and the cathode distance. The hatched line is the space-charge saturation for the feed beam.

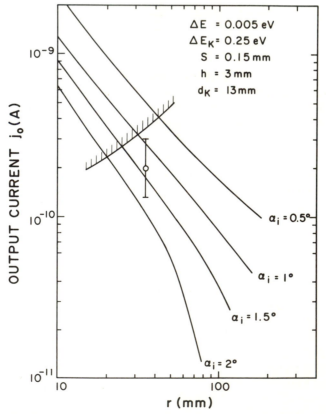

Fig. 2.19 Monochromatic current as a function of the cylinder mean radius *r*.

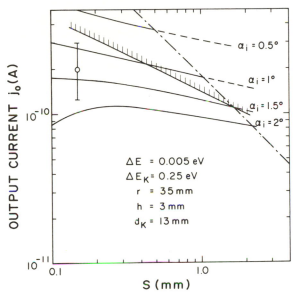

Fig. 2.20 Monochromatic current versus the slit width s. For large s the extension of the space-charge zone is no longer determined by α_i which sets another upper bound to the current (dash–dotted line).

The resolution of 0.005 eV is a typical value in most applications. The half-width of the input energy distribution is smaller than the Maxwellian distribution of a cathode. By tuning a spectrometer to the maximum output current for a certain cathode temperature, a relatively sharp angular distribution of the feed beam is required. The chromatic aberration of the lens system between cathode and the entrance slit then typically brings about a narrower energy distribution of the feed beam, which is a favorable effect. The monochromatic current as a function of r, s, ΔE_K, and d_K each for a set of $\alpha_i = 0.5$, $1°$, $1.5°$, and $2°$ is shown in Figs. 2.18–2.20. In the derivation of the expression for the extension of the space-charge zone x_s, we had assumed $\alpha_i x_s \gg \frac{1}{2}s$. With smaller α_i the extension of the space-charge zone reduces [Eq. (2.66)]; it remains constant with a reduction of α_i below the limit $\alpha_i = s/2x_s$. The general increase in the output current with smaller α_i in Figs. 2.18–2.20 is therefore realistic only in so far as the extension of the space-charge zone according to Eq. (2.66) is larger than that found from Eq. (2.68). The space-charge extension according to Eq. (2.68) sets an upper bound on the increase of the monochromatic current with smaller α_i, in particular for large s (see Fig. 2.20). The estimated "cathode limit" is shown as a hatched line. The experimentally observed output current is also shown with an "error bar" to indicate the approximate extent to which data on the monochromatic current are reproducible.

Inspection of Figs. 2.18–2.20 indicates that the parameter reference frame we have chosen is relatively close to the optimum combination. The functional dependence of the current on ΔE_K, d_K, and s is relatively weak anyway. A reduction in r which might be promising is prevented by the cathode limit. Although the cathode limit was only an estimate, the experimental data indicate that the estimate is realistic. Also, reducing r and keeping the slit width constant requires the pass energy to be rather low [Eq. (2.15)]. This may have adverse effects on the smallest possible ΔE because of the problem of patchy work functions.

The monochromatic current increases linearly with the slit height h. There the limitation is more in the engineering and the required precision of parallel alignment of the slits. With $h = 3$ mm, the height may meet the average manufacturing performance. It is also on the safe side with respect to the danger of employing nonplanar beams to an extent that they might lower the resolution [Eq. (2.14)].

2.6   OTHER ELECTROSTATIC DEFLECTOR GEOMETRIES

2.6.1   The Spherical Deflector

The spherical deflector has frequently been used as the energy-dispersive element in electron spectrometers. A detailed description has been given by Kuyatt and Simpson [8]. By using the principles of Section 2.5, one may also estimate the monochromatic current of the spherical deflector with curved slits. The basic equation for the resolution (Table 2.1) is approximately

$$\Delta E/E = (s/2r) + \tfrac{1}{2}\alpha^2, \tag{2.87}$$

which leads to the transmission function T_α,

$$T_\alpha = 1 - \tfrac{2}{3}(r/s)\alpha_i^2 a^2, \qquad \alpha_i a \leq (s/2r)^{1/2}. \tag{2.88}$$

Table 2.1 also shows that there is no loss in transmission for beams with nonzero angle with respect to the deflecting plane which is an advantage compared to the cylindrical deflector. The energy dispersion in the vicinity of the entrance slit is the same as with the cylindrical deflector [Eq. (2.55)]. The extension of the space-charge zone is then

$$x_s = x_0 + \int_{x_0}^{\pi r} \frac{x_0}{x}\,dx = x_0\left(1 + \ln\frac{\pi r}{x_0}\right) \quad \text{with} \quad x_0 = \frac{4\alpha_i E_0 r}{\Delta E_K} \tag{2.89}$$

for $x_s\alpha \gg \tfrac{1}{2}s$, or

$$x_s = \tfrac{2}{3}(8E_0 rs/\Delta E_K)^{1/2} \tag{2.90}$$

for $x_s\alpha \ll \frac{1}{2}s$. Using the same criterion for the maximum acceptable angle as in Section 2.5.6, we obtain for the resolution of the spherical deflector

$$\frac{\Delta E}{E} = \tfrac{3}{4}(s/r), \tag{2.91}$$

and for the monochromatic current in the two limits $x_s\alpha \gg \frac{1}{2}s$ and $x_s\alpha \ll \frac{1}{2}s$, respectively,

$$j_o = \left(\frac{4}{3}\right)^{5/2} \frac{1}{\sqrt{2}} \frac{\Delta E^{5/2}}{\Delta E_K} \frac{rhk}{sx_s}\left(1 - \alpha_i\left(\frac{2r}{s}\right)^{1/2}\right), \tag{2.92}$$

$$j_o = \left(\frac{4}{3}\right)^{5/2} \frac{1}{\sqrt{2}} \frac{\Delta E^{5/2}}{\Delta E_K} \frac{rhk}{sx_s}, \tag{2.93}$$

with the appropriate expression for x_s. In both cases the condition $\Delta E \ll \Delta E_k$ must be fulfilled in order to have $x_s \ll \pi r$, which is a prerequisite of the model. When the same parameters as those used in Section 2.65 are inserted into Eq. (2.92), one calculates a monochromatic current for the spherical deflector approximately three times larger than that found earlier for the cylindrical deflector. This result is not unexpected, since the spherical deflector provides the same resolution at a higher pass energy. On the other hand, the use of the same parameter set as used for the cylindrical deflector may be somewhat questionable, since the spherical deflector requires the proper imaging of a curved slit onto the analyzer. This may be more difficult to achieve, compared to imaging a rectangular slit. When the slit height is reduced, so that rectangular slits can be used instead of curved slits, the monochromatic currents of the two devices become about equal.

In practical designs, round apertures have been used mostly for the spherical deflector. We have assumed the space charge to be constant up to a certain path length x_s, which was determined by the energy dispersion. The distortion of the trajectories due to space charge, in the case of a beam of round cross section is also approximately represented by our model. In the limit of $x_s\alpha \ll \frac{1}{2}s$, the remarkably simple formula

$$j_o = (\tfrac{8}{9})k[(\Delta E)^2/(\Delta E_K)^{1/2}], \qquad \Delta E \ll \Delta E_K \tag{2.94}$$

is obtained. This formula should represent the monochromatic current of those spherical deflectors which are used in the "virtual aperture" mode. Here the slit or apertures effective for the monochromator are images of real apertures at a higher potential. This design [8] was proposed with the intention of avoiding secondary electrons and charging. In the process of optimizing the transmission of the spherical deflector, the image of the apertures will automatically be made in such a way as to have the angular aperture very small, and correspondingly the image size of the aperture

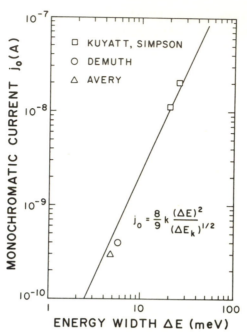

Fig. 2.21 The universal second-power law for the monochromatic current of a spherical deflector with circular apertures and a small aperture angle α_i. In this limit, the monochromatic current becomes totally independent of all design parameters. Experimental data are from three different sources: Kuyatt and Simpson [8], Demuth [10], and Avery [11], used with permission.

relatively large [Eq. (2.1)]. This is because larger α elongate the space-charge zone with adverse effects on the transmission, while the aperture size s cancels out in Eqs. (2.92) and (2.93) ($h = s$). The monochromatic current for the virtual circular aperture spectrometer therefore becomes independent of all design parameters. A comparison of Eq. (2.94) with some experimental data from different sources is shown in Fig. 2.21.

2.6.2 Mirror Deflectors

The virtue of mirror deflectors, the $30°$ plane mirror and the $42°$ cylindrical mirror, is the second-order focusing in α which allows these devices to accept substantially larger angles α. Despite this fact, both mirrors are not very effective as monochromators in terms of producing high currents. As seen from Eqs. (2.86) and (2.92), the monochromatic current is inversely proportional to the extension of the space-charge zone within the deflector. For small energy width ΔE, the space-charge zone in the cylindrical and spherical deflector can become quite small. For the mirror deflectors there is no energy dispersion on an appreciable fraction of the total path length (Fig. 2.3). Furthermore, after the beam enters the deflecting zone, the beam is already so broad that the dispersion does not effectively reduce the space charge for electrons having the pass energy. Space charge therefore affects the trajectories of electrons with the pass energy along almost the entire path length. Obviously this is not a favorable situation for the production of high monochromatic currents. In fact, monochromatic currents were found to be lower than for cylindrical or spherical deflectors by a factor of ~ 40 [12].

While mirror analyzers are not the best monochromators, their large acceptance angle can make the use of mirrors as analyzers beneficial, in particular when an efficient detection of inelastically scattered electrons with a broad angular distribution is desired. Thus the most effective energy-loss spectrometers may feature a combination of cylindrical deflector–plane mirror and spherical deflector–cylindrical mirror as monochromator and analyzer, respectively. So far, only symmetrical layouts have been realized.

2.7 BACKGROUND

In electron energy loss spectroscopy, the range of intensities between strong elastic and weak inelastic signals can be as high as 10^6. Therefore the resolution measured as the full width at half-maximum of the energy distribution is only one parameter to characterize the performance of the spectrometer. Quite frequently, weak loss features are of interest, such as those due to phonon excitations of the substrate in the regime of small energy losses. Then the tail of the energy distribution function is of more importance than the width at half-maximum. Such tails, which can extend quite far into the loss regime, originate from electrons passing through the analyzers with large angles α, as we have seen. These electrons may have been produced by the cathode emission system, but also by scattering from the slits. Machining the slits with sharp edges helps to keep this scattering small. Frequently, however, the slits act as unwanted lenses due to charging of the electrode surfaces. This charging is also likely to be rather inhomogeneous. The use of the "virtual slit" concept reduces these disturbing effects substantially. Virtual slits, i.e., images of real slits at a higher potential, are particularly important for circular apertures where the current load per surface area is higher than with elongated slits. As the adverse effects of charging become more severe at low pass energies, occasionally a trade-off of some resolution versus improved tails is beneficial.

Besides the tails of the energy distribution, additional structures, occasionally referred to as "ghost peaks," are observed in the loss spectrum even when the electrons are not scattered from a sample. These "ghost peaks" are produced by electrons scattered from the deflection plates. When they appear in the energy loss regime (rather than on the energy gain side) scattering is either from the inner electrode of the monochromator or the outer electrode of the analyzer. Electrons which scatter from the inner deflection plate of the monochromator originate from the low-energy part of the energy distribution fed into the monochromator. After being reflected from the inner deflection plate, a fraction of these electrons, nevertheless, may pass through the exit slit and appear as spurious energy loss structures. *Vice versa*, in observing the loss spectrum, the potential of the analyzer is raised in order to accelerate those electrons, having experienced a loss up to the appropriate pass energy. The elastically scattered electrons are then faster and eventually may hit the outer deflection plate and be reflected such as to pass through the exit slit and appear as a ghost peak. The "loss energy" where ghost peaks occur is approximately estimated, using the trajectory equations and assuming a mirror reflection from the deflecting plates midway between the entrance and the exit slit. For the cylindrical deflector, one obtains from Eq. (2.6)

$$\Delta E_{\text{loss}} = 2E_0 \left(\frac{r_0 - r}{r} \right) \tag{2.95}$$

with r_0 the radius of the outer plate of the analyzer, or

$$\Delta E_{\text{loss}} = 2E_0 \left(\frac{r - r_i}{r} \right) \tag{2.96}$$

with r_i the radius of the inner plate of the monochromator, respectively. In deriving the relations from Eq. (2.6), we have assumed α to be small, otherwise different solutions are possible in addition to those displayed in Eqs. 2.95 and 2.96 [13]. By using the basic resolution equation, we may derive the relation between the ghost peak position and the energy width for a cylindrical deflector,

$$\Delta E_{\text{loss}} \approx 1.5 \, \Delta E (r_0 - r)/s. \tag{2.97}$$

This relation shows that it is possible to design the spectrometer such that the ghost peak appears at a fairly high "energy loss" by using small slits. For the monochromator, this places the energy where the ghost peak occurs outside the energy distribution supplied by the cathode emission systems. For the analyzer, a high value of ΔE_{loss} is also preferred, since the electron optics between the sample and the analyzer will be tuned to optimize the focus for those electrons having experienced a real energy loss at the sample. The elastically scattered electrons which produce the ghost peak after being reflected by the outer deflection plate, will be defocused, and this reduces the number entering the analyzer as one scans through the loss spectrum.

Consequently the intensity of a ghost peak will be smaller, the larger the energy ΔE_{loss}, where it appears. A further reduction of the ghost peak can be achieved by an appropriate corrugation of the deflection plate [13] such as to avoid the possibility of mirror reflections from the plates. A third way to further reduce the intensity of ghost peaks exploits the fact that electrons which produce the ghost peaks have a different kinetic energy than electrons that have experienced an energy loss. One does this by putting a lens with strong chromatic aberration between the analyzer and the detector. This lens then acts as an additional analyzer with low resolution. By using all three means, the intensity of the ghost peaks can be reduced to roughly 10^{-6} of the intensity of the elastically scattered beam.

A complete removal of the background is achieved by employing two energy dispersion systems, one for the monochromator and one for the analyzer. Since the transmission functions of the two stages multiply, the tails of the energy distributions are also reduced substantially. Furthermore, the ultimate resolution may be higher since the second monochromator is fed by a relatively small input current, which reduces work function inhomogeneities insofar as they are caused by charging or electron-beam-assisted deposition of contaminants on the electrodes.

2.8 SOME FURTHER TECHNICAL ASPECTS

2.8.1 Magnetic Fields

The small deflecting fields in high-resolution spectrometers require magnetic shielding, although in general relatively moderate measures will suffice, since a variation of the electrostatic field can partly compensate for the influence of a homogeneous magnetic field. In order to estimate an upper bound on the residual magnetic field that can be tolerated, we consider the deviation of an electron of pass energy E_0 from its path by a magnetic flux B. This is

$$y = [e/2(2mE_0)^{1/2}]x^2B, \qquad (2.98)$$

with x the path length.

For the cylindrical deflector, this deviation should be smaller than the height of the slits in order to avoid heavy distortion of the trajectories within the cylindrical deflector. This limits the magnetic field to

$$B < (h/r^2)1.37\sqrt{E_0}, \qquad (2.99)$$

where h and r are in centimeters, E_0 is in electron volts, and B in gauss. With $r = 35$ mm, $h = 3$ mm, and $E_0 = 1$ eV, one calculates that the static field should be smaller than 30 mG.

More critical than static fields are ac magnetic fields. Here the deviation *y* must be definitely smaller than the slit width in order to avoid a reduction of resolution. We have here

$$B_{ac} \ll (s/r^2)1.37\sqrt{E_0},\qquad(2.100)$$

with the same units as in Eq. (2.99).

Finally, there is a requirement on the homogeneity of the magnetic field, in that the magnetic field effective for any two different trajectories of electrons having the pass energy E_0 must not differ by an amount that would let the image points of those trajectories differ by more than the slit width. Otherwise the resolution also would be spoiled. This leads to the following requirement on the homogenity of the field:

$$dB/dz \ll (s/wr^2)1.38\sqrt{E_0},\qquad(2.101)$$

where *w* is the maximum of any distance between the two trajectories, in centimeters, *s* and *r* are in centimeters, E_0 in electron volts, and dB/dz in gauss per centimeter. This relation leads to a homogenity requirement of about 1 mG/cm for typical dimensions of the spectrometer. Such homogenity is not difficult to achieve inside a magnetic shield. Care must be taken, however, to avoid the use of magnetic materials (most stainless steels included) in the construction of those spectrometer parts which are close to the beam.

2.8.2 Construction Materials

Most parts of electron energy loss spectrometers frequently have been machined from copper or copper–beryllium/silicon alloys. Molybdenum and titanium have also been used. Even totally nonmagnetic stainless steel, if available, is appropriate except for the slits, which are mostly made from molybdenum. The choice between these materials seems to be more or less a matter of convenience, and seems not to affect the spectrometer performance. An appropriate coating of all surfaces which face the electron beam in order to minimize the effect of charging and the inhomogeneity of the surface potentials is, however, necessary when high resolution is desired. The most effective coating, which is also easily accomplished, is made by dipping all parts into a suspension of colloidal graphite in isopropanol. The residues of isopropanol are readily removed by baking the parts at several hundred degrees centigrade in air. Surfaces coated by this procedure not only seem to have fairly homogeneous surface potentials, but the potentials also exhibit a remarkable long-term stability even after intermediate exposure to air. Electron spectrometers coated with graphite typically require only minor adjustments of the potentials of the order of a few millielectron volts after exposure to air, and subsequent bake-out of the vacuum system.

2.8.3 The Cathode System

As we have seen in our discussion of the transmission of the monochromator, the cathode system should supply an intense feed beam at a small voltage and with a small angular aperture of only a few degrees. The available feed current for the monochromator is limited by the space charge rather than by the emissivity of the cathode. Small angular apertures of the feed beam are achieved by either a small emission area of the cathode or a small emission angle. This may be seen from Abbe's sine law [Eq. (2.11)] which here assumes the form

$$k_B T A_k \sin^2 \theta_k = (E_0 + k_B T) A_s \sin \alpha \sin \beta, \qquad (2.102)$$

where A_k is the cathode area, A_s the area of the entrance slit, $k_B T$ the energy of the thermally emitted electrons, and θ_k, α, β the angular aperture for the cathode (circular symmetric) and the entrance slit, respectively. Because it is easier to reduce the cathode area than to limit the angle of emission accepted by the lens system, and also to avoid the generation of excess electrons which might cause a background noise problem, the cathode area should be kept small. Upon taking $\sin \beta = h/2d_k$ for the cylindrical deflector and allowing $\sin \theta_k$ to be 1, the necessary cathode area becomes

$$A_k = \frac{E_0 + kT}{kT} \frac{h^2 s}{2d_k} \sin \alpha, \qquad (2.103)$$

which for the set of parameters used in Section 2.5.5 and a tungsten cathode ($T = 2700°$K) becomes $A_k \approx 10^{-5}$ cm^2 for $E_0 = 1$ eV. Therefore, a small tungsten or LaB$_6$-tip cathode is most suitable. The LaB$_6$ cathode has the additional advantage of a smaller energy distribution.

The use of a small-area cathode and the formation of the beam in a single step has, besides being a relative-straightforward lens system, the advantage of placing the cathode relatively close to the entrance aperture of the monochromator. Magnetic fields which originate from the cathode supply current may be shielded by having a repeller around the cathode made from μ metal. The extra heating of the entrance slit keeps this slit, which has to take the highest current load, permanently at a higher temperature [14]. This reduces charging of the entrance slit appreciably. The reason is that charging of the surface is caused by the low conductance of deposits on the metal surfaces, and the conductance of such layers typically increases exponentially with the temperature.

REFERENCES

1. See, e.g., O. Klemperer, "Electron Optics." Cambridge Univ. Press, London and New York, 1971; A. B. El-Kareh and F. C. F. El-Kareh, "Electron Beams, Lenses, and Optics," Vols. 1 and 2. Academic Press, New York, 1970; P. Grivet, "Electron Optics." Pergamon, Oxford, 1965.
2. H. Boersch, J. Geiger, and W. Stickel, *Z. Phys.* **180**, 415 (1964).
3. D. Roy and J. D. Carette, *in* "Electron Spectroscopy for Surface Analysis" (H. Ibach, ed.) (Topics in Current Physics 4). Springer-Verlag, Berlin, Heidelberg, New York, 1977.
4. D. Roy and J. D. Carette, *Appl. Phys. Lett.* **16**, 413 (1970): *Canad. J. Phys.* **49**, 2138 (1971).
5. R. Herzog, *Z. Physik* **97**, 556 (1935); **41**, 18 (1940).
6. H. Wollnik and H. Ewald, *Nuclear Instr. Methods* **36**, 93 (1965).
7. H. Boersch, *Z. Phys.* **139**, 115 (1954).
8. C. E. Kuyatt and J. A. Simpson *Rev. Sci. Instr.* **38**, 103 (1967).
9. W. Knauer, *Optik* **54**, 211 (1979).
10. J. E. Demuth, private communication.
11. N. Avery, private communication.
12. S. Andersson, private communication.
13. H. Froitzheim, H. Ibach, and S. Lehwald, *Rev. Sci. Instr.* **46**, 1325 (1975).
14. H. Ibach, Auslegeschrift 2851 743, Deutsches Bundespatentamt.

CHAPTER 3

BASIC THEORY OF ELECTRON
ENERGY LOSS SPECTROSCOPY

3.1 GENERAL REMARKS

In Chapter 1, we presented a brief discussion of the interaction between electrons and the vibrational motion of atoms in the surface, or molecules adsorbed on it. As these entities vibrate, intense, small-angle scatterings are produced by their oscillating electric dipole moment. This dipolar lobe, when present, appears superimposed on a broad, angular distribution of electrons scattered inelastically by the impact mechanism. In this chapter, we discuss the techniques by which these contributions to the cross section may be calculated, along with application of the results to a number of specific cases.

In Chapter 1, we discussed dipole scattering from a qualitative point of view, along with scattering through large-angle deflections, where a fully microscopic description of the interaction of electron with the vibrating substrate is required. This last regime was described as the impact-scattering regime. As we shall illustrate by a simple example, a complete analysis of the scattering of electrons from a vibrating entity will incorporate both the dipolar and "impact" scattering contributions into a single expression for the cross section. In practice, it is extremely difficult to carry through a realistic calculation that does this, even for a small molecule in free space. Thus, the theorist prefers to develop distinctly different techniques for the two regimes. Since, as we saw in Chapter 1, the electrons inelastically scattered by the long-range dipole part of the oscillating potential set up by the vibrational motion suffer only small-angle deflections, to the experimentalist the "dipole lobe" often appears as a clear and distinct feature in the angular distribution of inelastically scattered electrons.

We can appreciate the principal elements of the theory from the following simple argument. Suppose we consider the scattering of an electron from a small molecule. Let the position $\mathbf{R}_l$ of each nucleus be written $\mathbf{R}_l = \mathbf{R}_l^{(0)} + \mathbf{u}_l$, where $\mathbf{R}_l^{(0)}$ is a vector to the equilibrium position of nucleus l, and $\mathbf{u}_l$ is the amplitude of the vibrational motion about its equilibrium position. Suppose the electron interacts with the molecule through a superposition of two-body potentials

$$V(\mathbf{x}) = \sum_l V_l(\mathbf{x} - \mathbf{R}_l). \tag{3.1}$$

If we consider the scattering of an electron with wave vector $\mathbf{k}^{(I)}$ into a final state $\mathbf{k}^{(S)}$, then in the first Born approximation (kinematical theory), the matrix element for scattering from the molecule is

$$V(\mathbf{Q}) = \int \frac{d^3x}{V} e^{+i\mathbf{Q} \cdot \mathbf{x}} V(\mathbf{x}),$$

where V is a basic quantization volume, and $\mathbf{Q} = \mathbf{k}^{(I)} - \mathbf{k}^{(S)}$ is the change in wave vector of the electron. From Eq. (3.1), we have

$$V(\mathbf{Q}) = \sum_l V_l(\mathbf{Q}) \exp(i\mathbf{Q} \cdot \mathbf{R}_l),$$

with $V_l(\mathbf{Q}) = \int d^3x \, V_l(\mathbf{x}) \exp(i\mathbf{Q} \cdot \mathbf{x})/V$. If by $V^{(1)}(\mathbf{Q})$ we denote the contribution to $V(\mathbf{Q})$ first order in the nuclear displacements $\mathbf{u}_l$, we have

$$V^{(1)}(\mathbf{Q}) = i \sum_l V_l(\mathbf{Q}) \exp(i\mathbf{Q} \cdot \mathbf{R}_l^{(0)})\mathbf{Q} \cdot \mathbf{u}_l. \tag{3.2}$$

If we consider very small angle scatterings, where $|\mathbf{Q}|d_0 \ll 1$ with d_0 a measure of the size of the molecule, then we may set the factors $\exp(i\mathbf{Q} \cdot \mathbf{R}_l^{(0)})$ to unity. Furthermore, if Z_l is the net charge which surrounds the nucleus located at $\mathbf{R}_l^{(0)}$, we have

$$\lim_{|\mathbf{Q}| \to 0} V_l(\mathbf{Q}) = 4\pi Z_l e^2/Q^2,$$

so we have

$$\lim_{|\mathbf{Q}| \to 0} V^{(1)}(\mathbf{Q}) = i4\pi(\mathbf{Q}/Q^2) \cdot \sum_l eZ_l\mathbf{u}_l,$$

or

$$\lim_{|\mathbf{Q}| \to 0} V^{(1)}(\mathbf{Q}) = i4\pi e\mathbf{Q} \cdot \mathbf{p}/Q^2,$$

where

$$\mathbf{p} = \sum_l eZ_l\mathbf{u}_l \tag{3.3}$$

is the oscillating portion of the electric dipole moment of the molecule.

From this argument, we can see that in the limit of very small wave vector transfers, the cross section for inelastic scattering of the electron from the vibrational motion of the molecule is controlled by only one parameter, the component $\mathbf{p}$ of the electric dipole moment which oscillates at the frequency ω_0 of the normal mode responsible for the scattering. As we shall see shortly, once we know the symmetry of a particular normal mode of an adsorbed molecule, or of the substrate from group theory, we may deduce the nonzero components of $\mathbf{p}$. The electron scattering cross section can then be cast entirely in terms of these components, which may be regarded as parameters to be deduced from the experimental data.

The point of the remarks above is that we can carry through with a discussion of dipole scattering without the need for a fully microscopic treatment of the interaction between the electron and the vibrating nuclei. In fact, we shall see that the cross section can be cast in a form considerably more general than the foregoing description suggests. Oscillating electric dipole moments in crystals are produced not only by the vibrational motion of nuclei, but also by other excitations such as fluctuations in the electronic degrees of freedom. Examples are fluctuations in the free carriers in metals or doped semiconductors, and interband transitions in insulators. The latter have an electric dipole moment given by $e\langle\psi_c|\mathbf{x}|\psi_v\rangle$, with $|\psi_c\rangle$ and $|\psi_v\rangle$ the wave functions for the valence and conduction bands, respectively, and the frequency of the virtual fluctuations always present is $(E_c - E_v)/\hbar$. In this chapter, we cast the dipole scattering formula in language sufficiently general that the result may be applied to inelastic scattering by fluctuations in the electronic degrees of freedom, as well as the vibrational modes.

From Eq. (3.2), we can also see that when $|\mathbf{Q}|d_0$ is not small, the calculation of the electron scattering cross section requires detailed and microscopic knowledge of the nature of the electron/molecule interaction, as well as the geometry. That is, even to carry through a calculation of the cross section within the simple kinematical description already outlined, we require the detailed form of $V_i(\mathbf{Q})$ as well as the equilibrium positions $\mathbf{R}_l^{(0)}$ of the nuclei. This is the impact scattering regime, where the cross section for inelastic scattering cannot by expressed simply in terms of a small number of phenomenological parameters.

It is useful to note that in the impact scattering regime, the angular and energy variation of the energy-loss cross section contains detailed structural information on the entity which produces the scattering. In electron energy loss spectroscopy, with electron energy in the range 1–100 eV, the surface is illuminated with waves that have a wavelength comparable to the internuclear separations. This is in contrast to the optical spectroscopies, which in all cases employ radiation with wavelength very long compared to the internuclear separation. Through electron energy loss studies in the impact scattering regime, one has the possibility, in principle, of obtaining detailed structural information from the loss cross sections, while this is not possible with optical methods.

The foregoing discussion outlines some key issues in the theory of electron energy loss from surfaces. The reader must keep the pedagogical nature of the presentation in mind. In any low-energy electron–surface scattering study, a kinematical description (use of the first Born approximation) is inadequate, and multiple scattering effects enter importantly. Any proper theory placed in contact with real data must recognize this. Also, we assumed that as the nucleus is displaced from its equilibrium position $\mathbf{R}_l^{(0)}$ to $\mathbf{R}_l$, the potential $V_i(\mathbf{x} - \mathbf{R}_l)$ remained unchanged in shape. Thus, we were lead to the particularly simple expression for the moment $\mathbf{p}$ given in Eq. (3.3). In fact, as the nuclei shift away from their equilibrium position, charge flows within the molecule, so that the potential of the distorted entity is not simply a superposition of potentials subjected to a rigid displacement. It is necessary to understand the influence of the nuclear displacement on the electron charge distribution in a microscopic fashion before one may calculate $\mathbf{p}$.

Despite the simplifications we have made, the conclusions we have reached remain unaltered within the framework of a complete analysis: in the dipole dominated regime, the cross section may be described accurately by a phenomenological theory, while in the impact scattering regime, a fully microscopic approach is required. A proper analysis would treat both regimes in a single stroke, but to both the theorist and the experimentalist, it is useful to treat each regime as distinctly identifiable. In this chapter, we first discuss the dipole scattering theory, then turn to a discussion of impact scattering.

3.2 SMALL-ANGLE INELASTIC SCATTERING BY DIPOLE FIELDS

We have seen that strong, small-angle scattering is produced by long-ranged dipolar fields generated by the molecular vibration, when the dipole selection rule allows these to be nonzero. As remarked in the previous section, dipolar scattering is produced not only by vibrational motions of adsorbed molecules and atoms in the crystal surface, but by any elementary excitation of the sample accompanied by a fluctuation in charge density. In this section, we summarize the theoretical description of this scattering. The expression for the cross section can be cast into a rather general form, and this can be applied to a diverse array of physical situations, including the description of scattering from dipole-active vibrations of adsorbed molecules.

We begin by considering a semi-infinite crystal, possibly with an adsorbed layer present. In the ground state, $\rho_0(\mathbf{x})$ is the charge density in the system. We include in $\rho_0(\mathbf{x})$ the contribution to the charge density from the electrons, and also from the nuclei. If the system is disturbed in some fashion, possibly by a thermal fluctuation, the charge density becomes $\rho_0(\mathbf{x}) + \rho_1(\mathbf{x}, t)$, where considerations of electrical neutrality require

$$\int d^3x\, \rho_1(\mathbf{x}, t) = 0, \tag{3.4}$$

that is, the fluctuation simply rearranges charge already present, so the system is always electrically neutral.

At the moment, we need say little about the precise nature of $\rho_1(\mathbf{x}, t)$. We can imagine it is nonzero by virtue of the vibrational motion of the nuclei of an adsorbed molecule, fluctuations in density of free carriers, or a virtual fluctuation in which an electron in a valence band of an insulator makes a transition into a conduction band, and back. At a later stage in the analysis, we shall see that the fluctuation–dissipation theorem allows us to relate the amplitude $\rho_1(\mathbf{x}, t)$ of the charge-density fluctuation back to the dielectric response function of the crystal.

We imagine the surface of the sample lies in the xy plane, with the sample in the half space $z < 0$ and vacuum above. An electron which approaches the surface from above then sees the electrostatic potential

$$\varphi(\mathbf{x}, t) = e \int_{z' < 0+} \frac{d^3x'\, \rho_1(\mathbf{x}', t)}{|\mathbf{x} - \mathbf{x}'|}, \tag{3.5}$$

where the integral in Eq. (3.5) extends over the sample, and e is the charge of the electron. In the case where scattering by bulk excitations is considered, it is important to include in the potential any charge density induced on the surface by the charge fluctuation inside the crystal. Thus, the integral in Eq. (3.5) is extended into the vacuum above the crystal a bit, to ensure such contributions are included fully. Note that the neutrality condition stated in Eq. (3.4) requires that at large distances from the surface, the leading contribution to $\varphi(\mathbf{x}, t)$ is dipolar in character.

It should be noted that in Eq. (3.5), retardation effects are ignored, and we assume that information about the charge fluctuation at $\mathbf{x}'$ is transmitted instantly to the point $\mathbf{x}$ where the electron is located. In all the applications we shall consider this approximation is fully justified. The criterion for its validity is the following: If we consider scattering by an excitation with frequency ω_0, and if the electron "sees" the excitation when it is a distance l_0 above the surface, then when the inequality $l_0 \ll c/\omega_0$ is satisfied with c the velocity of light, retardation effects can be ignored. In effect, information from the crystal is transmitted to the electron on a time scale very short compared to the period of the excitation, so the electron senses the instantaneous position of the charges in the medium. In Chapter 1, we saw $l_0 \cong Q_{\|}^{-1}$, where $\mathbf{Q}_{\|} = \mathbf{k}_{\|}^{(I)} - \mathbf{k}_{\|}^{(S)}$ is the wave vector transfer experienced by the electron, projected onto a plane parallel to the surface. Thus, the criterion may also be written $\omega_0/Q_{||} \ll c$. For the example in Chapter 1, where we consider scattering of an electron from a molecular vibration, with a loss of 150 meV, we have $\omega_0 = 2.29 \times 10^{14}$ sec^{-1}, and if we consider an electron with impact energy of 5 eV which is incident along the normal and suffers an angular deflection of $1°$, we have $Q_{||} = 2.01 \times 10^6$ cm^{-1}. Hence $\omega_0/Q_{||} = 1.14 \times 10^8$ cm/sec, which is more than two orders of magnitude smaller than the velocity of light $c = 3 \times 10^{10}$ cm/sec.

Through use of the identity

$$\frac{1}{|\mathbf{x} - \mathbf{x}'|} = 2\pi \int \frac{d^2Q_{||}}{Q_{||}} e^{i\mathbf{Q}_{||} \cdot \mathbf{x}_{||}} e^{-Q_{||}|z - z'|},$$

with the subscript $||$ used to denote a vector that lies in the xy plane, the potential seen by the electron outside the crystal may be written in the form

$$\varphi_1(\mathbf{x}, t) = 2\pi e \int \frac{d^2Q_{||}}{Q_{||}} e^{i\mathbf{Q}_{||} \cdot \mathbf{x}_{||}} e^{-Q_{||}z} \int_{z' < 0+} \rho_1(\mathbf{Q}_{||}z'; t) e^{+Q_{||}z'} dz' \quad (3.6)$$

where we have introduced

$$\rho_1(\mathbf{Q}_{||}z'; t) = \int d^2x_{||} e^{-i\mathbf{Q}_{||} \cdot \mathbf{x}_{||}} \rho_1(\mathbf{x}', t').$$

The next step in the formal development is to insert the potential in Eq. (3.6) into the Schrödinger equation, and use perturbation theory to calculate the scattering cross section. Since the details of this procedure have been given elsewhere [1], we shall omit the derivation. However, it is useful to comment on some important aspects of the problem of general interest.

When we consider an incident electron with wave vector $\mathbf{k}^{(I)}$, and calculate the probability that it scatters to a final state with wave vector $\mathbf{k}^{(S)}$, then this scattering event is produced by the contribution to the right-hand side of Eq. (3.6) with $\mathbf{Q}_{||} = \mathbf{k}_{||}^{(I)} - \mathbf{k}_{||}^{(S)}$. Components of wave vector parallel to the surface are conserved in the scattering process, but there is no conservation relation on the perpendicular component of the wave vector. If the electron loses energy $\hbar\omega$ in the scattering process, then it emerges with energy $E_S = E_I - \hbar\omega$. Thus, for a given value of $\mathbf{Q}_{||}$ and the energy transfer ω, the direction of the outgoing electron is determined uniquely.

If we consider the contribution to the potential in Eq. (3.6) with wave vector $\mathbf{Q}_{||}$ parallel to the surface, then we see that the potential extends into the vacuum a distance roughly equal to $Q_{||}^{-1}$. Precisely the same behavior was obtained for the dipole potential considered in Chapter 1; as remarked there, this behavior of the potential follows simply because it must obey Laplace's equation in the vacuum above the crystal. Any contribution to the potential which exhibits the spatial variation $\exp(i\mathbf{Q}_{||} \cdot \mathbf{x}_{||})$ in the directions parallel to the surface, no matter what its physical origin, must necessarily decay as $\exp(-Q_{\parallel}z)$ as we move away from the substrate. Thus, any fluctuation in charge density in or on the crystal, whether it is caused by the motion of adsorbate nuclei, or electronic fluctuations in the bulk, produces a long-ranged disturbance in the vacuum that gives rise to small-angle scatterings. We may determine the range of the potential simply from the scattering kinematics, i.e., from the value of $\mathbf{Q}_{\parallel}$ associated with a given event.

From Eq. (3.6), and the form of the integral on z', it also follows that the contribution to the potential proportional to $\exp(i\mathbf{Q}_{\parallel} \cdot \mathbf{x}_{\parallel})$ is produced by charge fluctuations that extend down to a distance $l_0(Q_{\parallel}) \approx Q_{\parallel}^{-1}$ *below* the crystal surface. Thus, an electron which suffers a scattering that changes its wave vector components parallel to the surface by the amount $\mathbf{Q}_{\parallel}$ "looks down" into the medium roughly the distance $Q_{\parallel}^{-1}$, which again is controlled only by the kinematics of the scattering process. If one wishes to probe not only the outermost atomic layer or two, but explore the near vicinity of the surface, this provides us with a simple rule for determining the depth probed by the electron scattering event.

Once again, the next step is to insert the potential $\varphi_1(\mathbf{x}, t)$ into the Schrödinger equation, and calculate the amplitude of the scattered wave first order in $\varphi_1(\mathbf{x}, t)$. For the moment, we shall be content with perturbation theory, and later on, we shall comment on higher-order scatterings. If we consider an incoming electron with wave vector $\mathbf{k}^{(I)}$, and calculate the probability amplitude for scattering to a final state $\mathbf{k}^{(S)}$, then four distinct scattering processes contribute [1]. These are illustrated in Fig. 3.1. The

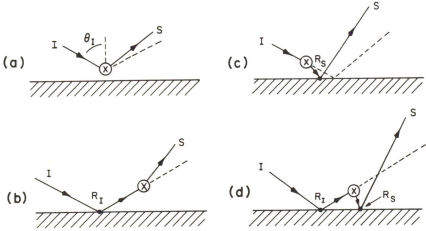

Fig. 3.1 The four scattering processes which take an electron from an initial state $\mathbf{k}^{(I)}$ to a final state $\mathbf{k}^{(S)}$ near the specular direction. The encircled crosses denote the point where the electron scatters from the fluctuating electric field in the vacuum above the crystal.

electron may approach the surface, then at some point in the vacuum above the crystal, it may scatter from the fluctuating electric field without ever striking the crystal surface. This process is illustrated in Fig. 3.1a. Except for the special case where the angle of incidence θ_I is very close to 90°, if the final state $\mathbf{k}^{(S)}$ lies very close to the specular direction, the scattering event involves a large momentum transfer in the direction perpendicular to the surface. In contrast, the two events in Figs. 3.1b and 3.1c involve only small-angle scatterings. One finds [1, 2] that the event in Fig. 3.1a contributes to the scattering amplitude weighted by the matrix element

$$M^{(a)}(\mathrm{I} \to \mathrm{S}) = \int_0^\infty dz \, \exp[-(Q_{||} + ik_\perp^{(I)} + ik_\perp^{(S)})z]$$

$$= (Q_{||} + ik_\perp^{(I)} + ik_\perp^{(S)})^{-1},$$

where $k_\perp^{(I)}$ and $k_\perp^{(S)}$ are the magnitudes of the wave vector components of the incident and scattered electron normal to the surface. In a similar fashion, the scatterings shown in Figs. 3.1b–3.1d are weighted by

$$M^{(b)}(\mathrm{I} \to \mathrm{S}) = \int_0^\infty dz \, \exp[-(Q_{||} - ik_\perp^{(I)} + ik_\perp^{(S)})z]$$

$$= (Q_{||} - ik_\perp^{(I)} + ik_\perp^{(S)})^{-1},$$

$$M^{(c)}(\mathrm{I} \to \mathrm{S}) = \int_0^\infty dz \, \exp[-(Q_{||} + ik_\perp^{(I)} - ik_\perp^{(S)})z]$$

$$= (Q_{||} + ik_\perp^{(I)} - ik_\perp^{(S)})^{-1},$$

and

$$M^{(d)}(\mathrm{I} \to \mathrm{S}) = \int_0^\infty dz \, \exp[-(Q_{||} - ik_\perp^{(I)} - ik_\perp^{(S)})z]$$

$$= (Q_{||} - ik_\perp^{(I)} - ik_\perp^{(S)})^{-1}.$$

From earlier discussions, we have seen that in typical small-angle scattering events, $Q_{||} \cong 10^6$ cm^{-1}, while $k_\perp^{(I)}$ and $k_\perp^{(S)}$ are both the order of 10^8 cm^{-1}. Hence, in this regime, where the *difference* $k_\perp^{(I)} - k_\perp^{(S)}$ is a small fraction of either $k_\perp^{(I)}$ or $k_\perp^{(S)}$, the processes in Figs. 3.1b and 3.1c make the dominant contribution to the scattering cross section. Hence, we obtain an accurate description of the scattering amplitude by retaining only these two terms, and this will be done in all that follows.

A convenient quantity to calculate is the scattering efficiency S per unit solid angle $d\Omega(\hat{k}_S)$ per unit energy $dh\omega$. We denote this quantity by $d^2S/d\Omega(\hat{k}_S) \, dh\omega$, and the combination $(d^2S/d\Omega(\hat{k}_S) \, dh\omega) \, d\Omega(\hat{k}_S) \, dh\omega$ is dimensionless, and gives the probability that an electron will scatter from its initial state into a final state in the solid angle $d\Omega(\hat{k}_S)$, and in the energy range $dh\omega$. Thus, it emerges with energy $E_S = E_I - \hbar\omega$, and its final state wave vector is $\mathbf{k}^{(S)}$, where $\mathbf{k}_{||}^{(S)} = \mathbf{k}_{||}^{(I)} - \mathbf{Q}_{||}$. The scattering efficiency just described is given by [1]

$$\frac{d^2S}{d\Omega(\hat{k}_S)\,dh\omega} = \frac{m^2 e^2 v_\perp^2}{2\pi^2 \hbar^5 \cos\theta_I} \left(\frac{k_S}{k_I}\right) \frac{P(\mathbf{Q}_{||},\omega)}{Q_{||}^2}$$

$$\times \frac{|v_\perp Q_{||}(R_S + R_I) + i(R_I - R_S)(\omega - \mathbf{v}_{||} \cdot \mathbf{Q}_{||})|^2}{[v_\perp^2 Q_{||}^2 + (\omega - \mathbf{v}_{||} \cdot \mathbf{Q}_{||})^2]^2}. \quad (3.7)$$

In this expression, θ_I is the angle of incidence measured with respect to the crystal normal, as illustrated in Fig. 3.1a, k_I and k_S are the magnitudes of the wave vector of the incident and scattered electron, respectively, while $v_\perp$ is the magnitude of the incoming electron velocity normal to the surface, and $\mathbf{v}_{||}$ the velocity parallel to it. Finally, the quantity $P(\mathbf{Q}_{||},\omega)$ is defined as

$$P(\mathbf{Q}_{||},\omega) = \int d^2 x_{||} \int_{-\infty}^{+\infty} dt \exp(i\mathbf{Q}_{||} \cdot \mathbf{x}_{||} - i\omega t)$$

$$\times \int_{-\infty}^{0^+} dz' \int_{-\infty}^{0^+} dz'' \exp(Q_{||}[z' + z'']) \langle \rho_1(\mathbf{x}_{||}z'',t)\rho_1(0z',0)\rangle_T, \quad (3.8)$$

where by $\rho_1(\mathbf{x}_{||}z,t)$ we denote the amplitude of the fluctuating charge density of the medium at the position $\mathbf{x} = \mathbf{x}_{||} + \hat{z}z$, with $\mathbf{x}_{||}$ the projection of the position vector $\mathbf{x}$ onto the xy plane. The angular brackets with subscript T denote the average of the quantity enclosed over the appropriate statistical ensemble at the temperature T.

We shall shortly cast the result in Eq. (3.7) in a form more suitable for direct evaluation. Before we do this, some general comments are useful. According to Eq. (3.7) the inelastic intensity diverges when θ_I approaches $90°$. This divergence simply results from the facts that within the framework of our model, the electron spends an infinite amount of time near the surface. The divergence is removed in any more realistic model, e.g., when the influence of the image potential is taken into consideration.

The quantities R_I and R_S that appear in Eq. (3.7), and which are illustrated in Figs. 3.1b and 3.1c, are the probability *amplitudes* that describe specular reflection of the incident and scattered electron from the surface. If we consider the incident electron with energy E_I and wave vector $\mathbf{k}^{(I)}$, then the intensity of the specularly reflected, elastic beam normalized to the incident intensity, is $|R_I|^2$. The amplitudes R_I and R_S rather than the intensities $|R_I|^2$ and $|R_S|^2$ enter Eq. (3.7) because the two processes illustrated in Figs. 3.1b and 3.1c interfere coherently; their respective contributions are added to the inelastic scattering *amplitude*, and this (complex) quantity is then squared to form the expression for the *intensity* of the inelastically scattered electrons. Both R_I and R_S are complex numbers with an amplitude and phase; a measurement of the intensity of low-energy electrons elastically diffracted from the surface (a LEED measurement) provides one with information on only

$|R_I|^2$ and $|R_S|^2$, and no information on the phase of their quantities is obtained from LEED data. We know of no experimental method that may be used to determine both the amplitude and phase of R_I and R_S.

Thus, as it stands, Eq. (3.7) is a most cumbersome expression to utilize, even if we know the form of $P(\mathbf{Q}_{||}, \omega)$. In fact, in many experimental studies of small-angle inelastic electron scattering from surfaces, the energy transfer $\hbar\omega$ is a small fraction of the incident energy E_I. When this is the case, and the angular deflection experienced by the electron is small, then we may then suppose that $R_S \cong R_I$. Equation (3.7) then becomes

$$\frac{d^2S}{d\Omega(\hat{k}_S)\,dh\omega} = \frac{2m^2e^2v_\perp^4}{\pi\hbar^5\cos\theta_I}\left(\frac{k_S}{k_I}\right)\frac{|R_I|^2P(\mathbf{Q}_{||}, \omega)}{[v_\perp^2Q_{||}^2 + (\omega - \mathbf{v}_{||}\cdot\mathbf{Q}_{||})^2]^2}. \tag{3.9}$$

The expression in Eq. (3.9) is a great simplification when compared to Eq. (3.7), since all the factors which enter, save for $P(\mathbf{Q}_{||}, \omega)$, are fully determined by either the scattering geometry, or are directly measureable. The quantity $P(\mathbf{Q}_{||}, \omega)$ contains information on the physics of the surface region of the crystal, and it is upon this object that we shall focus our attention shortly.

In practice, one must take care that it is reasonable to suppose that $R_S \cong R_I$ before Eq. (3.9), or one of the special forms subsequently deduced from it, are applied to analyze experimental data. Even though $\hbar\omega$ may be a small fraction of E_I, and the angular deflections from the specular small, at low energies very fine scale resonances can appear in $|R_I|^2$ under certain experimental conditions [3]. These fine structure resonances, whose origin is intimately related to the image potential felt by the incoming electron, have widths the order of 100 meV. Thus, if an electron scattering experiment is carried out in a regime of energy and angle where the fine structure resonances are present in the specular reflectivity $|R_I|^2$, the assumption $R_S \cong R_I$ required to obtain Eq. (3.9) from Eq. (3.7) may break down badly, so that Eq. (3.9) and the results below deduced from it will not describe the data quantitatively.

The result in Eqs. (3.7) and (3.9) is quite general, and forms for the theory of small-angle electron scattering from surfaces an expression analogous to the well known Van Hove formula for the scattering of thermal neutrons from solids. We require the form of $P(\mathbf{Q}_{||}, \omega)$, and to obtain this we must turn to a model description of the surface of the sample.

We shall proceed as follows: First, the charge fluctuations that produce $\rho_1(\mathbf{x}, t)$ necessarily generate an electric field $\mathbf{E}(\mathbf{x}, t)$ related to $\rho_1(\mathbf{x}, t)$ via

$$\rho_1(\mathbf{x}, t) = (4\pi)^{-1}\nabla\cdot\mathbf{E}(\mathbf{x}, t). \tag{3.10}$$

Thus, through use of Eq. (3.10), we may eliminate $\rho(\mathbf{Q}_{||}, \omega)$ in favor of the electric field $\mathbf{E}(\mathbf{x}, t)$. When this is done, we are then led to require correlation functions for the electric field fluctuations, $\langle E_\alpha(\mathbf{x}t)E_\beta(\mathbf{x}'t')\rangle_T$, where α and β refer to Cartesian components of the electric field. These may be related to the dielectric response functions of the substrate through use of the fluctuation–dissipation theorem. Thus, ultimately we may relate $P(\mathbf{Q}_{||},\omega)$ to the dielectric properties of the substrate.

The details of this procedure are described in Appendix A. There we resort to a specific model which can be used to describe several different physical situations, through appropriate choice of constants. We consider a semi-infinite substrate with a complex, but isotropic dielectric constant $\varepsilon_b(\omega)$. Upon the substrate is a layer of thickness d characterized by a dielectric constant $\varepsilon_s(\omega)$. The geometry is illustrated in Fig. 3.2. Quite clearly, as $d \to 0$, we describe the scattering of electrons from a semi-infinite dielectric substrate; with appropriate choice of $\varepsilon_b(\omega)$, we may discuss scattering from a metal or semiconductor with free carriers present, or from an insulator where interband electronic transitions make the dominant contribution to $\varepsilon_b(\omega)$. With the surface layer added, we may describe a layer of adsorbate molecules, or we may model a semiconductor with depletion or accumulation layers near the surface. We shall consider these various cases in turn in the discussion that follows.

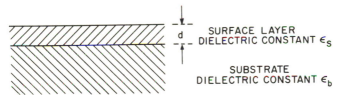

Fig. 3.2   The model used to calculate the response function $P(\mathbf{Q}_{||},\omega)$ which appears in Eq. (3.9). We have a semi-infinite substrate with complex dielectric constant $\varepsilon_b(\omega)$, and a surface layer of thickness d with dielectric constant $\varepsilon_s(\omega)$.

For the two-layer model just described, and illustrated in Fig. 3.2, and from Appendix A, one finds the result (quoted in an earlier paper [5])

$$P(Q_{||},\omega) = \frac{2\hbar Q_{||}}{\pi}\left[1 + n(\omega)\right]\operatorname{Im}\left\{\frac{-1}{\bar{\varepsilon}(Q_{||},\omega) + 1}\right\}, \qquad (3.11)$$

where

$$\bar{\varepsilon}(Q_{||},\omega) = \varepsilon_s(\omega)\left[\frac{1 + \Delta(\omega)\exp(-2Q_{||}d)}{1 - \Delta(\omega)\exp(-2Q_{||}d)}\right], \qquad (3.12)$$

where $\Delta(\omega) = [\varepsilon_b(\omega) - \varepsilon_s(\omega)]/[\varepsilon_s(\omega) + \varepsilon_b(\omega)]$ and $n(\omega) = [\exp(\hbar\omega/k_B T) - 1]^{-1}$.

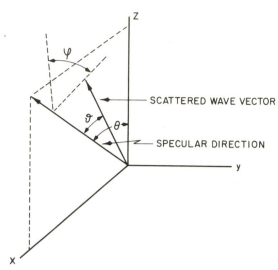

Fig. 3.3   The scattering geometry that enters the discussion of dipole scattering and energy loss near the specular direction.

Then Eq. (3.9) becomes

$$\frac{d^2S}{d\Omega(\hat{k}_S)\,d\hbar\omega} = \frac{4m^2e^2v_\perp^4}{\pi^2\hbar^3\cos\theta_I}\left(\frac{k_S}{k_I}\right)\frac{|R_I|^2Q_{||}}{[v_\perp^2Q_{||}^2 + (\omega - \mathbf{v}_{||}\cdot\mathbf{Q}_{||})^2]^2}$$

$$\times[1 + n(\omega)]\,\mathrm{Im}\left\{\frac{-1}{\bar{\varepsilon}(Q_{||},\omega) + 1}\right\}. \tag{3.13}$$

The prefactors in Eq. (3.13) may be expressed conveniently in terms of two angles, ϑ and φ, which measure the angular deviation of the electron from the specular direction. The scattering geometry we use is illustrated in Fig. 3.3. The angle between the wave vector $\mathbf{k}^{(S)}$ of the scattered electron, and that of the specular beam is ϑ, while φ measures the azimuthal orientation of $\mathbf{k}^{(S)}$, as one looks "down" the specular beam. The regime $\vartheta \ll 1$, very small deflections from the specular, are of primary interest and the expressions to be quoted make use of this assumption. Finally, the angle between the specular beam and the normal to the crystal is equal to the angle of incidence θ_I.

For the geometry in Fig. 3.3, when $\vartheta \ll 1$, the following expressions apply, with $\vartheta_E = \hbar\omega/2E_I$,

$$v_\perp^2Q_{||}^2 + (\omega - \mathbf{v}_{||}\cdot\mathbf{Q}_{||})^2 = 4E_I^2(\vartheta^2 + \vartheta_E^2)\cos^2\theta_I \tag{3.14}$$

and

$$Q_{||} = k_I[(\vartheta_E\sin\theta_I - \vartheta\cos\theta_I\cos\varphi)^2 + \vartheta^2\sin^2\varphi]^{1/2}. \tag{3.15}$$

With these expressions, Eq. (3.13) may then be arranged to read

$$\frac{d^2 S}{d\Omega(\hat{k}_S)\,dh\omega} = \frac{2|R_I|^2}{\pi^2 a_0 k_I E_I \cos\theta_I} (1 - 2\vartheta_E)^{1/2}$$

$$\times \frac{[(\vartheta_E \sin\theta_I - \vartheta\cos\theta_I \cos\varphi)^2 + \vartheta^2 \sin^2\varphi]^{1/2}}{[\vartheta^2 + \vartheta_E^2]^2}$$

$$\times [1 + n(\omega)] \operatorname{Im}\left\{\frac{-1}{1 + \tilde{\varepsilon}(Q_{||},\omega)}\right\}, \tag{3.16}$$

where a_0 is the Bohr radius.

The result in Eq. (3.16) is the principal result of the present section. Shortly we shall apply it to a variety of physical situations. We conclude with one more step: in the limit $Q_{||}d \ll 1$, which is well satisfied for $\vartheta \cong \vartheta_E$ in typical energy loss studies of adsorbate layers, the expression for $d^2 S/d\Omega(\hat{k}_S)\,dh\omega$ may be decomposed into two terms. The first, $d^2 S_b/d\Omega(\hat{k}_S)\,dh\omega$, survives in the limit $d \to 0$ and describes electron scattering by electric field fluctuations in the vacuum above the crystal produced by excitations in the bulk. The second, $d^2 S_s/d\Omega(\hat{k}_S)\,dh\omega$, is proportional to d and describes scattering by electric fields produced by fluctuations within the surface layer. We have

$$\frac{d^2 S}{d\Omega(\hat{k}_S)\,dh\omega} = \frac{d^2 S_b}{d\Omega(\hat{k}_S)\,dh\omega} + \frac{d^2 S_s}{d\Omega(\hat{k}_S)\,dh\omega}, \tag{3.17}$$

where

$$\frac{d^2 S_b}{d\Omega(\hat{k}_S)\,dh\omega} = \frac{2|R_I|^2}{\pi^2 a_0 k_I E_I \cos\theta_I} (1 - 2\vartheta_E)^{1/2}$$

$$\times \frac{[(\vartheta_E \sin\theta_I - \vartheta\cos\theta_I \cos\varphi)^2 + \vartheta^2 \sin^2\varphi]^{1/2}}{[\vartheta^2 + \vartheta_E^2]^2}$$

$$\times [1 + n(\omega)] \operatorname{Im}\left\{\frac{-1}{1 + \varepsilon_b(\omega)}\right\}, \tag{3.18}$$

and

$$\frac{d^2 S_s}{d\Omega(\hat{k}_S)\,dh\omega} = \frac{2|R_I|^2 d}{\pi^2 a_0 E_I \cos\theta_I} (1 - 2\vartheta_E)^{1/2}$$

$$\times \frac{[(\vartheta_E \sin\theta_I - \vartheta\cos\theta_I \cos\varphi)^2 + \vartheta^2 \sin^2\varphi]^{1/2}}{[\vartheta^2 + \vartheta_E^2]^2}$$

$$\times [1 + n(\omega)] \operatorname{Im}\left\{\frac{-1}{\varepsilon_s(\omega)}\left(\frac{\varepsilon_b^2(\omega) - \varepsilon_s^2(\omega)}{[\varepsilon_b(\omega) + 1]^2}\right)\right\}. \tag{3.19}$$

From Eq. (3.19), the presence of the factor $\varepsilon_b(\omega)$ inside the braces shows the fluctuations in the adsorbed layers excite the substrate, so that the field seen by the electron is that produced by both the excitation in the adsorbed layer, and its image in the substrate. We shall comment on the significance of this later.

3.3 BULK LOSSES IN THE REGIME OF SMALL-ANGLE SCATTERING

3.3.1 The Intensity

In Eq. (3.18), we have given the expression for inelastic scattering of electrons from bulk excitations, via the dipole mechanism. In this section, we apply this form to a number of specific situations. First, it is convenient to rewrite the expression a bit. Let

$$\hat{\vartheta} = \vartheta/\vartheta_E \qquad (3.20)$$

be a reduced scattering angle, and note that with the polar axis of a spherical coordinate system aligned along the specular direction, we have $d\Omega(\hat{k}_s) = \vartheta\,d\vartheta\,d\varphi$. In an actual experiment, it is not $d^2S/d\Omega(\hat{k}_s)\,d\hbar\omega$ that is measured,

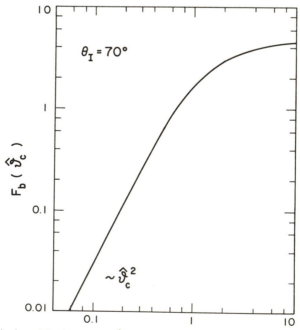

Fig. 3.4 A plot of the function $F_b(\hat{\vartheta}_c)$ that describes the number of inelastically scattered electrons collected by a spectrometer that picks up all electrons with scattering angles $0 \leq \vartheta \leq \vartheta_c$. Recall $\hat{\vartheta}_c = \vartheta_c/\vartheta_E$.

but rather this quantity integrated over the solid angle subtended by the entrance slit of the analyzer. We suppose this to be circular, centered on the specular direction, and that it catches all electrons scattered into the angular range $0 \leq \vartheta \leq \vartheta_c$. We then have

$$\frac{dS_b}{d\hbar\omega} \equiv \int d\Omega(\hat{k}_S) \frac{d^2S}{d\Omega(\hat{k}_S)\,d\hbar\omega}$$

$$= \frac{4|R_I|^2(1 - 2\vartheta_E)^{1/2}}{\pi^2 a_0 k_I \cos\theta_I} \frac{1 + n(\omega)}{\hbar\omega} \operatorname{Im}\left\{\frac{-1}{1 + \varepsilon_b(\omega)}\right\} F_b(\hat{\vartheta}_c), \quad (3.21)$$

where

$$F_b(\hat{\vartheta}_c) = \int_0^{2\pi} d\varphi \int_0^{\hat{\vartheta}_c} d\hat{\vartheta}\,\hat{\vartheta}\, \frac{[(\sin\theta_I - \hat{\vartheta}\cos\theta_I\cos\varphi)^2 + \hat{\vartheta}^2\sin^2\varphi]^{1/2}}{[1 + \hat{\vartheta}^2]^2}. \quad (3.22)$$

In Fig. 3.4, we present a plot of $F_b(\hat{\vartheta}_c)$ which is also tabulated in Table 3.1.

TABLE 3.1

$F_b(\hat{\vartheta}_c)$

$\hat{\vartheta}_c$ \ θ_I	70°	60°	50°	40°
0.1	0.0293	0.0270	0.0239	0.0201
0.2	0.1142	0.1053	0.0933	0.0786
0.3	0.2467	0.2279	0.2024	0.1711
0.4	0.4158	0.3846	0.3425	0.2912
0.5	0.6093	0.5646	0.5045	0.4318
0.6	0.8159	0.7576	0.6796	0.5858
0.7	1.0265	0.9553	0.8604	0.7475
0.8	1.2342	1.1512	1.0412	0.9121
0.9	1.4343	1.3410	1.2181	1.0764
1.0	1.6240	1.5219	1.3884	1.2383
1.2	1.9674	1.8521	1.7043	1.5476
1.4	2.2622	2.1388	1.9852	1.8301
1.6	2.5128	2.3854	2.2326	2.0828
1.8	2.7260	2.5978	2.4494	2.3062
2.0	2.9082	2.7812	2.6392	2.5032
3.0	3.5132	3.4080	3.2976	3.1938
4.0	3.8470	3.7624	3.6746	3.5928
5.0	4.0560	3.9860	3.9140	3.8470
6.0	4.1986	4.1392	4.0782	4.0216
7.0	4.3016	4.2502	4.1976	4.1486
8.0	4.3796	4.3342	4.2880	4.2448
9.0	4.4406	4.4000	4.3588	4.3204
10.0	4.4894	4.4530	4.4156	4.3810

Note that we have the limiting behavior

$$\lim_{\hat{\vartheta}_c \to 0} F_b(\hat{\vartheta}_c) = \pi \sin \theta_I \, \hat{\vartheta}_c^2 \tag{3.23a}$$

and also

$$\lim_{\hat{\vartheta}_c \to \infty} F_b(\hat{\vartheta}_c) = \tfrac{1}{2}\pi^2. \tag{3.23b}$$

From Eq. (3.23b), for a spectrometer which samples a sufficiently wide acceptance angle, the expression for the loss intensity becomes remarkably simple. We have for the scattering efficiency $dS_b^{(\infty)}/dh\omega$ when $\hat{\vartheta}_c \gg 1$ the form

$$\frac{dS_b^{(\infty)}}{dh\omega} = \frac{2|R_I|^2(1 - 2\vartheta_E)^{1/2}}{a_0 k_I \cos \theta_I} \frac{1 + n(\omega)}{h\omega} \mathrm{Im}\left\{\frac{-1}{1 + \varepsilon_b(\omega)}\right\}. \tag{3.24}$$

When $h\omega \ll 2E_I$, so the factor $(1 - 2\vartheta_E)^{1/2}$ is approximated by unity, the energy variation is controlled by the factor k_I in the denominator, so the integrated loss intensity falls off as $E_I^{-1/2}$. For a spectrometer with finite acceptance angle, $F_b(\hat{\vartheta}_c)$ decreases as the energy decreases, yielding an optimum value for the scattering efficiency. The reflectivity $|R_I|^2$ is a most important factor to consider, since the absolute intensity of a given loss is controlled by this factor and it is the absolute intensity that is the key quantity from the experimental point of view. Thus, one tends to search for impact energies where the reflectivity $|R_I|^2$ is high in order to optimize the loss intensity. From Eq. (3.19), we see that the same observation holds for surface losses.

We see that Eq. (3.24) relates the inelastic intensity to the dielectric properties of the material. Just as in an optical experiment, one may therefore recover $\varepsilon_b(\omega)$ from $\mathrm{Im}\{-1/(\varepsilon_b(\omega) + 1)\}$ by a Kramers–Kronig analysis of the data collected over a wide spectral range. In fact, electron energy loss spectroscopy (in transmission at high energies, not low-energy reflection spectroscopy considered here) has been used to determine the optical properties of metals in the ultraviolet [4].

Here our concern is with the infrared regime. In the infrared, three elementary excitations can make a major contribution to the loss function: in ionic, nonconducting crystal lattices, we have infrared active optical phonons. In semiconductors, we have the free carriers which give rise to plasmons. Also, in small-gap materials, direct interband transitions contribute to $\varepsilon_b(\omega)$, and these may contribute to the energy loss spectrum. These cases will be discussed more thoroughly at a later stage.

In a number of circumstances, such as scattering from the surface of a metal or doped semiconductor in the relevant range of concentrations, the free carriers provide a nonresonant contribution to the loss function $\mathrm{Im}\{-1/(1 + \varepsilon_b(\omega))\}$, and this appears as a background that varies smoothly as a function of frequency. We turn to this situation first.

3.3.2 Free Carrier Contributions to the Loss Function

In a doped semiconductor or a metal, the dielectric function may be described by supplementing the contribution ε_∞ from interband transitions by the Drude term contributed by the free carriers. We have

$$\varepsilon_b(\omega) = \varepsilon_\infty - \frac{\omega_p^2}{\omega(\omega + [i/\tau(\omega)])} \equiv \varepsilon_\infty + \frac{4\pi i\sigma(\omega)}{\omega}. \tag{3.25}$$

Here $\omega_p^2 = 4\pi ne^2/m^*$ is the plasma frequency of the carriers, $\tau(\omega)$ a possibly frequency-dependent relaxation time, and we take the background dielectric constant ε_∞ real for the present discussion. In the second step in Eq. (3.25), we write $\varepsilon_b(\omega)$ in terms of the frequency-dependent conductivity $\sigma(\omega)$. The loss function $\text{Im}\{-1/(1 + \varepsilon_b(\omega))\}$ becomes

$$\text{Im}\left\{\frac{-1}{1 + \varepsilon_b(\omega)}\right\} = \frac{\omega\omega_{sp}^2}{(1 + \varepsilon_\infty)\tau(\omega)}\left[(\omega_{sp}^2 - \omega^2)^2 + \omega^2/\tau^2(\omega)\right]^{-1}. \tag{3.26}$$

Here $\omega_{sp} = \omega_p/(1 + \varepsilon_\infty)^{1/2}$ is the frequency of the surface plasmons which propagate on the metal–vacuum interface.

We see that Eq. (3.26) describes the inelastic scattering of electrons from the surface plasmons. If the relaxation time $\tau(\omega)$ is sufficiently long, the expression describes a loss peak centered at the surface plasmon frequency ω_{sp}. In a metal, the bulk plasma frequency ω_p lies in the range 10–15 eV. Then for energy transfers that correspond to frequencies in the infrared, one is sampling the very low frequency wing of the surface plasmon loss feature. Here we have $\omega \ll \omega_{sp}$, so that Eq. (3.26) reduces to

$$\text{Im}\left\{\frac{-1}{1 + \varepsilon_b(\omega)}\right\} = \frac{\omega}{\omega_p^2}\frac{1}{\tau(\omega)} \equiv \frac{\omega}{4\pi\sigma_R(\omega)}, \tag{3.27}$$

where $\sigma_R(\omega)$ is the real part of the frequency-dependent conductivity of the metal. Since the conductivity of a metal is inversely proportional to the temperature T the background should be proportional to T. This temperature dependence is actually found in experiments (Fig. 3.5).

The result in Eq. (3.27) shows that from the information provided by the bulk losses, one can determine dielectric properties of metals in the infrared. For spectrometers with moderate energy resolution, $\hat{\vartheta}_c$ can be made larger than unity (see Chapter 2), so that $F_b(\hat{\vartheta}_c)$ becomes insensitive to the actual value of $\hat{\vartheta}_c$. The loss function $\text{Im}\{-1/(1 + \varepsilon_b(\omega))\}$ can then be determined to high accuracy. In fact, the use of electron energy loss spectroscopy to measure the frequency-dependent conductivity $\sigma_R(\omega)$ is in some senses superior to infrared spectroscopy, where it is the reflectivity $|R(\omega)|^2$ that is obtained. The reflection amplitude $R(\omega)$ for infrared radiation is very close to unity in the infrared, and one has to measure $|R(\omega)|^2$ with very high accuracy to obtain

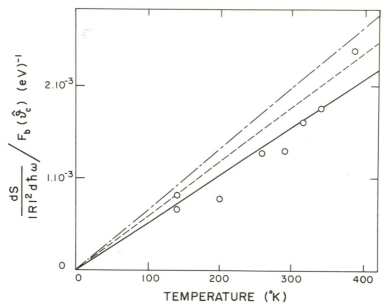

Fig. 3.5   The temperature variation of the loss function for scattering of electrons from a clean Ni surface. The experimental data (—) are compared with the theory, with Eq. (3.21) combined with Eq. (3.27). In Eq. (3.27), the conductivity is taken from dc measurements (----), and that from infrared optical studies (–––).

reliable values of the quantity $1 - |R(\omega)|^2$, which contains the information on the dielectric response of the metal. In essence, electron spectroscopy provides a direct measurement of $1 - R(\omega)$ itself. For $R(\omega)$ near unity, one has the relation

$$\mathrm{Im}\left\{\frac{-1}{1 + \varepsilon_b(\omega)}\right\} = \tfrac{1}{2}\,\mathrm{Im}[1 - R(\omega)]. \tag{3.28}$$

The proportionality of the loss function to $[1 - R(\omega)]$ can therefore be exploited to obtain a sensitive and accurate measurement of the deviation of $R(\omega)$ from unity. Figure 3.5 shows a comparison between an absolute measurement the temperature-dependent background and theory, for the case of a clean Ni crystal surface.

We now turn from metals to a discussion of the background from scattering off free carriers in semiconductors. The dielectric function given in Eq. (3.25) continues to apply, except the plasma frequency ω_p typically lies in the infrared, rather than in the ultraviolet as in the case of metals. This is because the free-carrier concentration n is smaller than in metals by at least four orders of magnitude, and sometimes by very much more. The effective mass

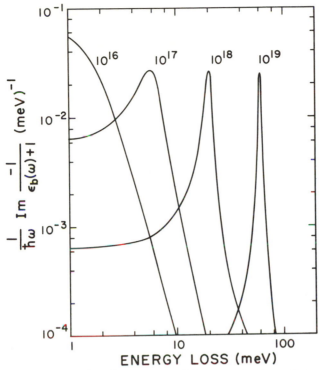

Fig. 3.6 A plot of $(\hbar\omega)^{-1} \, \mathrm{Im}\{-1/(1 + \varepsilon_b(\omega))\}$ for scattering from electrons in n-type silicon at room temperature. The loss function is given for several free carrier concentrations.

m^* is typically much smaller in semiconductors than in metals, but the combined effect of the decrease in carrier concentration with that of the large value of ε_∞ encountered in many semiconductors pulls the surface plasmon frequency ω_{sp} down into the infrared. At the same time, the dc mobility $\mu = e\tau(0)/m^*$ can be two orders of magnitude larger than in metals, since the time $\tau(0)$ between collisions is very much larger, and the effective mass is smaller. These considerations lead us into a parameter regime very different than that encountered in our discussion of metals.

In Fig. 3.6, we have plotted $(\hbar\omega)^{-1} \, \mathrm{Im}\{-1/(1 + \varepsilon_b(\omega))\}$ for a semiconductor with the free carrier concentration as a parameter. The rest of the parameters, ε_∞, m^*, and the mobility, have been chosen to represent silicon at room temperature, and we have supposed that the relaxation time $\tau(\omega)$ is frequency independent. Note that the loss function has a maximum at approximately the surface plasmon frequency $\omega_{sp}^2 = 4\pi n e^2/m^*(1 + \varepsilon_\infty)$. For small carrier concentrations, however, the loss structure is rather broad and featureless, since the relaxation rate τ^{-1} is no longer small compared to the plasma frequency ω_p.

In order to make a rough estimate of the integrated intensity of these loss features, we suppose we are in the limit $\omega_{sp}\tau \gg 1$. The loss function may then be represented by the delta function form

$$\frac{1}{\hbar\omega}\,\text{Im}\left\{\frac{-1}{1 + \varepsilon_b(\omega)}\right\} \cong \frac{\pi}{2(1 + \varepsilon_\infty)}\,\delta(\hbar\omega - \hbar\omega_{sp}) \tag{3.29}$$

so that if

$$S_b^{(tot)} = \int d\hbar\omega\,\frac{dS_b}{d\hbar\omega} \tag{3.30}$$

is the integrated intensity of the surface plasmon feature in the spectrum, we have for $\vartheta_E \ll 1$ and $n(\omega) \approx 0$,

$$\frac{S_b^{(tot)}}{|R_I|^2} = \frac{2}{\pi}\,\frac{1}{a_0 k_I \cos\theta_I}\,\frac{F_b(\hat{\vartheta}_c)}{(1 + \varepsilon_\infty)} \tag{3.31}$$

$$= \frac{\pi}{a_0 k_I \cos\theta_I}\,\frac{1}{(1 + \varepsilon_\infty)},$$

where the last line follows when $\hat{\vartheta}_c \gg 1$.

The right-hand side of Eq. (3.31) is quite close to unity under conditions encountered in high-resolution electron spectroscopy. This means that the broad and featureless distributions in Fig. 3.6 will result in a broadened energy distribution of electrons reflected from semiconductor surfaces. This may make a high-resolution analysis of vibrations difficult if not impossible. To illustrate this point further, we have calculated an energy distribution for electrons that strike the surface with energy $E_I = 5$ eV, and an initially Gaussian distribution of width $\Delta E = 5$ meV. We then show the energy distribution after reflection from a silicon surface with electron concentration $n = 3 \times 10^{16}$ cm^{-3}. The results are given in Fig. 3.7. The additional broadening provided by inelastic scattering from the free carriers is quite significant. The situation is even worse with semiconductors of higher carrier concentration, when the mobility is low. Therefore, to carry out vibrational spectroscopy on semiconductor surfaces requires consideration of the doping and temperature range to be used, in order to avoid this broadening effect.

As in the case of metals, this background is controlled by the same dielectric response function as encountered in infrared spectroscopy. Nevertheless, there are again interesting differences between the two techniques. For example, in the calculation so far, we have assumed that the electron concentration does not depend on the distance from the surface, which corresponds to a flat band condition in the semiconductor. Both clean

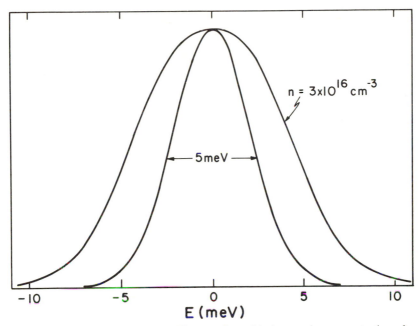

Fig. 3.7   For scattering from a silicon surface with free carrier concentration of $n = 3 \times 10^{16}$ cm^{-3}, we show the broadening of the energy distribution of the reflected electrons as a consequence of scattering from free carriers. The electron beam that strikes the surface has energy $E_I = 5$ eV and width 5 meV, and scattering from free carriers produces substantial additional broadening.

and adsorbate covered surfaces typically show band bending, with the consequence that the free carrier concentration is a function of distance from the surface. In Chapter 1, and again in the present chapter, we saw that as the electron scatters inelastically from the surface, it "looks" into the crystal roughly the distance $Q_\parallel^{-1}$ given by

$$Q_\parallel^{-1} \cong \frac{1}{k_1 \vartheta_E} = \frac{2}{k_I} \frac{E_I}{\hbar\omega}. \tag{3.32}$$

This probing depth may be compared with the thickness of the space-charge layer, which in a simple Schottky model is

$$d_s = \left(\frac{\varepsilon(0)\phi_s}{2\pi e N_n} \right)^{1/2}, \tag{3.33}$$

where $\varepsilon(0)$ is the static dielectric constant of the material, ϕ_s the surface potential, and N_n the donor concentration which, at room temperature, is roughly equal to the electron concentration n in the bulk. If the surface plasmon frequency ω_{sp} is inserted into Eq. (3.32), and the result rearranged a bit, we have

$$Q_{\|}^{-1} \cong d_s[(1 + \varepsilon_\infty)/\varepsilon(0)]^{1/2}(m^*/m)^{1/2}(E_I/e\phi_s)^{1/2}. \qquad (3.34)$$

With typical band bending potentials $e\phi_s$ a few tenths of a volt, E_I in the range of a few electron volts, and the small values of m^*/m typical of most semiconductors, from Eq. (3.34) we have $Q_{\|}^{-1} \cong d_s$. This means a quantitative analysis of electron energy loss data is complicated in this situation, and our simple two-layer model represents the situation only crudely. On the other hand, we see that electron energy loss spectroscopy has the potential of providing depth resolved information on the dynamics of space charge and inversion layers, when spectrometers with small acceptance angle are employed, so one may select $\mathbf{Q}_{\|}$.

3.3.3   Interband Transitions

In the present discussion of the electronic contribution to the bulk and surface loss function, we have considered only scattering from free carriers. Nevertheless, it has been shown experimentally that interband transitions, such as those between the conduction and valence bands of semiconductors, make a substantial contribution, in the appropriate range of energy loss. In

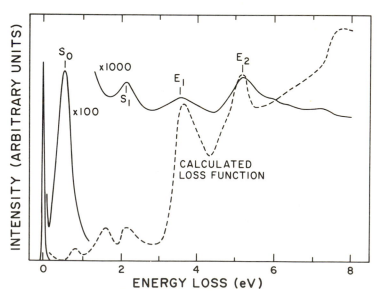

Fig. 3.8   An energy loss spectrum of a cleaved (111) (2 × 1) silicon surface at an angle of incidence $\theta_I = 78.7°$. The data are compared with the theory (---) using bulk optical constants. $E_0 = 50$ eV.

addition, transitions between surface states also can be seen in the loss spectrum. A good quantitative account of this data is provided by combining Eqs. (3.18) and (3.19) with the surface state contribution included in $\varepsilon_s(\omega)$. In Fig. 3.8, we show the energy loss spectrum of a clean, cleaved silicon surface, and compare this with that calculated from Eq. (3.18). The features labeled E_1 and E_2 are bulk interband transitions, and the strong feature at 500 meV is due to excitation of electrons from one subband of surface states to a second. The band gap of roughly 250 meV between the two subbands is intimately related to the 2 × 1 reconstruction of the surface. After annealing the crystal in ultrahigh vacuum, the clean surface exhibits the well-known 7 × 7 reconstructed surface. On the 7 × 7 reconstructed surface, the electronic structure and nature of the surface states is not well understood. Nevertheless, one would expect the surface band to split into a considerable number of subbands separated by very small gaps. In the loss spectrum of a clean silicon (7 × 7) surface, a continuous background is observed (Fig. 3.9). This is believed to have its origin in transitions between these many subbands. Also, a two-dimensional metallic band of surface states could account for the continuous loss feature. The background disappears when the surface states are removed by hydrogen adsorption (Fig. 3.9).

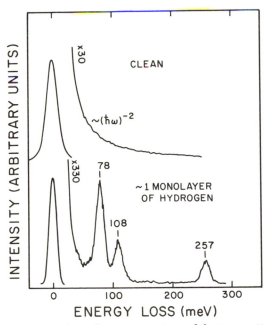

Fig. 3.9   The upper curve shows the energy spectrum of electrons scattered from the (111) surface of silicon, with the 7 × 7 reconstruction pattern present. One sees a background that varies roughly as $(\hbar\omega)^{-2}$. When hydrogen is adsorbed on the surface (lower curve), one sees loss peaks from the hydrogen vibrational modes, but the background disappears.

3.3.4   The Contribution from Surface Optical Phonons of Ionic Insulators

In Section 3.3.2, we discussed the contribution to the bulk loss function from the scattering produced by free carriers. This produces the features described by Eq. (3.26), where it is the surface plasmons that propagate on the metal vacuum interface that generate the electric fields in the vacuum responsible for the inelastic scatterings. This surface plasmon feature in the loss spectrum is broadened by the finite lifetime of the free carriers. In the case of metals, the low-frequency wing of the loss structure appears as the background described by Eq. (3.27), while in semiconductors with low carrier concentration we have $\omega_{sp}\tau(\omega) \ll 1$, with the consequence that we have a feature centered around zero energy loss, described in Fig. 3.7.

When electrons are scattered from the surface of an ionic insulator, a surface mode similar in character to the surface plasmon is found. This is the surface optical phonon with frequency ω_s that lies between that ω_{TO} of the bulk long-wavelength transverse optical phonon, and the frequency ω_{LO} of the bulk longitudinal optical phonon. We may describe the contribution to the loss cross section from this mode with the same expression, Eq. (3.21), used in our discussion of scattering from free carriers and from electric field fluctuations associated with interband electronic transitions.

For a cubic material with one infrared active bulk transverse optical (TO) phonon, the dielectric function may be written

$$\varepsilon_b(\omega) = \varepsilon_\infty + \frac{4\pi n e^{*2}}{M_r} \frac{1}{\omega_{TO}^2 - \omega^2 - i\omega\gamma(\omega)}, \qquad (3.35)$$

where n is the number of unit cells per unit volume, e^* the Born or transverse effective charge, and M_r the reduced mass of the unit cell. If $\varepsilon(0) = \varepsilon_\infty + 4\pi n e^{*2}/M_r\omega_{TO}^2$ is the static dielectric constant of the material, then the loss function $\mathrm{Im}\{-1/[1 + \varepsilon_b(\omega)]\}$ has a peak at the frequency

$$\omega_s = \omega_{TO}\left(\frac{\varepsilon(0) + 1}{\varepsilon_\infty + 1}\right)^{1/2}, \qquad (3.36)$$

providing the damping function $\gamma(\omega)$ is sufficiently small. The frequency ω_s necessarily lies above ω_{TO}, but below the longitudinal optical phonon frequency $\omega_{LO} = \omega_{TO}(\varepsilon(0)/\varepsilon_\infty)^{1/2}$.

In the limit of small damping, we have in place of Eq. (3.29)

$$\mathrm{Im}\left\{\frac{-1}{1 + \varepsilon_b(\omega)}\right\} = \frac{\pi}{2}\frac{\varepsilon(0) - \varepsilon_\infty}{[\varepsilon(0) + 1][\varepsilon_\infty + 1]}[\delta(\hbar\omega_s - \hbar\omega) - \delta(\hbar\omega_s + \hbar\omega)], \qquad (3.37)$$

where here we allow for the possibility of inelastic scattering with phonon absorption, as well as that with phonon emission. Then the probability, normalized to that for specular reflection, that an electron will scatter from the surface with emission of a surface optical phonon is given by

$$\frac{S_b^{(tot+)}}{|R_1|^2} = \frac{2}{\pi} \frac{1 + \overline{n}_s}{a_0 k_1 \cos \theta_1} \left(\frac{\varepsilon(0) - \varepsilon_\infty}{\varepsilon(0) + 1} \right) \frac{F_b(\hat{\vartheta}_c)}{(1 + \varepsilon_\infty)}, \tag{3.38}$$

where $\overline{n}_s = [\exp(\hbar\omega_s/k_B T) - 1]^{-1}$. Similarly, the probability for scattering with phonon emission is

$$\frac{S_b^{(tot-)}}{|R_1|^2} = \frac{2}{\pi} \frac{\overline{n}_s}{a_0 k_1 \cos \theta_1} \left(\frac{\varepsilon(0) - \varepsilon_\infty}{\varepsilon(0) + 1} \right) \frac{F_b(\hat{\vartheta}_c)}{(1 + \varepsilon_\infty)}. \tag{3.39}$$

The first experimental study of the inelastic scattering of low-energy electrons from a crystal surface explored scattering produced by a surface optical phonon on the ionic semiconductor ZnO [6] (see also Fig. 4.14). The one-phonon loss feature was observed on both the energy gain and loss side of the energy spectrum of the scattered electrons, with the energy $\hbar\omega_s$ of 69 meV, in excellent accord with the prediction of Eq. (3.36). In this experiment, a number of features of the theory were verified explicitly. Among these are the variation of the loss intensity with $(\cos \theta_1)^{-1}$ over a rather wide range of incident angles, the proportionality to $E_1^{-1/2}$ [the factor k_1 in the denominator of Eq. (3.39) is proportional to $E_1^{1/2}$], and the expectation that ratio of the energy gain scattering to energy loss is, for one-phonon events, equal to $\exp(-\hbar\omega_s/k_B T)$. By this time, scattering from surface optical phonons has been observed for a number of ionic compounds.

3.3.5 Multiquantum Scattering Processes in the Dipolar Regime

All the discussions in the previous subsections explore small-angle dipole scattering of electrons from the surface by various mechanisms. This scattering is quite strong. One may see that, for either emission of a plasmon or surface optical phonon at the absolute zero of temperature, the scattering probability integrated over solid angle and normalized to that of the specular beam is given by

$$\Gamma = \frac{\pi}{a_0 k_1 \cos \theta_1} \left(\frac{\varepsilon(0) - \varepsilon_\infty}{\varepsilon(0) + \varepsilon_\infty} \right) \frac{1}{1 + \varepsilon_\infty}, \tag{3.40}$$

where the formula applies to the metal upon letting the static dielectric constant $\varepsilon(0)$ become infinite.

It is easy to see that under conditions commonly encountered, Γ may become comparable to unity, as remarked earlier. Under these conditions, the perturbation theory upon which Eq. (3.40) is based breaks down. It is difficult to obtain formulas for the loss probability in this strong coupling regime in general, but for the special case where we consider scattering from either surface optical phonons or surface plasmons, simple closed-form expressions may be obtained [2].

Suppose we let $S_n^{(+)}$ be the total probability the electron excites n quanta as it passes by the surface, losing the energy $n\hbar\omega_s$ in the process. We then have, for scattering from thermal excitations,

$$S_n^{(+)} = \Gamma_n |R_1|^2, \tag{3.41}$$

where, with $I_n(x)$ the modified Bessel function,

$$\Gamma_n = \left(1 + \frac{1}{\bar{n}_s}\right)^{n/2} I_n[2\Gamma\{\bar{n}_s(1 + \bar{n}_s)\}^{1/2}] \exp[-\Gamma(1 + 2\bar{n}_s)]. \tag{3.42}$$

In the limit as the temperature and $\bar{n}_s$ vanish, this reduces to the well-known Poisson distribution

$$\Gamma_n = (\Gamma^n/n!) \exp(-\Gamma). \tag{3.43}$$

For the case where the electron gains the energy $n\hbar\omega_s$ by absorbing n quanta, we have for the scattering efficiency $S_n^{(-)}$

$$S_n^{(-)} = \exp(-n\hbar\omega_s/k_B T)S_n^{(+)}. \tag{3.44}$$

Several features of the preceding results are interesting. First of all, when Γ is not small compared to unity, the intensity of the specular beam is reduced from the value $|R_1|^2$ characteristic of the rigid crystal to the smaller value (at $T = 0$) $\exp(-\Gamma)|R_1|^2$. Thus, for scattering of electrons from an ionic crystal, or from a metal under circumstances that the dimensionless coupling constant Γ is appreciable, before the measured specular intensity may be compared with the results of a LEED calculation, one should recognize the correction factor $\exp(-\Gamma)$.

As the coupling constant Γ increases, and the integrated strength of the loss structures does also, the specular intensity is reduced simply because a nonnegligible fraction of the electrons are scattered out of the specular beam. One has the sum rule, with $S_0|R_1|^2$ the specular reflectivity with depletion of the incident beam by inelastic events described by S_0 [Eq. (3.42) with $n \equiv 0$],

$$|R_1|^2 S_0 + |R_1|^2 \sum_{n=1}^{\infty} (S_n^{(+)} + S_n^{(-)}) \equiv |R_1|^2, \tag{3.45}$$

which shows that when the coupling constant $\Gamma \neq 0$, the number of electrons that emerges from the surface is the same as when $\Gamma = 0$. They are redistributed among the various inelastic beams, however.

The experimental study [6] of the scattering by surface optical phonons on the surface of ZnO observed multiquantum losses (up to $n = 5$), in addition to the one-quantum loss described in the previous subsection. The intensities of the various features in the loss cross section are nicely accounted for by the Poisson distribution (see Fig. 4.14).

The results just quoted apply to the case where the electron scatters from thermal excitations on the surface. If instead we have a coherently generated wave of macroscopic amplitude, the expression for the multiquantum loss is distinctly different from Eq. (3.42). Consider, for example, a coherently generated optical phonon with amplitude

$$\mathbf{u}(\mathbf{x}, t) = \exp(-Q_{||}z)[\mathbf{u}_0 \exp(i\mathbf{Q}_{||} \cdot \mathbf{x}_{||} - i\omega_s t) + \text{c.c.}].$$

Here the probability the electron is reflected from the surface with energy transfer $\pm n\hbar\omega_s$ is given by [2]

$$S_n^{(\pm)} = \tilde{\Gamma}_n |R_1|^2, \tag{3.46}$$

with

$$\tilde{\Gamma}_n = |J_n(\tilde{\Gamma})|^2$$

where $J_n(x)$ is the ordinary Bessel function, and

$$\tilde{\Gamma} = \frac{16\pi v_\perp nee^* |u_0|^2}{(1 + \varepsilon_\infty)[(\omega_s - \mathbf{v}_{||} \cdot \mathbf{Q}_{||})^2 + v_\perp^2 Q_{||}^2]}.$$

3.4 SURFACE LOSSES

3.4.1 Interpretation of the Loss Function in Terms of the Eigenmodes of a Slab

In Section 3.2, we have developed a formalism which allows us to describe electron energy losses from charge density fluctuation within a semi-infinite dielectric medium, or a dielectric covered by a thin dielectric overlayer. The formulas for the inelastic cross section in this scheme become independent of the nature of the excitation, and in the last section we have already discussed the application to scattering by both intraband electronic transitions (Figs. 3.5–3.7), and interband transitions of the bulk (Fig. 3.8). We have also found that the probing depth of the electron was of the order of 100 Å. Therefore energy losses of that kind might be called "surface losses" in some sense, and the pole in the loss function $\text{Im}(-1/(\varepsilon_b(\omega) + 1))$ for a free electron gas is in fact called the surface plasmon, since the surface is a prerequisite for the excitation. Nevertheless, the surface plasmon and other elementary excitations on a depth scale of 100 Å do not represent surface losses to the extent that they are characteristic of the outermost surface layer of the substrate, or an adsorbate layer to which we wish to address ourselves now. A different excitation spectrum for the surface atoms or an adsorbed layer can be accounted for within the inelastic scattering formalism by using the approximation of a thin surface layer, i.e., $Q_{||}d \ll 1$, where $\hbar Q_{||}$ is the parallel momentum transfer and d the thickness of the layer. For this situation we have found the loss spectrum to be proportional to [Eq. (3.19)]

$$I_{inel} \sim \frac{\varepsilon_b^2}{(\varepsilon_b + 1)^2} \text{ Im} \frac{-1}{\varepsilon_s(\omega)} + \frac{1}{(\varepsilon_b + 1)^2} \text{ Im } \varepsilon_s(\omega) \qquad (3.47)$$

when ε_b, the bulk dielectric constant, is assumed to be real and frequency independent for the moment. The poles of these two contributions for the cross section occur at the frequencies of the longitudinal and transverse *bulk* modes of the thin layer. This, at first glance, is a somewhat puzzling result since the electric field of a true bulk excitation is necessarily zero outside the material itself, as we shall see, and cannot lead to inelastic electron scattering as long as the electron does not penetrate into the material. We have assumed this in the derivation of the scattering formalism. For the purpose of a better understanding of Eq. (3.47), we therefore briefly discuss the eigenmodes of slab in the following.

For an isotropic dielectric medium with a purely real dielectric function $\varepsilon_b(\omega)$, bulk eigenmodes are obtained by letting either div $\mathbf{P} = 0$ or curl $\mathbf{P} = 0$, where $\mathbf{P}$ is the polarization. This leads to transverse and longitudinal waves, respectively. Here again we assume $c \to \infty$, i.e., we are considering unretarded solutions only, as already discussed. Surface modes of the dielectric are obtained by letting div $\mathbf{P} =$ curl $\mathbf{P} = 0$ and div $\mathbf{E} =$ curl $\mathbf{E} = 0$. Surface solutions therefore satisfy the Laplace equation for the potential $\varphi(\mathbf{x}, t)$. A particular solution for the slab geometry is

$$\varphi(x, z, t) = e^{i(Q_{||}x - \omega t)}(Ae^{-Q_{||}(z - d/2)} + Be^{+Q_{||}(z + d/2)}). \qquad (3.48)$$

The two constants A and B must be chosen to satisfy the boundary condition for the normal component of $\mathbf{D} = \mathbf{E} + 4\pi\mathbf{P}$ at the interfaces. This leads to a secular equation of the form

$$\frac{\varepsilon_s(\omega) - 1}{\varepsilon_s(\omega) + 1} \frac{\varepsilon_s(\omega) - \varepsilon_b}{\varepsilon_s(\omega) + \varepsilon_b} = e^{-2Q_{||}d}. \qquad (3.49)$$

This condition is equivalent to the condition for a pole in the loss function for the layer–substrate system [Eq. (3.16)]. The dispersion of the solutions of Eq. (3.49) when $\varepsilon_s(\omega)$ has the typical dispersion of an harmonic oscillator are depicted in Figs. 3.10a and b for the cases $\varepsilon_b > 0$ and $\varepsilon_b < 0$, respectively. In the limit $Q_{||}d \gg 1$, the coupling between each of the surface waves localized on the two interfaces vanishes, and the frequencies approach the frequencies of the surface waves on the boundary between two semi-infinite materials,

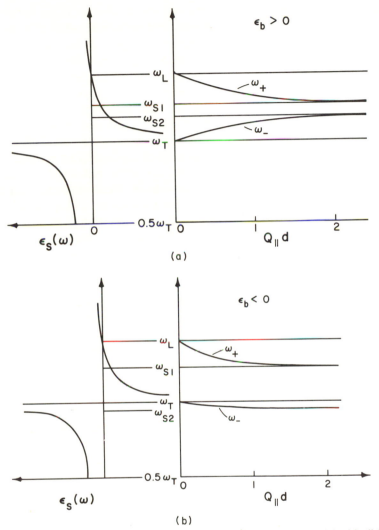

Fig. 3.10 Surface eigenmode dispersion relations for a slab of material with dielectric function $\varepsilon_s(\omega)$ are shown when the slab is placed on a substrate with dielectric constant ε_b. We show the dispersion relation for two cases, (a) $\varepsilon_b > 0$ and (b) $\varepsilon_b < 0$.

which is given by the condition

$$\varepsilon_s(\omega) = -\varepsilon_{\text{outside}}, \tag{3.50}$$

where $\varepsilon_{\text{outside}}$ is the appropriate dielectric constant of the adjacent medium, i.e., 1 or ε_b in our case. In the limit $Q_{\|}d \ll 1$, the frequencies of the surface waves approach the frequencies of the bulk excitations ω_L and ω_T in the slab, respectively. Nevertheless, they remain surface waves when $Q_{||}d$ is finite, even though it is small, with an electric field outside the slab. The field, however, reduces in intensity as $Q_{||}d$ approaches zero. This is the origin of the extra factor of $Q_{||}d$ included in the prefactor of Eq. (3.19), and makes the intensity of the surface losses decrease less rapidly than bulk losses when observed progressively farther from the specular direction $(Q_{||} \neq 0)$. It is also illuminating to consider the electric field of the two branches of surface modes (Fig. 3.11). The branch which approaches ω_T for small $Q_{||}d$ in this limit represents a polarization of the surface layer entirely parallel to the surface while the branch which approaches ω_L is polarized entirely perpendicular to the slab surfaces in the same limit. The two terms in Eq. (3.47) therefore represent losses from these two excitations. For a metal surface (and many semiconductors where $\varepsilon_b \gg 1$), only the first term with the perpendicular polarization survives. Thus, from a very different viewpoint, we recover the screening of parallel dipole moments which we have considered in Chapter 1, and Fig. 1.3. By comparing the two situations $\varepsilon_b = 1$ and $\varepsilon_b \to \infty$ we also see that the intensity of inelastic scattering from perpendicular polarizations is quadrupled as a result of the image of the dipole in the substrate, just as the local semimicroscopic picture predicts.

We finally note that in the limit where $|\varepsilon_b| \gg 1$, the loss function from Eq. (3.19)

$$\mathrm{Im} \ \frac{-1}{\varepsilon_s(\omega)} \frac{\varepsilon_b^{\,2}(\omega) - \varepsilon_s^2(\omega)}{[\varepsilon_b(\omega) + 1]^2}$$

reduces to simply $\mathrm{Im}[-1/\varepsilon_s(\omega)]$, even when $\varepsilon_b(\omega)$ is complex and frequency dependent. In the same limit, the ω_- branch of the surface waves degenerates with the transverse "bulk" frequency ω_T of the slab. No field is associated with this mode. If the dielectric layer is considered to consist of an ensemble of harmonic oscillators of dipolar character, ω_T is therefore equivalent to the mechanical eigenfrequency ω_0 of the oscillator. This observation will help to make contact with a microscopic description of the surface layer.

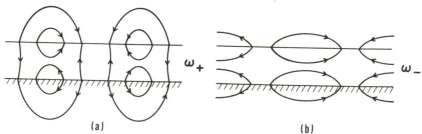

Fig. 3.11   The electric fields present, when the (a) ω_+ and (b) ω_- modes of Figs. 3.10(a) and (b), respectively, are excited.

3.4.2 Infrared Reflection Spectroscopy

As already mentioned in the Introduction, infrared reflection spectroscopy has also become an important tool for studies of the vibrations of adsorbed molecules. Compared to electron spectroscopy, the resolution is substantially better. On the other hand, it is difficult to cover a wide range of frequencies. As a consequence, up to now most experimental work has concentrated on the molecular vibrations of adsorbed CO. It has been found that the frequencies observed with infrared and electron spectroscopies compare quite well. In fact it can be shown that for a metal substrate, electron spectroscopy and infrared reflection studies probe the same aspects of the dielectric response of the surface layer. Even intensities observed with the two techniques can be related to each other [7].[1]

In infrared reflection spectroscopy the signal is detected as a *reduction* of the intensity R reflected from a metal surface. This reduction can be calculated by considering once again a layered system in the limit where the active surface layer is thin, and the reduction in intensity is small [8]. The result for light polarized perpendicularly and parallel to the plane of incidence is [9], with θ_1 the angle of incidence,

$$\frac{\Delta R_\perp}{R_\perp} = \frac{8\pi d \cos \theta_1}{\lambda} \operatorname{Im} \frac{\varepsilon_s(\omega) - \varepsilon_b}{1 - \varepsilon_b}, \tag{3.51}$$

$$\frac{\Delta R_{||}}{R_{||}} = \frac{8\pi d \cos \theta}{\lambda} \operatorname{Im} \frac{\varepsilon_s(\omega) - \varepsilon_b}{1 - \varepsilon_b} \frac{1 - [(\varepsilon_b + \varepsilon_s(\omega))/\varepsilon_b \varepsilon_s] \sin^2 \theta_1}{1 - [(1 + \varepsilon_b)/\varepsilon_b] \sin^2 \theta_1}. \tag{3.52}$$

For a metal where $|\varepsilon_b| \gg 1$ the component of the electric field in the incident light parallel to the surface is nearly zero and $\Delta R_\perp$ is consequently very small. For $\Delta R_{||}$ we obtain in the same limit after some algebra,

$$\frac{\Delta R_{||}}{R_{||}} = \frac{8\pi d}{\lambda} \sin^2 \theta_1 \left[\cos \theta_1 \left(1 + \frac{1}{|\varepsilon_b|^2} \frac{\sin^2 \theta_1}{\cos^4 \theta_1} \right) \right]^{-1} \operatorname{Im} \frac{-1}{\varepsilon_s(\omega)}. \tag{3.53}$$

Apart from kinematical factors, infrared reflection and electron spectroscopy probe the same type of surface loss. These are characterized by the pole of $\operatorname{Im}(-1/\varepsilon_s(\omega))$. This effect was first noticed when thin alkali halide layers deposited on silver seemed to show an absorption line near the longitudinal optical phonon frequency, rather than at the transverse optical phonon

[1] In this paper, the prefactor in the expressions for both the electron scattering intensity and the infrared reflectivity are in error by a factor of 2. The error in the expression for the electron scattering intensity evidently appears earlier in Ref. [2]; the 8 in Eq. (3.28) of the first paper cited in Ref. [2] should be changed to a 4, and the remaining results corrected. In addition, in the expression for both the electron scattering intensity and the infrared reflectivity, the factor ε_∞ should appear as indicated here, if e^* is to be interpreted as the dynamic charge of a single, isolated entity adsorbed on the surface. Table 1 of the paper cited in the present footnote thus really tabulates the quantity $e^*/\sqrt{2}\varepsilon_\infty$.

frequency (Berreman effect [10]). We understand from our previous discussion, however, that it is not the bulk mode but a surface mode of the layer which is actually observed in the experiment [11].

The prefactor in Eq. (3.53) is also quite interesting. The reduction in reflectivity has a maximum near grazing incidence at approximately, for $|\varepsilon_b| \gg 1$,

$$\theta = \tfrac{1}{2}\pi - (3^{1/4}/|\varepsilon_b|^{1/2}),\tag{3.54}$$

at which point $\Delta R_\parallel / R_\parallel$ becomes

$$\frac{\Delta R_\parallel}{R_\parallel} = \frac{8\pi d}{\lambda}\frac{3^{3/4}}{4}\sqrt{|\varepsilon_b|}\operatorname{Im}\frac{-1}{\varepsilon_s(\omega)}.\tag{3.55}$$

The reduction in the reflectivity as a function of the angle of incidence is plotted in Fig. 3.12, for a platinum surface with $\varepsilon = -120 + 340i$ and the integrated absorption strength is matched to represent the observed value for a monolayer of CO on platinum.

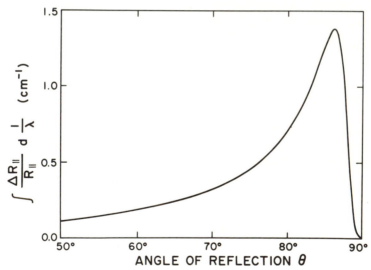

Fig. 3.12 The reflectivity change, integrated over frequency, as a function of reflected angle θ for p-polarized light incident on an adsorbate-covered CO surface. The data represent CO on Pt(111) (H. Ibach [7] and further references therein, used with permission).

3.4.3   Application to Vibration of Adsorbed Molecules

While up to now the formal description of energy losses or reflectivity change produced by the surface layer did not specify the nature of the elementary excitation responsible, we now orient the discussion to vibrations of adsorbed layers on metals or semiconductors with $|\varepsilon_b| \gg 1$. The simplest model of the dielectric function for an adsorbed layer with a single (perpendicular) vibrational frequency is

$$\varepsilon_s(\omega) = \varepsilon_\infty + \frac{\omega_p^2}{\omega_T^2 - \omega^2 - i\gamma\omega}, \tag{3.56}$$

where ε_∞ is high-frequency limit of $\varepsilon_s(\omega)$ above the vibrational resonance and ω_p is the ion-plasma frequency. For adsorbed diatomic molecules with no mechanical coupling, ω_p is proportional to the density of molecules n

$$\omega_p^2 = 4\pi e^{*2} n/M_r, \tag{3.57}$$

where e^* is an effective ionic charge and M_r the appropriate reduced mass. The transverse eigenfrequency ω_T is equal to the mechanical eigenfrequency of the adsorbed molecules (with no local field corrections in this case of a layer on a metal surface). With this notation the loss function $\mathrm{Im}(-1/\varepsilon_s(\omega))$ assumes the form, in the limit of small γ,

$$\mathrm{Im}(1/\varepsilon_s(\omega)) = (\pi\omega_p^2/2\omega_s\varepsilon_\infty^2)\{\delta(\omega_s - \omega) + \delta(\omega_s + \omega)\}. \tag{3.58}$$

The resonance frequency is given by

$$\omega_s^2 = \omega_T^2 + (\omega_p^2/\varepsilon_\infty) = \omega_T^2 + (4\pi e^{*2} n/M_r\varepsilon_\infty), \tag{3.59}$$

which we have already shown to be equivalent to the bulk longitudinal frequency of the layer. We note that we shall have to link up our macroscopic description with a proper microscopic theory before we have a precise definition of ω_T, e^*, and ε_∞. In the case of a very dilute adsorbate system, we may assume that ω_T and e^* are independent of n, and $\varepsilon_\infty = 1$. Then the intensity becomes proportional to the concentration n. We shall come back to this problem in more detail shortly.

With the help of Eqs. (3.19) and (3.58), the integrated intensity of an electron loss line divided by the elastic intensity becomes, in the limit $\hbar\omega_s \gg k_B T$,

$$\frac{I_{\text{inel}}}{I_{\text{el}}} = \frac{1}{|R_I|^2} S = \frac{2\pi\hbar}{a_0 E_I \cos\theta_I} (1 - 2\vartheta_E)^{1/2} F_s(\hat{\vartheta}_c) \frac{e^{*2} n_s}{M_r\omega_s\varepsilon_\infty^2}, \tag{3.60}$$

where we have replaced $d \cdot n$ by n_s, the surface concentration, and where $F_s(\hat{\vartheta}_c)$ is

$$F_s(\hat{\vartheta}_c) = \frac{2}{\pi} \int_0^{2\pi} d\varphi \int_0^{\hat{\vartheta}_c} \hat{\vartheta}\, d\hat{\vartheta}\, \frac{(\sin\theta_1 - \hat{\vartheta}\cos\theta_1\cos\varphi)^2 + \hat{\vartheta}^2\sin^2\varphi}{[1 + \hat{\vartheta}^2]^2}$$

$$= (\sin^2\theta_I - 2\cos^2\theta_I)\frac{\hat{\vartheta}_c^2}{1 + \hat{\vartheta}_c^2} + (1 + \cos^2\theta_I)\ln(1 + \hat{\vartheta}_c^2). \tag{3.61}$$

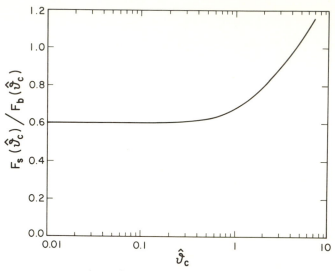

Fig. 3.13 The ratio $F_s(\hat{\vartheta}_c)/F_b(\hat{\vartheta}_c)$ for the case where the angle of incidence $\theta_1 = 70°$.

TABLE 3.2

The Function $F_s(\hat{\vartheta}_c)$

$\hat{\vartheta}_c$ \ θ_1	70°	60°	50°	40°
0.1	0.0175	0.0149	0.0117	0.0083
0.2	0.0688	0.0586	0.0462	0.0330
0.3	0.150	0.128	0.102	0.0740
0.4	0.255	0.220	0.177	0.131
0.5	0.379	0.329	0.267	0.202
0.6	0.515	0.451	0.371	0.287
0.7	0.659	0.581	0.485	0.383
0.8	0.806	0.716	0.606	0.488
0.9	0.953	0.854	0.731	0.601
1.0	1.10	0.991	0.860	0.720
1.2	1.38	1.26	1.12	0.967
1.4	1.64	1.52	1.37	1.22
1.6	1.89	1.77	1.62	1.47
1.8	2.11	2.00	1.86	1.71
2.0	2.32	2.21	2.08	1.95
3.0	3.16	3.10	3.04	2.97
4.0	3.78	3.78	3.78	3.78
5.0	4.26	4.31	4.37	4.44
6.0	4.66	4.76	4.87	4.99
7.0	5.01	5.14	5.29	5.46
8.0	5.30	5.46	5.66	5.88
9.0	5.56	5.76	5.99	6.24
10.0	5.80	6.02	6.28	6.57

Similar to the bulk loss intensities [Eq. (3.23)], $F_s(\hat{\vartheta}_c)$ is proportional to $\hat{\vartheta}_c^2$ for small $\hat{\vartheta}_c$,

$$\lim_{\hat{\vartheta}_c \to 0} F_s(\hat{\vartheta}_c) = 2\hat{\vartheta}_c^2 \sin^2 \theta_I. \tag{3.62}$$

However, $F_s(\hat{\vartheta}_c)$ does not approach a finite upper limit as the corresponding function for the bulk losses $F_b(\hat{\vartheta}_c)$ does. The ratio of the two functions is plotted in Fig. 3.13. We have also tabulated $F_s(\hat{\vartheta}_c)$ in Table 3.2.

We can cast Eq. (3.69) into another form using the expectation value of the perpendicular component of the dipole moment,

$$\hbar e^{*2}/2M_r\omega_s = |\langle 0|\mu_\perp|v\rangle|^2. \tag{3.63}$$

Then the inelastic intensity due to a transition from the vibrational ground state to the vth vibrational state becomes

$$\frac{I_{\text{inel}}}{I_{\text{el}}} = \frac{4\pi(1 - 2\vartheta_E)^{1/2}n_s}{a_0 E_I \cos\theta_I \varepsilon_\infty^2} |\langle 0|\mu_\perp|v\rangle|^2 F_s(\hat{\vartheta}_c). \tag{3.64}$$

In this form the equation for the loss intensity is applicable to adsorbed molecules of arbitrary structure. The same equation can also be derived more directly [2, 12] by considering the interaction of electrons with dipole active harmonic oscillators. The advantage of the presentation we have chosen is that it shows the intimate relations between various spectroscopies as well as between the vibration spectroscopy and spectroscopy of electronic transitions.

We note that Eq. (3.64) still contains ε_∞ of the adsorbed layer, a quantity which can be defined only if the layer may be considered as an entity separable from the substrate. In fact, Eq. (3.64) should only be used as long as the dipole moment matrix element can be identified with properties of the molecule alone. As soon as the chemical interaction between the molecule and the surface causes a significant redistribution of charge between the surface and the molecule, the model of the dielectric layer with image charge in the substrate breaks down. Instead of considering the dipole moment of the molecule and its image, one then should consider the dipole moment of an entire surface complex, including the metal atoms. Then of course the dipole moment operator includes the normal coordinates of the metal atoms as well. One must also bear in mind that in this description a factor of 4 should be eliminated in Eq. (3.64), since the image effect is now included in the matrix element. We also have to find a proper definition of ε_∞, which will be discussed in the following subsection.

The results of this subsection have been developed for application to the scattering of electrons from an adsorbed layer of molecules whose vibrational motion leads to the appearance of a dipole moment. If we consider a perfect crystal, with no adsorbates present, the atoms in the outermost one or two atomic layers may behave in a fashion similar to adsorbed molecules, in that their vibrational motion may generate macroscopic fields in the vacuum above the crystal. The atoms in the surface layer necessarily sit at a site which lacks an inversion center. If such atoms are displaced from their equilibrium position, a local dipole moment is generated in the surface, and this produces a field in the vacuum above the crystal. In a metal, if an atom deep inside the crystal is displaced, a local dipole moment is produced, but the electric field produced by it is screened to zero in a distance the order of 1 Å. If the atom is in the surface, there are no electrons in the vacuum above it, and an incoming particle sees the whole dipole moment. In a semiconductor such as silicon, where the infinitely extended crystal contains an inversion center, a symmetry argument requires displacements of the atoms by a long-wavelength optical phonon to generate no macroscopic electric field. A surface optical phonon may generate an electric field, however, since the symmetry argument breaks down there. In both of these examples, one may view the outermost one or two atomic layers as "dipole active," in a manner very similar to an adsorbed layer of molecules. In fact, with parameters interpreted properly, the formulas of the present section apply to this case.

One of the early studies of low-energy electron energy loss spectroscopy explored the (2 × 1) reconstructed (111) surface of silicon [13]. Here one finds scattering from a surface optical phonon with $\hbar\omega_s = 56$ meV. The mode is clearly microscopic in origin, i.e., it is an eigenmode in which the atomic displacements are confined to the outermost one or two atomic layers (see also Section 5.3.2). The scattering is clearly dipolar in nature, and interpretation of the data within the present model leads to the estimate $e^*/\varepsilon_\infty \approx 0.5e$ to $1.0e$ [2], where e is the magnitude of the electron charge.

3.4.4   Frequency Shifts and the Coverage Dependence of Intensities

It is a commonplace and frequently discussed observation in infrared reflection spectroscopy and electron energy loss spectroscopy that the frequencies of adsorbed molecules shift with increasing coverage. Furthermore, the intensities are found to be not proportional to the coverage, but rather seem to saturate at higher coverages. The frequency shifts and the nonlinearity of the intensity was found to be particularly large for adsorbed CO, and they are thus readily observable even at relatively small coverages. On the other hand, frequency shifts observed with hydrocarbons were found to be generally much smaller. This suggests that a long-range dipole coupling between the adsorbed CO molecules causes the frequency shifts, and not so much a direct chemical interaction or a substrate mediated interaction. Several attempts have been made to calculate the shifts in a microscopic model for the dipole–dipole interactions. Despite the experimental hints suggesting dipole coupling to be the major reason for the frequency shift, Mahan and Lucas [14] concluded that the effect was too small. In their calculation, however, the authors used the oscillator strength of gas phase CO, which is indeed much too small to produce the experimentally observed intensities. In fact it is easy to understand that the oscillator strength for adsorbed CO should be larger, since an additional contribution to the dipole moment must come from the charge fluctuations produced by the carbon atom moving against the metal surface. Later on, this increased oscillator strength was modeled by considering the "self-image" of the adsorbed CO [15]. In this model, however, not only the far field of the imaged dipole but also the field at nearest neighbor distances enters. At this point, the image model becomes questionable as a realistic representation of the system.

In a microscopic description of a chemically strongly bound atom or molecule (such as CO on a transition metal surface), it is more natural to consider the adsorbed species together with its nearest-neighbor surface atoms as a unit, and describe this unit by an effective ionic charge q^* and an electronic polarizability α_e. Then the question arises how the quantities e^* and ε_∞ and the frequency shift of the macroscopic model described earlier are related to the newly defined q^* and α_e.

In the limit of very small coverage, the contact between the two descriptions is easily made, since then $\omega_s \to \omega_0$ and $\varepsilon_\infty \to 1$. As we have already discussed, the effective charge e^* is to be replaced by $\frac{1}{2}q^*$, since we are now including the image in our q^*, while we have seen that in the macroscopic picture the image added an extra factor of 4 to the cross section from this source.

The effect of finite coverages can be studied for the simple case of adsorbed atoms or diatomic molecules like CO. Since we are interested mainly in the eigenmodes for $Q_\parallel \cong 0$, we can assume all surface units to move in phase. The equation of motion in this case is then, with u the normal coordinate of a simple entity,

$$M_r \ddot{u} + M_r \omega_0^2 u = q^*(E_{ext} + E_{dip}),\qquad (3.65)$$

where M_r is the appropriate reduced mass, E_{ext} an external field, and E_{dip} the dipole field of all other units, which is

$$E_{dip} = -p \sum_{i \neq j} (1/r_{ij}^3) \qquad (3.66)$$

with

$$p = q^*u + \alpha_e(E_{ext} + E_{dip}). \qquad (3.67)$$

These equations can be rearranged to yield

$$M_r \ddot{u} + M_r \left(\omega_0^2 + \frac{q^{*2}\Sigma}{M_r(1 + \alpha_e\Sigma)} \right) u = \frac{q^*}{1 + \alpha_e\Sigma} E_{ext}. \qquad (3.68)$$

From this equation we see that the eigenfrequency of the system is now

$$\omega_s^2 = \omega_0^2 + \frac{q^{*2}\Sigma}{M_r(1 + \alpha_e\Sigma)}, \qquad (3.69)$$

which may be compared to the macroscopic relation in Eq. (3.59). The interaction with an external field is now determined by a factor $q^*/(1 + \alpha_e\Sigma)$ rather than by e^*/ε_∞. We therefore are to replace $2e^*/\varepsilon_\infty$ in Eq. (3.60), by $q^*/(1 + \alpha_e\Sigma)$ to obtain

$$\frac{I_{iael}}{I_{el}} = \frac{\pi h}{2a_0 E \cos\theta_I} (1 - 2\vartheta_E)^{1/2} F_s(\hat{\vartheta}_c) \frac{q^{*2}}{M_r \omega_s} \frac{n_s}{(1 + \alpha_e\Sigma)^2}. \qquad (3.70)$$

We may also at this point transform the relation for the infrared reflectivity change to

$$\int \frac{\Delta R_\parallel}{R_\parallel} d\, 1/\lambda$$

$$= \frac{\pi}{c^2} \sin^2\theta_I \left[\cos\theta_I \left(1 + \frac{1}{|\varepsilon_b|^2} \frac{\sin^2\theta_I}{\cos^4\theta_I} \right) \right] \frac{q^{*2}}{M_r} \frac{n_s}{(1 + \alpha_e\Sigma)^2}. \qquad (3.71)$$

In both equations the denominator $(1 + \alpha_e\Sigma)^2$ introduces a nonlinearity of the intensity with increasing coverage. In Eqs. (3.70) and (3.71), the effective charge may be replaced by an ionic polarizability α_v

$$\alpha_v = q^{*2}/M_r\omega_0^2, \qquad (3.72)$$

where ω_0^2 is the (unperturbed) eigenfrequency, or the eigenfrequency in the low coverage limit. We note again however, that α_v and α_e are not the gas phase values for an adsorbed molecule, but the corresponding quantities for the adsorbate complex. If one wishes to make quantitative estimates of the nonlinearity of the intensities and of the expected frequency shift, values of α_v and α_e have to be taken from experimental data on the adsorbed layer. For instance, a careful measurement of the intensity versus the coverage plus a reasonable model for the dipole sum should provide the data for predicting the frequency shift. One has examples of particular interest, where the nonlinearity of intensity with coverage is very pronounced. Such a case was reported by Pfnür et al. [16] for CO on Ru(001), where the intensity actually passes through a maximum (Fig. 3.14) as a result of the screening effect produced by the electronic polarizability α_e. From the absolute intensity of the reflectivity change in Ref. [16], one obtains $\alpha_v/(1 + \alpha_e\Sigma)^2 \sim 0.25 \text{ Å}^3 \pm 20\%$ at the maximum. In order to obtain the functional dependence

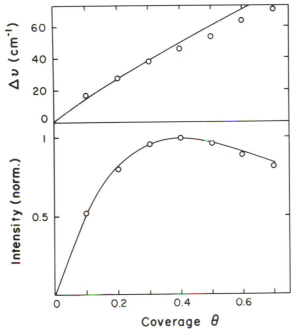

Fig. 3.14 The intensity and frequency shift as a function of coverage for CO adsorbed on Ru(001). The experimental points are from Ref. [16], and the solid lines are from the theory described in the text. $\alpha_e = 3.7 \text{ Å}^3$, $\alpha_v = 0.56 \text{ Å}^3$.

of the intensity on the coverage Θ, one has to consider the lattice sum $\sum_{i \neq j} 1/r_{ij}^3$ in more detail. Most of the depolarizing effect occurs at higher coverage, where the lateral interaction between the CO molecules is repulsive. Then it is reasonable to use the lattice sum for a uniformly compressed structure. The lattice sum, then, is

$$\sum_{i \neq j} 1/r_{ij}^3 = 8.9 n_s^{3/2} \tag{3.73}$$

for a hexagonal, close-packed lattice structure [14]. The prefactor is not very structure sensitive. With this lattice sum the value of α_e can be taken from the coverage where the maximum in the intensity occurs and one now obtains $\alpha_e = 3.7 \text{ Å}^3$ and $\alpha_v \sim 0.56 \text{ Å}^3$ from the absolute intensity. With these two values taken from essentially one data point, a good overall description of the intensity versus coverage curve *and* the frequency shift is obtained as shown in Fig. 3.14. The solid lines are the theoretical predictions, once the information from the single data point has been used. We conclude, therefore, that the major contribution to the frequency shifts observed with adsorbed CO results from dipole–dipole coupling. The same conclusion was reached by Andersson and Persson [17] in their study of the frequency shift versus momentum transfer $Q_{||}$ for CO on Cu(001). We note that the values for α_v and α_e we deduce from the fit in Fig. 3.14 are substantially larger than those for gaseous CO (0.057 and 2.5). It is therefore no surprise that Mahan and Lucas [14] in their earlier discussion of the same problem had calculated a much lower frequency shift when they used these gas phase values. It is, however, rather obvious that $\alpha_v = 0.057$ Å^3 would be too low by an order of magnitude to account for the actually observed intensities.

3.5   THE IMPACT SCATTERING REGIME

Except for the introductory remarks in Section 3.1, our treatment of inelastic electron scattering has been confined to the regime of very small-angle scattering by excitations of the bulk and surface regions which generate dipole fields in the vacuum above the crystal surface. From the opening remarks in the chapter, we have seen that at large scattering angles, simple phenomenology cannot be used to describe the electron scattering processes. One must resort to a microscopic theory instead. Also, even near the specular, certain vibrational motions of adsorbates, such as that of an adsorbate atom parallel to the surface, fail to generate an oscillating dipole normal to the surface, and are hence "dipole forbidden." Here again, one must resort to a microscopic theory to obtain a full description of the scattering cross section.

Even for dipole-allowed modes, examination of the cross section in the regime of large-angle deflections shows the cross section does not fall off, but instead levels off to a value substantially lower than found near the specular, but the scattering is nonetheless detectable. This is again impact scattering. Until recently, rather little attention from either theorists or experimentalists has been devoted to large-angle inelastic scattering of electrons from adsorbate vibrations. The topic has been under active discussion in the recent literature, and a number of basic features of the impact scattering mechanism have been outlined. The area is potentially very rich, in that the energy and angle variation of the loss cross section contain detailed information on the surface geometry, as we shall see.

In Fig. 3.15, we show data on the angular variation of the energy loss cross section for scattering from the vibrational motion of hydrogen adsorbed on the surface of W(100) [18]. The hydrogen forms an ordered monolayer (the β phase in this case), with a hydrogen atom sitting on a bridge site between two tungsten atoms. The site has C_{2v} symmetry. Thus, if we assume only the hydrogen atom moves and the tungsten atoms remain at rest when the hydrogen vibrates, there are three normal modes of vibration each with a distinctly different frequency. One vibrational mode is the vibration normal to the surface. We then have two normal modes in which

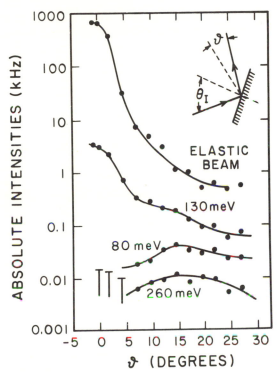

Fig. 3.15   The angular profile of electrons scattered inelastically from an ordered mono-layer of hydrogen adsorbed on the W(100) surface. The figure, taken from the work of Ho, *et al.* [18], shows the angular profile of the elastic beam, that for scattering from the 130-meV vibration (normal to the surface), the 80-meV vibration (parallel to the surface), and the overtone of the 130-meV mode at 260 meV.

the hydrogen moves parallel to the surface. One involves motion of the hydrogen perpendicular to the triangle formed by it and the two tungstens involved in the bridge bond, and in the second the hydrogen rocks back in forth parallel to the plane of the triangle.

Figure 3.15 first shows the angular profile of the specular (elastic) beam. The intensity scale is logarithmic, so the intensity of the specular beam drops off by two orders of magnitude, when the angular deflection is 5°. Also shown is the angular profile of the cross section for scattering from the mode at 130 meV. At large deflection angles, the scattering cross section varies smoothly with scattering angle, to assume a value some four orders of magnitude smaller than the intensity of the elastic beam on specular. As one moves in toward the specular direction, for small values of the deviation away from specular, the intensity of scattering from the 130-meV mode increases dramatically. Near the specular, the scattering intensity from the 130-meV mode is larger by roughly a factor of 50 than it is at large deflection angles. The 130-meV mode is the vibration of the hydrogen normal to the surface, and the dramatic increase in intensity as one moves toward the specular direction is the dipole scattering. The angular width of the dipole lobe is roughly the same as that of the elastic beam; this suggests that the characteristic angle ϑ_E is smaller than the angular resolution of the apparatus. The impact energy of the electrons is 9.65 eV for the data shown in Fig. 3.15 so we have $\vartheta_E = \hbar\omega/2E_I = 0.370°$, which is indeed smaller than the half-width of the specular beam.

Also shown in Fig. 3.15 is the angular profile for scattering from a mode with $\hbar\omega = 80$ meV. We see no rise in the scattering intensity as the specular direction is approached, so this is a mode in which the hydrogen moves parallel to the surface. This low-frequency mode, in fact, is the rocking of the hydrogen perpendicular to the plane it forms with the two tungsten atoms which participate in the bridge bond. By our previous criterion, this is a "dipole forbidden" mode. Not only do we see no increase in the cross section as the specular direction is approached, but the scattering cross section drops substantially as ϑ becomes small. We shall see shortly that in the impact scattering regime, we again encounter selection rules [19] which complement the dipole selection rule introduced in the early Literature [2]. These require the impact cross section for a parallel vibration such as this to *vanish* as the specular direction is approached, and the behavior of the 80-meV mode displayed in Fig. 3.15 is consistent with the selection rule. Not given in Fig. 3.15 is the cross section for scattering from the 160-meV mode, which is the motion of the hydrogen atom parallel to the plane of the triangle. The 260-meV feature is the overtone of the 130-meV mode.

We next turn to a description of the theory of inelastic electron scattering by adsorbate vibrations. We can see the main outlines from the remarks in Section 3.1, supplemented by the following.

If we consider the angular regime in Fig. 3.15 where ϑ is in the range 20–30°, then $Q_{\parallel} \approx 10^8$ cm^{-1}, and the length $l_0(Q_{\parallel})$ that describes, in the dipole theory, the distance the electron is above the surface when it scatters, becomes $l_0(Q_{\parallel}) \cong Q_{\parallel}^{-1} = 1$ Å. Hence the physical picture of the scattering event differs qualitatively from that discussed in the small-angle regime. The electron does not "see" the electric field until it is 1 Å from the surface. Since the electron typically penetrates into the material a distance the order of 10 Å, the inelastic event is initiated primarily while the electron is *inside* the structure, not when it is in the vacuum far above it. To describe the impact scattering regime, we require a microscopic theory capable of representing the electron wave function near the surface of the material. We now turn to a discussion of how such calculations may be carried out.

3.5.1 Basic Theory of Inelastic Scattering in the Impact Regime

In quantum-mechanical scattering theory, the basic quantity that enters is the scattering amplitude $f(\mathbf{k}^{(S)}, \mathbf{k}^{(I)})$ that describes the amplitude of the wave that radiates outward from the scattering center. If we send a plane wave with wave vector $\mathbf{k}^{(I)}$ into the potential responsible for the scattering, then the outgoing wave has the form $f(\mathbf{k}^{(S)}, \mathbf{k}^{(I)}) \exp(ik^{(S)}r)/r$. Once an expression for the scattering amplitude is obtained, it is a straightforward matter to form an expression for the cross section.

In the discussion of inelastic electron scattering by surface vibrations, the formal discussion may also be framed in terms of a certain scattering amplitude. One begins by invoking the adiabatic approximation central to most analyses in solid state physics. That is, we consider a macroscopic but nonetheless finite solid, and suppose the various nuclei are frozen in positions $\mathbf{R}_1, \mathbf{R}_2, \ldots, \mathbf{R}_N$. The collection of position vectors to the various nuclear sites will be denoted by $\{\mathbf{R}\}$. The Schrödinger equation for the electron is then solved, and the relevant wave functions are obtained. In our case, we consider a finite crystal with surface and possibly adsorbates present, and send an electron with wave vector $\mathbf{k}^{(I)}$ into the surface. Just as in quantum mechanical theory, the electron scatters from the solid, to emerge from it with wave vector $\mathbf{k}^{(S)}$. The probability amplitude that it emerges with this wave vector is again described by a scattering amplitude, just as in elementary quantum mechanics. This amplitude now depends parametrically on the nuclear positions, so we refer to it as $f(\mathbf{k}^{(S)}, \mathbf{k}^{(I)}; \{\mathbf{R}\})$.

If the nuclei are fixed at the lattice sites of a perfect semi-infinite crystal, with ordered layer of adsorbates present, then the calculation of the scattering amplitude already outlined is that encountered in the theory of low-energy electron diffraction (LEED) [20]. If we consider the two-dimensional unit cell appropriate to the surface region of the structure, and denote by $\mathbf{G}_{\parallel}$ the reciprocal lattice vectors associated with this structure, then the scattering amplitude is nonzero only when $\mathbf{k}_{\parallel}^{(S)} = \mathbf{k}_{\parallel}^{(I)} + G_{\parallel}$, i.e., from the perfectly periodic structure, we have only Bragg scattering. This scattering is elastic, so the energy E_S of the scattered electron equals that E_I of the incident electron. In the presence of the surface, the wave vector is conserved within a reciprocal lattice vector for wave vector components parallel to the surface but there is complete breakdown of wave vector conservation for components normal to the surface, because of the lack of translational symmetry in this direction.

When the electron encounters the solid, the nuclei are not fixed in their equilibrium positions at each instant of time, but thermal vibrations shift them away. The solid is in fact disordered, with nucleus i at the position

$\mathbf{R}_i = \mathbf{R}_i^{(0)} + \mathbf{u}_i$, where $\mathbf{R}_i^{(0)}$ is the vector to the equilibrium site, and $\mathbf{u}_i$ the displacement away produced by the thermal vibration. Because of the thermal disorder, not all the electrons emerge along the Bragg beam directions as they would if the solid were perfectly periodic at each instant of time. A certain fraction are scattered away from the Bragg directions to form a background referred to as the thermal diffuse background. The electron deflected away from the Bragg direction is in fact scattered by emission or absorption of a vibrational quantum. The scattering is thus not elastic, but the electron emerges with energy $E_1 \pm \hbar\omega_s$, with ω_s the frequency of the vibrational motion.

The basic issue addressed in the theory of impact scattering from surface vibrations can then be viewed as the calculation of the energy spectrum of the electrons which contribute to the thermal diffuse background, for angular deflections sufficiently large that dipole scattering may be supposed small (or for scatterings produced by dipole forbidden modes). We may proceed as follows: As we have seen, the basic scattering amplitude $f(\mathbf{k}^{(S)}, \mathbf{k}^{(I)}; \{\mathbf{R}\})$ depends parametrically on the nuclear positions. The amplitude of the thermal vibrations is small, so we may expand the scattering amplitude in powers of the displacement $\mathbf{u}_i$ from equilibrium and keep only the leading term:

$$f(\mathbf{k}^{(S)}, \mathbf{k}^{(I)}; \{\mathbf{R}\}) = f(\mathbf{k}^{(S)}, \mathbf{k}^{(I)}; \{\mathbf{R}^{(0)}\}) + \sum_{i\alpha} \left(\frac{\partial f}{\partial R_{i\alpha}} \right)_0 u_{i\alpha} + \cdots , \quad (3.74)$$

where $u_{i\alpha}$ is the αth Cartesian component of the displacement vector $\mathbf{u}_i$.

Now let s refer to a particular vibrational normal mode of the system, and let $\xi_{i\alpha}^{(s)}$ denote the amplitude of the displacement of nucleus i in Cartesian direction α when a quantum of vibrational motion is excited. Then we may write the displacement $u_{i\alpha}$ in Eq. (3.74) as an operator expressed in terms of the annihilation and creation operators a_s and a_s^+ of the vibrational quanta. If the eigenvectors $\xi_{i\alpha}^{(s)}$ are normalized so that

$$\sum_{i\alpha} |\xi_{i\alpha}^{(s)}|^2 = 1, \quad (3.75)$$

then we have

$$u_{i\alpha} = \sum_s \left(\frac{\hbar}{2\omega_s M_i} \right)^{1/2} \xi_{i\alpha}^{(s)} (a_s + a_s^+), \quad (3.76)$$

where N is the number of unit cells in the system.

It is now clear that the terms proportional to $\mathbf{u}_i$ in Eq. (3.74) describe scatterings in which a single vibrational quantum is either absorbed or emitted. If we consider events in which a particular vibrational quantum is emitted, then it is convenient to introduce the matrix element

$$M(\mathbf{k}^{(I)}, \mathbf{k}^{(S)}; +s) = \langle n_s + 1 | f(\mathbf{k}^{(S)}, \mathbf{k}^{(I)}; \{\mathbf{R}\}) | n_s \rangle$$

$$= (n_s + 1)^{1/2} \left(\frac{\hbar}{2N\omega_s} \right)^{1/2} \left(\frac{\partial f}{\partial Q_s} \right), \qquad (3.77)$$

where

$$\left(\frac{\partial f}{\partial Q_s} \right) = \sum_{i\alpha} \left(\frac{\partial f}{\partial R_{i\alpha}} \right)_0 \frac{\xi_{i\alpha}^{(s)}}{\sqrt{M_i}}, \qquad (3.78)$$

N is the number of unit cells in the system, and n_s the number of vibrational quanta present. From the Bose–Einstein statistics, one has $n_s = [\exp(\hbar\omega_s / k_B T) - 1]^{-1}$.

The cross section for inelastic scattering from the surface, with emission of a vibrational quanta of frequency ω_s is conveniently expressed in terms of the matrix element defined in Eq. (3.77). Before we give the result, it is useful to resort to a more explicit description of the vibrational normal modes of the system. So far, we have said little about the collection of quantum numbers that describe the vibrational excitations, except to use the symbols to denote the collection of them required to label each normal mode uniquely. We shall always confine our attention to structures periodic in the two dimensions parallel to the surface. The surface of the crystal may be reconstructed, but for the purposes of the present discussion we require the reconstructed surface configuration be commensurate with the underlying crystal. If an adsorbate layer is present, we assume it ordered again with structure commensurate with that of the underlying lattice planes.

We then have a two-dimensional unit cell in the plane parallel to the surface that is characteristic of the geometry, and associated with this is a reciprocal lattice in the space of allowed wave vectors. The presence of translational symmetry ensures that the set of vibrational normal modes of the structures may be labeled by a wave vector $\mathbf{Q}_{\parallel}$ that lies within the first Brillouin zone of the two-dimensional reciprocal lattice. Then by s, we mean the value of $\mathbf{Q}_{\parallel}$ that is associated with the mode, along with any other quantum numbers or labels required to identify it. As this volume progresses, we shall discuss in detail the normal modes of vibration characteristic of the surface region. If we have a layer of adsorbates on the crystal, then there will be for fixed $\mathbf{Q}_{\parallel}$ a set of vibrational modes with displacement largely, but not entirely confined to the adsorbate layer itself. To first approximation, such modes may be viewed as normal modes of the adsorbate layer, though we shall appreciate that a proper description of them requires the inclusion of motions in the substrate necessarily always present. Also, associated with a particular $\mathbf{Q}_{\parallel}$, we may have vibrational modes localized to the near vicinity of the surface (surface phonons), even if an adsorbate layer is not present. Finally, there are bulk vibrational quanta that may propagate up to and reflect off the surface, to excite vibrational motion there. In our subsequent discussions we shall consider in detail the scattering produced by each of these types of normal model. For the moment, we replace the symbol s by the combination $Q_{||}\alpha$, where the label α contains all indices other than the wave vector parallel to the surface.

Now in a scattering event, the existence of translational symmetry in the two directions parallel to the surface ensures that in the emission (or absorption) of a phonon with wave vector $Q_{||}$, wave vector components parallel to the surface are conserved to within a reciprocal lattice vectors. For a given scattering geometry, if one searches for electrons that emerge with a particular energy E_S, the difference $k_{||}^{(S)} - k_{||}^{(I)}$ is fixed. If $G_{||}$ is the particular reciprocal lattice vector (necessarily unique, and possibly zero) which when subtracted from $(k_{||}^{(S)} - k_{||}^{(I)})$ yields a vector $Q_{||}$ that lies within the first Brillouin zone, then the vibrational quanta that may scatter the electrons in the direction determined by $k^{(S)}$ are only those with wave vector $Q_{||}$. This follows from the requirement of wave vector conservation parallel to the surface, and outlines how one may isolate the particular phonon modes responsible for scattering electrons into a particular direction. If $(dS_\alpha(k^{(I)}, k^{(S)})/d\Omega)\,d\Omega$ is the probability that the vibrational quanta $(Q_{||}\alpha)$ scatters the electron into the solid angle $d\Omega$, then one may show that, with A the surface area of the crystal,

$$\frac{dS_\alpha(k^{(I)}, k^{(S)})}{d\Omega} = \frac{mE_I \cos^2\theta_S}{2\pi^2\hbar^2 \cos\theta_I} A\left|M(k^{(I)}, k^{(S)}; Q_{||}\alpha)\right|^2. \tag{3.79}$$

In the end, when the normalization condition in Eq. (3.75) is noted, the right-hand side of Eq. (3.79) is independent of A, as it must be. We have considered only inelastic scatterings accompanied by emissions of vibrational quanta. The corresponding results for scatterings by absorption of vibrational quanta are found from the results above, upon replacing the factor $(1 + n_s)^{1/2}$ in Eq. (3.77) by $n_s^{1/2}$, then reversing the sign of $Q_{||}$ everywhere in the foregoing discussion.

First, we need two distinctly different sets of quantities. We require the vibrational frequencies ω_s and the eigenvectors $\xi_{i\alpha}^{(s)}$, and these must come from a theory of lattice vibrations of the surface region. Then we must calculate the derivatives $(\partial f/\partial R_{i\alpha})_0$ of the scattering amplitude with atomic displacement. This portion of Eq. (3.78) requires a detailed theory of how the electron interacts with the atoms of the crystal, when they are displaced from their equilibrium position by the thermal vibrations in the material. Our attention in the remainder of the present section will focus on the properties of the scattering amplitude derivatives $(\partial f/\partial R_{i\alpha})_0$, and the resulting scattering probabilities from models of adsorbate vibrations sufficiently simple that the eigenvector amplitudes $\xi_{i\alpha}^{(s)}$ may be calculated easily. Later in the volume, we shall turn to more realistic descriptions of lattice vibrations in the surface.

So far, the description of inelastic electron scattering from adsorbate vibrations is quite general and applicable to the small angle regime, as well as impact scattering. We now turn to a specific procedure that generates expressions for $(\partial f/\partial R_{i\alpha})_0$, with emphasis on the regime where electron penetration into the crystal is of central importance.

Calculations have been carried out recently of the energy and angle variation of the impact scattering cross section within the following framework [19]. One can imagine each nucleus surrounded by a spherically symmetric potential $V(\mathbf{x} - \mathbf{R}_i)$ which is nonzero only within a sphere of some radius R_0 that surrounds each lattice site. The potential associated with the substrate and any adsorbate layer is regarded as a linear superposition of these nonoverlapping "muffin tins." In between these, the potential is constant, with a value below that of the vacuum by an amount equal to the inner potential. This physical picture is in fact identical to that employed in theoretical analyses of low-energy electron diffraction from crystals, and also of angular resolved photoemission [20]. Each of these invokes the evaluation of certain scattering amplitudes, which are calculated with the muffin tin potentials centered about the equilibrium lattice positions.

In the problem of present interest, we shall require in addition the change produced in the crystal potential when a particular nucleus is displaced. If we set this question aside for the moment, and just refer to the derivative of the crystal potential with nuclear displacement as $\partial V(\{\mathbf{R}\})/\partial R_{i\alpha}$, the analysis of the multiple scattering series provides the following expression for the scattering amplitude derivative [19]:

$$\left(\frac{\partial f}{\partial R_{i\alpha}}\right)_0 = \left\langle \mathbf{k}^{(S)} \left| (G + GT_0G) \frac{\partial V(\{\mathbf{R}\})}{\partial R_{i\alpha}} (1 + GT_0) \right| \mathbf{k}^{(I)} \right\rangle. \quad (3.80)$$

In Eq. (3.80), $|\mathbf{k}^{(I)}\rangle$ and $|\mathbf{k}^{(S)}\rangle$ are plane waves that describe the electron as it approaches and exits the crystal, G is the free space Green's function for the electron, and T_0 the t matrix for (multiple) scattering of the electron from the crystal, with all nuclei on their equilibrium lattice sites. Equation (3.80) is written symbolically in operator form, and each term involves integrations over coordinate space, so one includes each possible path of the electron as it approaches or exits from the unit cell that contains the displaced nucleus.

The expression in Eq. (3.80) may be also written in the form

$$\left(\frac{\partial f}{\partial R_{i\alpha}}\right)_0 = \left\langle \mathbf{k}^{(S)} \left| g_{PE}^{(F)} \frac{\partial V(\{\mathbf{R}\})}{\partial R_{i\alpha}} \right| \psi_{LEED}^{(I)} \right\rangle, \quad (3.81)$$

where $|\psi^{(I)}_{LEED}\rangle$ is the same wave function as that encountered in the theory of low-energy electron diffraction, while $g^{(F)}_{PE}$ is the Green's function that describes the emitted electron in the theory of photoemission. Thus, as the electron approaches the unit cell which contains the displaced nucleus, it engages in multiple scattering off the crystal potential. After it strikes the unit cell ($i\alpha$), it is deflected away from the path followed by a LEED electron, and the calculation of its subsequent path is performed as in the theory of photoemission; the electron engages in multiple scattering off the crystal once again as it leaves.

As remarked earlier, a crucial ingredient in the calculation is $\partial V(\{\mathbf{R}\})/\partial R_{i\alpha}$. In the calculations carried out to date, it has been assumed that the spherically symmetric muffin-tin potential follows the displaced nucleus unchanged in form and shape. This goes beyond the adiabatic approximation, which allows the electron charge cloud that surrounds a nucleus to deform continuously as the nucleus shifts position. However, with methods presently available, it is difficult to construct an accurate description of the change in crystal potential with nuclear displacement, for an atom on or near the crystal surface. It is thus difficult to see how to improve this approximation, in a computationally tractable manner. If we accept it, then all quantities in Eq. (3.80) may be expressed in terms of certain matrix elements of the free electron Green's function G, the phase shifts $\delta_l(E)$ for scattering off the various muffin tin potentials, and the radial solution of Schrödinger's equation within that muffin tin which contains the displaced nucleus [19].

While it may be difficult to improve on the description of $\partial V(\{\mathbf{R}\})/\partial R_{i\alpha}$ just outlined, the approximation has one crucial limitation. If a nucleus is displaced, and the electron charge cloud follows it without distortion, then no electric dipole moment is generated. To obtain a nonzero electric dipole moment, it is necessary for the electronic charge that surrounds the nucleus to distort in shape, or there must be charge transfer between nearby unit cells. Hence, within the framework of the muffin tin model, combined with the approximate description of the electron–phonon interaction outlined in the previous paragraph, we are unable to obtain a description of large-angle inelastic electron scattering that can be brought into direct contact with the dipole scattering theories outlined earlier in the present chapter. We regard this as a major difficulty with the theory at its present state of development, but it is also a difficulty that will be hard to remove in any satisfactory manner. For example, a more realistic model of the electron–phonon interaction would allow an electric dipole moment to be generated within a given muffin tin, when its nucleus is displaced. To calculate the electron scattering amplitude produced by such an improved description of $\partial V(\{\mathbf{R}\})/\partial R_{i\alpha}$ would then require understanding of how this dipole moment is screened by the electrons in the metal. We did not need to confront this issue in the theory of dipole scattering presented earlier, because the dipole moment effective charge e^* was viewed as a phenomenological parameter to be obtained from experimental data, and the screening process need not be described in microscopic terms. Finally, dipole moments produced by displacement of nuclei near the surface are only partially screened by the electrons (this is why an incoming electron sees an electric field in the vacuum above the surface), and generates a long-range potential as a consequence. Then the potential between the muffin tins can no longer be treated as a constant unaffected by small nuclear displacements, and one of the key assumptions in the standard multiple scattering theory of electron/solid interactions breaks down.

Within the scheme just outlined, and with use of the simple form of the electron phonon interaction described after Eq. (3.81), a series of studies of large-angle electron scattering from a $c(2 \times 2)$ CO layer on the Ni(100) surface have been carried out [19]. We conclude this section with a summary of some results of this investigation.

In Fig. 3.16, for a particular scattering geometry, we show the energy variation and magnitude of the scattering efficiency ($dS/d\Omega$) for scattering from the CO stretch vibration. In the calculation, the wave vectors $\mathbf{k}^{(I)}$, $\mathbf{k}^{(S)}$ and the normal to the surface lie in a plane, the scattering plane, aligned along the (100) direction. The angle of incidence $\theta_I = 30°$, and the scattered electron

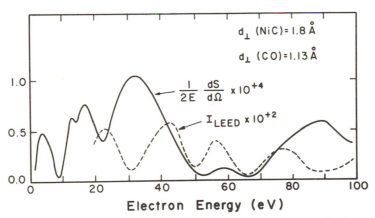

Fig. 3.16 The solid line is a plot of the quantity ($dS/2E_I d\Omega$) for scattering from the CO stretch vibration, with a $c(2 \times 2)$ layer on the Ni(100) surface. The energy E_I is expressed in hartrees. The angle of incidence $\theta_I = 30°$, and the electrons emerges with the angle $\theta_S = 35°$ relative to the surface normal, and the scattering plane is aligned along Ni(100). The dotted line is a calculation of the energy variation of the (00) LEED beam for scattering from the same structure.

emerges with the angle $\theta_S = 35°$, about $5°$ off the specular direction. Several general aspects of scattering in the impact regime may be appreciated from the figure.

An important result of the dipole scattering analysis, displayed first in Eq. (3.9), is that the scattering cross section is proportional to the intensity $|R_1|^2$ of the specular beam. Certain assumptions, evidently comfortably satisfied in many experiments, are required for this result to hold. This allows one to bring data taken in the dipole dominated regime in quantitative contact with the theory, without the need for a detailed quantitative description of the scattering of the electron from the substrate. One may measure the intensity $|R_1|^2$ of the specular beam, divide this quantity out of the data, and the combination $|R_1|^{-2}(d^2S/d\Omega\,d\hbar\omega)$ may then be compared with the relevant version of the dielectric model, when this is appropriate. We saw a number of such comparisons earlier in the present chapter. In the impact scattering regime, it is no longer true that the cross section for vibrational excitation is proportional to the intensity of the specular beam, modulated by a smooth and simply describable function of energy. This is illustrated in Fig. 3.16 for the geometry described in the previous paragraph. We give the probability per unit solid angle $dS/d\Omega$ for exciting the CO stretch vibration, and compare the energy variation of this quantity with calculations $|R_1|^2$, the intensity of the specular beam. The inelastic cross section $dS/d\Omega$ has been divided by $2E_I$, where E_I is the electron impact energy expressed in rydbergs. The energy variation of $(dS/d\Omega)$ and $|R_1|^2$ differ qualitatively, for all energies considered. Thus, in presenting data taken in the impact scattering regime, the experimentalists should not divide the inelastic scattering cross section by the intensity of the specular beam, since this will introduce spurious and unphysical structure into the former.

If one examines the expressions for the scattering efficiency in the dipole scattering regime, the excitation cross section decreases as the impact energy E_I increases, provided the impact energy is sufficiently large that the characteristic deflection angle ϑ_E is small compared to the spectrometer acceptance angle ϑ_c. The dipole scattering cross section decreases for two reasons. As the energy is raised, on the average $|R_1|^2$ falls, and the dipole excitation cross section $|R_1|^{-2}\,d^2S/d\Omega\,d\hbar\omega$ also decreases with energy when $\vartheta_E < \vartheta_c$. The latter behavior is a characteristic feature of Coulomb scattering. From Fig. 3.16, one sees that in the impact scattering regime, the total excitation efficiency $(dS/d\Omega)$ increases as the electron impact energy increases. It thus would be most favorable to study large-angle inelastic electron scattering at impact energies substantially larger than used in present generation experiments, if suitable spectrometers could be constructed (see also Section 2.6).

The fact that $(dS/d\Omega)$ increases with increasing electron energy is not a special feature of the scattering geometry used in Fig. 3.16, but is a general feature of the impact scattering regime. That this is so may be appreciated from the following simple argument. Suppose the intensity for specular reflection from a layer of atoms with all nuclei clamped in place is denoted by $|R_1|_0^2$. Then if the atoms are allowed to execute thermal vibrations, in the simplest description of the scattering process, the specular intensity is reduced from the value $|R_1|_0^2$ characteristic of the rigid lattice by the Debye–Waller factor. We have

$$|R_1|^2 = |R_1|_0^2 \exp[-\tfrac{1}{2}\langle(\mathbf{Q}\cdot\mathbf{u})^2\rangle_T],$$

where $\mathbf{Q} = \mathbf{k}^{(S)} - \mathbf{k}^{(I)}$, $\mathbf{u}$ is the displacement of an atom from its equilibrium position, and the angular brackets denote a thermal average. The specular intensity is reduced when the atoms are allowed to vibrate, because inelastic scattering of electrons by phonons deflects electrons away from the specular direction into the thermal diffuse background. The depletion of the specular beam from one-phonon processes is given by $\tfrac{1}{2}\langle(\mathbf{Q}\cdot\mathbf{u})_T^2\rangle|R_1|^2$ in this description, and one sees that since Q^2 is proportional to electron energy E_1 (when $E_1 \gg \hbar\omega$, as in Fig. 3.16), the cross section for one-phonon events scales with electron impact energy roughly as $E_1|R_1|^2$, while the dipole excitation cross section is given by $|R_1|^2$ multiplied by a function that *decreases* with impact energy. Since Fig. 3.16 shows that the structure in $dS/d\Omega$ for the example explored in the figure has structure at energies distinctly different from that in $|R_1|^2$, the foregoing kinematical description cannot be taken literally, though it allows one to appreciate the overall trends in the figure.

It is interesting to compare the magnitude of the impact scattering efficiency, as obtained from Fig. 3.16, with the scattering efficiency for dipole excitation. For electrons with impact energy in the range 5–20 eV, the calculations displayed in Fig. 3.16 show that for impact scattering, we have $(dS/d\Omega) \approx 10^{-4}$. Calculations for a variety of scattering geometries show that this is a typical number in this energy range, for impact excitation of the CO stretching vibration, for CO on the Ni(100) surface in the $c(2 \times 2)$ structure [19]. If the electrons are collected with a spectrometer that collects all electrons that lie within a cone of half-angle equal to $1°$, then the solid angle subtended by the spectrometer slits is roughly 10^{-3} steradians. Hence, roughly one electron in 10^7 contained in the *incident* beam is scattered into the spectrometer slits by the impact mechanism. For a strongly dipole-allowed mode such as the CO stretch vibration, at an impact energy of 5 eV about one electron in 10^5 from the incident beam is scattered into the near specular dipolar lobe, if $|R_1|^2 \cong 10^{-2}$. Thus, at such low beam energies, the cross section for large-angle impact scattering is about two orders of magnitude smaller than that in the near specular dipole lobe. At higher energies, say around 50 eV, the dipole cross section will be smaller than that at 5 eV by one order of magnitude or possibly more. At the same time, as we have seen, on the average the impact cross section increases. Hence, by operating at higher beam energies impact, rather than dipole scattering, is emphasized. We shall present a more complete analysis of the relative intensity of dipole and impact scattering later in this chapter, to take explicit account of the spectrometer geometry.

In Fig. 3.16, we see that the excitation efficiency $(dS/d\Omega)$, when considered a function of energy, contains peaks and structures at certain energies distinct from those where structure appears in the specular reflection coefficient. These structures arise from coherent interferences between the scattered beams, when the electron is on either the entrance or exit portion of its trajectory. As is the case also in low-energy electron diffraction and photoemission studies, the position of these structures is sensitive to the details of the geometry of the adsorbed layer and underlying substrate. We illustrate this in Fig. 3.17, where we show the energy variation of the scattering efficiency $(dS/d\Omega)$ for excitation of the CO stretching vibration, for two different choices of the carbon–oxygen bond length. We see pronounced differences between the structures present in the two cases. From Fig. 3.16, we see that the information contained in the energy variation of the excitation probability is distinctly different from that obtained from low-energy

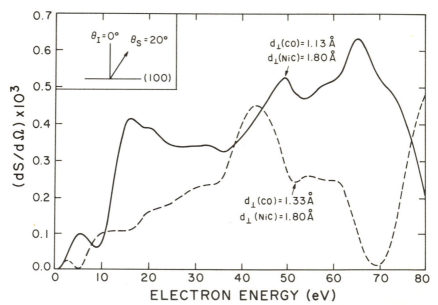

Fig. 3.17 The energy variation of the excitation efficiency $(dS/d\Omega)$ for scattering off the CO stretching vibration, for a $c(2 \times 2)$ layer of CO on the Ni(100) surface. The scattering geometry is illustrated in the inset.

electron diffraction. In the case of a $c(2 \times 2)$ layer of CO on a substrate such as Ni, the cross section for elastic backscattering from the absorbate layer is modest, so that a substantial fraction of the incoming electrons are transmitted through the CO layer, to backscatter from the substrate. The energy variation of the intensities seen in LEED studies is thus sensitive to the spacing between the adsorbate layer and the substrate. In the calculation of $(dS/d\Omega)$, the cross section for backscattering from the CO layer with excitation of a vibrational quanta is substantial, and much larger than that for forward scattering with vibrational excitation. The origin of this is the factor of $(\mathbf{k}^{(I)} - \mathbf{k}^{(S)})^2$ that entered the elementary discussion of one-phonon scattering given earlier. Thus, in the presence of the strong backscattering cross section for vibrational excitation, the structures in $(dS/d\Omega)$ are influenced most importantly by the carbon–oxygen spacing, rather than that between the adsorbate layer and the substrate. This is not true for all scattering geometries; a detailed study finds an appreciable range of incident and exit angles where the energy variation of $(dS/d\Omega)$ is influenced importantly by the distance between the carbon layer, and the outermost atomic layer of Ni [19].

The energy variation of the electron scattering efficiency in electron energy loss studies is thus a potentially rich source of information on surface structure. To obtain data over the wide energy range displayed in Figs. 3.16 and 3.17 will be most difficult, however, since an accurate calibration of the transmission of the spectrometer in a wide energy range would be required.

As we saw in the data on H adsorbed on the W(100) surface given in Fig. 3.15, outside the dipolar lobe, the cross section for scattering from modes polarized parallel to the surface is comparable to that for scattering from the perpendicular mode. In the case of CO adsorbed on a top-bonding geometry, i.e., with the carbon bonded directly on top of a Ni atom so the adsorption site has four fold symmetry, each mode with vibrational motion parallel to the surface is twofold degenerate. We thus have two distinct frequencies, one associated with a hindered rotation in which the motion of the carbon and oxygen is 180° out of phase, and the second is associated with a hindered translation, where the oxygen and carbon move in phase, so the whole molecule rocks back and forth in a wagging motion. In Fig. 3.18, for an electron with incident energy of 6.5 eV, we show calculations of the angular variation of the scattering efficiency $(dS/d\Omega)$ for excitation of the frustrated rotation (the $\omega_x^{(+)}$ mode) polarized parallel to the scattering plane, and the frustrated translation (the $\omega_x^{(-)}$ mode) also polarized parallel to the scattering plane, which is aligned along the (100) direction. We see the scattering efficiencies for exciting the parallel modes is found to be comparable in magnitude to that for exciting the CO stretching vibration, as in the case of H on W(100).

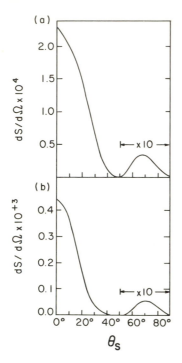

Fig. 3.18 The angular variation of the scattering efficiency $(dS/d\Omega)$ for scattering produced by the two normal modes, (a) $\omega_x^{(+)}$ and (b) $\omega_x^{(-)}$, of CO polarized parallel to the surface, and to the plane which contains the incident wave vector and the normal to the surface (the scattering plane); $\theta_I = 50°$. The wave vector of the scattered electron also lies in this plane, and the scattering plane is aligned along (100).

Note that the scattering efficiency $(dS/d\Omega)$ for excitation of each of these modes vanishes, when the scattered electron emerges right along the specular direction. This is not an accident, but rather a consequence of a selection rule which operates in the impact-scattering regime.

3.5.2 Selection Rules in the Impact Scattering Regime

In this section, we turn to a systematic discussion of selection rules that operate in the impact scattering regime. For this purpose, we employ the scattering geometry given in Fig. 3.19. The incident electron wave vector $\mathbf{k}^{(I)}$ lies in the xz plane, and makes an angle θ_I with the z axis, which is the normal to the surface. The plane which contains $\mathbf{k}^{(I)}$ and the z axis, the xz plane in Fig. 3.19, will be called the scattering plane. The scattered wave vector $\mathbf{k}^{(S)}$ makes the angle θ_S with the z axis, and φ is the angle between the (negative) x axis and the projection of $\mathbf{k}^{(S)}$ onto the xy plane. When $\varphi = 0$, $\mathbf{k}^{(S)}$ will lie in the scattering plane, and $\mathbf{k}^{(S)}$ lies along the specular direction when $\theta_S = \theta_I$, and $\varphi = 0$.

The central quantity in the analysis is the quantity $(\partial f/\partial Q_s)$ defined in Eq. (3.77). We shall examine this quantity, to see that for modes of selected

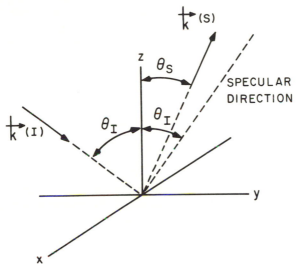

Fig. 3.19   The scattering geometry that forms the basis of the discussion of selection rules in the impact scattering regime. The xz plane is the scattering plane.

symmetry and certain scattering geometries, it will vanish. If we define

$$\frac{\partial V}{\partial Q_s} = \sum_i \frac{\xi_{i\alpha}^{(s)}}{\sqrt{M_i}} \frac{\partial V(\{\mathbf{R}\})}{\partial R_{i\alpha}}, \tag{3.82}$$

the incoming electron wave function $|\psi_{\mathbf{k}^{(I)}}^{(-)}\rangle = (1 + GT_0)|\mathbf{k}^{(I)}\rangle$ with a corresponding wave function for the outgoing wave, $|\psi_{\mathbf{k}^{(S)}}^{(+)}\rangle$, then we have for scattering from a particular mode Q_s,

$$\left(\frac{\partial f}{\partial Q_s}\right) = \left\langle \psi_{\mathbf{k}^{(S)}}^{(+)} \left| \frac{\partial V}{\partial Q_s} \right| \psi_{\mathbf{k}^{(I)}}^{(-)} \right\rangle. \tag{3.83}$$

Before we begin, we examine some properties of the wave functions $\psi_{\mathbf{k}}^{(+)}$ and $\psi_{\mathbf{k}}^{(-)}$. Both are solutions of the same Schrödinger equation, that for motion of an electron with the same energy $E_{\mathbf{k}} = \hbar^2 k^2/2m$ in the vacuum above the crystal. They are linearly independent, because they are subject to different boundary conditions. The function $\psi_{\mathbf{k}}^{(-)}$ consists of a single incoming wave, with wave vector $\mathbf{k}$, and this produces one or more Bragg beams reflected from it. Notice that $k_\perp < 0$ necessarily, in the present geometry with the crystal in the lower half-space. Far from the crystal $\psi_{\mathbf{k}}^{(+)}$ consists of a single outgoing beam (to construct $\psi_{\mathbf{k}}^{(+)}$ we must have $k_\perp > 0$), and to satisfy the boundary condition at the surface several incoming Bragg beams must be contained in it. Now, if we consider the complex conjugate $\psi_{\mathbf{k}}^{(-)*}$ of $\psi_{\mathbf{k}}^{(-)}$, we have another solution of the same Schrödinger equation with energy $E_{\mathbf{k}}$, in the vacuum above the crystal. Since the operation of complex conjugation changes $\exp(i\mathbf{k} \cdot \mathbf{x})$ to $\exp(-i\mathbf{k} \cdot \mathbf{x})$, $\psi_{\mathbf{k}}^{(-)*}$ consists of *one* outgoing beam of wave vector $-\mathbf{k}$, and several incoming beams, each oppositely directed to the outgoing beams in $\psi_{\mathbf{k}}^{(-)}$. This is a solution with the same form as $\psi_{-\mathbf{k}}^{(+)}$, and the uniqueness theorem in fact requires $\psi_{\mathbf{k}}^{(-)*}$ and $\psi_{\mathbf{k}}^{(+)}$ to be identical everywhere. Hence we have the identity

$$(\psi_{\mathbf{k}}^{(-)})^* = \psi_{-\mathbf{k}}^{(+)}, \tag{3.84}$$

and similar reasoning shows that

$$(\psi_{\mathbf{k}}^{(+)})^* = \psi_{-\mathbf{k}}^{(-)}. \tag{3.85}$$

With these identities, we may appreciate the origin of the selection rule that produces the zero for scattering along the specular direction, for the particular modes considered in the calculations carried out for Fig. 3.18. The crystal is invariant under reflection in the yz plane, while the parallel modes considered in the previous section are odd under this reflection. These symmetries combined with time-reversal invariance, and one approximation to be outlined, require $(dS/d\Omega)$ vanish when the scattered electron emerges along the specular. We now turn to the proof.

In quantum mechanics, if R denotes a symmetry operation and O an Hermitean operator, then application of R replaces O by the combination $O' = R^{-1}OR$, where for unitary operators $R^{-1} = R^+$. Now for the $c(2 \times 2)$ layer of CO on Ni(100), with the scattering plane aligned along (100), the crystal potential is unchanged by the operation of reflection through the yz plane. If we denote this by R_{yz}, then we have the identity $V(\{\mathbf{R}\}) = R_{yz}^{-1} V(\{\mathbf{R}\}) R_{yz}$. It follows that if we consider *any* normal mode odd under reflection through the yz plane, then

$$\left(\frac{\partial V}{\partial Q_s}\right) = -R_{yz}^{-1}\left(\frac{\partial V}{\partial Q_s}\right) R_{yz}, \tag{3.86}$$

i.e., the derivative $(\partial V/\partial Q_s)$ is odd under this reflection.

Now we calculate the matrix element

$$\left(\frac{\partial f}{\partial Q_s}\right) = \int d^3x\, \psi_{\mathbf{k}^{(S)}}^{(+)*}\left(\frac{\partial V}{\partial Q_s}\right)\psi_{\mathbf{k}^{(I)}}^{(-)} \tag{3.87}$$

for the case where the wave vector $\mathbf{k}^{(S)}$ lies along the specular direction. We do this after invoking the following approximation: we ignore the difference between $|\mathbf{k}^{(S)}|$ and $|\mathbf{k}^{(I)}|$, so in fact $\psi_{\mathbf{k}^{(S)}}^{(+)}$ is replaced by the wave function for an electron with the *incident* energy E_I rather than that E_S of the scattered wave function.

Now we have the identity, for any wave vector in the scattering plane,

$$R_{yz}\psi_{\mathbf{k}}^{(-)} = \psi_{\mathbf{k}'}^{(-)}, \tag{3.88}$$

where if $\mathbf{k} = \mathbf{k}_{\parallel} + \hat{z}k_z$, $\tilde{\mathbf{k}}_{\parallel} = -\mathbf{k}_{\parallel} + \hat{z}k_z$. Recall that the operator K that describes time-reversal simply replaces the wave function ψ by its complex conjugate, with spin ignored. Hence, upon invoking Eq. (3.84)

$$KR_{yz}\psi_{\mathbf{k}}^{(-)} = \psi_{-\tilde{\mathbf{k}}}^{(+)}, \tag{3.89}$$

or when applied to the calculation of Eq. (3.87) with the approximations just described,

$$\psi_{\mathbf{k}(S)}^{(+)} = R_{yz}\psi_{\mathbf{k}(I)}^{(-)*} \tag{3.90}$$

and

$$\psi_{\mathbf{k}(I)}^{(-)} = R_{yz}\psi_{\mathbf{k}(S)}^{(+)}, \tag{3.91}$$

so we have the identity, after minor rearrangement

$$\left(\frac{\partial f}{\partial Q_s}\right) = \int d^3x\, \psi_{\mathbf{k}(S)}^{(+)*} R_{yz}^{-1} \frac{\partial V}{\partial Q_s} R_{yz}\psi_{\mathbf{k}(I)}^{(-)}, \tag{3.92}$$

or upon invoking Eq. (3.86),

$$\left(\frac{\partial f}{\partial Q_s}\right) = -\int d^3x\, \psi_{\mathbf{k}(S)}^{(+)*} \frac{\partial V}{\partial Q_s} \psi_{\mathbf{k}(I)}^{(-)}. \tag{3.93}$$

A comparison between Eqs. (3.93) and (3.87) shows we have demonstrated that $(\partial f/\partial Q_s)$ and its negative are equal, so the only possibility is

$$\left(\frac{\partial f}{\partial Q_s}\right) \equiv 0. \tag{3.94}$$

We thus have the following statement, valid when little error is introduced into the matrix element by ignoring the difference between the actual energy E_S of the scattered electron and that E_I of the incoming beam. If the scattering plane is oriented so that the yz plane (Fig. 3.18) is a plane of reflection symmetry, and if we consider impact scattering by a mode with normal coordinate Q_s odd under R_{yz}, then this reflection symmetry combined with time-reversal invariance requires the scattering amplitude $(\partial f/\partial Q_s)$ to vanish when the wave vector of the scattered electron is directed along the specular direction. The statement has been proved for the case where the vibrational mode is nondegenerate. It applies also to degenerate modes, if they may be represented by eigenvectors with well-defined reflection symmetry.

It is this selection rule that is responsible for the zero along the specular in the scattering efficiency $(dS/d\Omega)$ displayed in Fig. 3.18. While the calculations summarized in the figure employed the scattered wave function at the actual scattered energy E_S, a zero is clearly evident.

We have considered the proof of the selection rule for the case where the adsorption site is such that R_{yz} is a good symmetry operation. If, instead, the z axis of Fig. 3.19 is an axis about which there is twofold symmetry, then this twofold symmetry combined with time-reversal symmetry allows us to prove once again that $(\partial f/\partial Q_s)$ vanishes, when the scattered wave vector is along the specular direction. This is so for modes with normal coordinate Q_s odd under the twofold rotation, and the proof once again requires one to ignore the difference between the energies of the scattered and incident electrons.

The scattering plane itself, the xz plane in Fig. 3.19, may also be aligned so it is a plane of reflection symmetry. Since both $\psi^{(-)}_{\mathbf{k}^{(0)}}$ and $\psi^{(+)}_{\mathbf{k}^{(s)}}$ are even under the operation R_{xz}, when this is a good symmetry operation, it follows that for *any* scattered wave vector $\mathbf{k}^{(S)}$ in the scattering plane, the scattering amplitude $(\partial f/\partial Q_s)$ must vanish identically, if the normal coordinate Q_s is odd under R_{xz}. The proof of this statement requires neither the presence of time-reversal symmetry, nor the assumption that $E_S \cong E_I$.

When the selection rules described in the present section are combined with the dipole selection rule discussed earlier in the present chapter, we have a means of uniquely identifying the normal modes of an adsorbate when the geometry is simple, or of placing constraints on the nature of the eigenvectors in more complex situations.

3.6   TECHNICAL REMARKS

In this section, we turn our attention to some technical aspects of electron energy loss spectroscopy when the results of the theory of inelastic scattering are combined with the properties of electron spectrometers, as outlined in Chapter 2. We consider more carefully than earlier the relative intensities of losses due to dipole and impact scattering, an estimate of the smallest amount of material from which a signal can be detected, and the problem of determining absolute cross sections of individual molecules.

3.6.1   The Intensities of Dipole and Impact Scattering

We begin by recalling some of the fundamental equations that allow us to relate the acceptance angle ϑ_c to the energy E_I and the energy width ΔE of the beam. In Section 2.5.4, we found the energy width related to the acceptance angle of the monochromator by

$$\Delta E \approx \alpha^2 E_0, \tag{3.95}$$

with E_0 the pass energy for a monochromator operating under space-charge conditions. For electrons which lose only a small fraction of their energy after scattering, the angular spread of the beam in the analyzer is also α, provided the spectrometer is symmetrically built and operated. With Abbe's sine law [Eq. (2.1)], we obtain

$$\Delta E = M_{opt}^2 E_I \vartheta_c^2, \tag{3.96}$$

where M_{opt} is the optical magnification between the exit aperture of the monochromator and its image on the sample, or likewise between the entrance slit of the analyzer and its time-reversed image on the sample. While this relation suggests that one should make $M_{opt} \ll 1$ in order to achieve large acceptance angles at the samples, most spectrometers have $M_{opt} > 1$, for reasons we have discussed in Chapter 2. In any case, for a given object–image distance, and a certain lens design, M_{opt} may be considered as approximately independent of the energy E_I. We therefore see from Eq. (3.96) that the acceptance angle ϑ_c becomes smaller the smaller we make ΔE, and becomes smaller also as the impact energy is increased.

We can now estimate the intensity of vibrational losses from dipole scattering, by using Eq. (3.96) to express ϑ_c in terms of ΔE, then using our earlier theoretical expressions for the dipole intensity. If we use Eq. (3.60), with $F_s(\hat{\vartheta}_c)$ approximated by Eq. (3.62), then eliminate ϑ_c through Eq. (3.96), we have

$$I_{inel} = I_I |R_I|^2 \left(\frac{16\pi n_s e^{*2} h^2 \sin^2 \theta_I}{a_0 M_r (\hbar \omega_s)^3 \varepsilon_\infty^2 \cos \theta_I} \right) \frac{\Delta E}{M_{opt}^2}, \tag{3.97}$$

where I_I is the monochromatic current incident on the sample.

Since I_I is itself proportional to $(\Delta E)^2$, the integrated loss intensity is proportional to $(\Delta E)^3$ in the limit where Eq. 3.97 is valid. This shows that one has to pay an extraordinarily high price for resolution in electron spectroscopy. Therefore, when two loss features are not well separated in a spectrum because of the limited resolution, the spectrum is not always improved by going to higher resolution since the increase in noise may degrade the gain from improved resolution. We also see from Eq. (3.97) that the dipole losses decrease in intensity with increasing $\hbar \omega_s$. One factor of $(\hbar \omega_s)^{-1}$ enters the right-hand side of Eq. (3.97) because increasing the frequency of the oscillator causes the mean square amplitude of vibration to decrease. The additional factor of $(\hbar \omega_s)^{-2}$ is a result of the increasing angular spread in the dipole losses for increasing $\hbar \omega_s$. Equation (3.97) may be used to estimate the intensity of losses relative to the elastically scattered peak. For a CO monolayer with $(e^*/\varepsilon_\infty) \approx 0.5e$, and for $\Delta E = 5$ meV, Eq. (3.97) gives a number the order of 10^{-3} for the ratio $I_{inel}/I_I |R_I|^2$. If the reflectivity were anywhere near unity, and the collection efficiency of the spectrometer also about one, absolute count rates of the order of 10^6 counts per second (CPS) could be achieved. In real situations, with $|R_I|^2 \ll 1$, and spectrometer efficiencies of the order of 10% a monolayer of CO gives count rates around 10^3 CPS.

We next turn to a simple estimate of the intensity of losses due to impact scattering. As argued earlier, one has the rough estimate, with $\mathbf{u}$ the vibration amplitude of the mode in question,

$$I_{\text{inel}} \approx \tfrac{1}{2}(\mathbf{Q} \cdot \mathbf{u})^2 I_{\text{el}}, \tag{3.98}$$

where $\mathbf{Q}$ is the wave vector transfer. For large-scattering angles well into the impact scattering regime, $|\mathbf{Q}| \cong |\mathbf{k}^{(1)}|$, the wave vector of the incident electron. The total elastic scattering cross section of a molecule is of the order of its van der Waals area [21]. For monolayer coverages, since a single inelastic scattering turns an electron through a large angle, we may replace I_{el} by I_1 rather than $|R_1|^2 I_1$, i.e., large-angle events such as that illustrated in Fig. 3.1a allow the electron to be directly scattered back into the vacuum by a single event, without the need for backreflection from the substrate. Since, in the impact scattering regime, the inelastically scattered current is distributed rather uniformly over the whole solid angle, the current picked up by the spectrometer is roughly $\pi \vartheta_c^2/4\pi$. If u^2 is replaced by the zero-point amplitude $h/2M_r \omega_s$, as is appropriate when $\hbar\omega_s \gg k_B T$, with $(\mathbf{Q} \cdot \mathbf{u})^2$ replaced by $k_1^2 u^2$, we obtain the estimate

$$I_{\text{inel}} = I_1 (mE_1/\hbar\omega_s M_r)(\vartheta_c^2/4). \tag{3.99}$$

Comparison of the prediction of Eq. (3.97) with the full multiple scattering analyses discussed earlier [19] shows that this simple estimate gives the correct order of magnitude as well as the correct trends with variations of E_1 and $\hbar\omega_s$, although it is equivalent to a simple kinematical single-scattering analysis and ignores all interference effects. If we eliminate ϑ_c^2 again with Eq. (3.96), we have

$$I_{\text{inel}} = I_1 \frac{\Delta E}{4\hbar\omega_s} \frac{m}{M_r} \frac{1}{M_{\text{opt}}^2}. \tag{3.100}$$

With a spectrometer collection efficiency of about 0.1, and also $1/M_{\text{opt}}^2 \approx 0.1$, one calculates inelastic currents of the order of 2 CPS, for a monolayer of CO with $\Delta E = 5$ meV. This is roughly two orders of magnitude lower than the intensity dipole scattering in specular reflection. For a CH stretching vibration, one would estimate count rates of roughly 10 CPS.

It is also illuminating to calculate the ratio of the impact scattering current to the dipole scattering current which, for the above numbers, is

$$I_{\text{impact}}/I_{\text{dipole}} \cong 10^{-4} |R_1|^{-2} (\hbar\omega_s \varepsilon_\infty/e^*)^2 \tag{3.101}$$

where $\hbar\omega_s$ is in electron volts, and e^* is in units of e. From this equation, we may again see that the count rate for impact scattering is much lower than for dipole scattering, for a strong dipole scatterer such as the CO stretching vibration. However, when vibrations of weak dipole scatterers such as H are considered, with $e^*/\varepsilon_\infty \cong 0.05e$, then impact and dipole scattering become equally intense features. One can further predict from Eq. (3.101) that impact scattering is more likely to prevail for the higher frequency modes, like the C—H stretching vibration of a hydrocarbon, as opposed to the CH bending vibrations which are typically a factor of 3 lower in frequency. We shall find these conclusions in accord with the general trends in the experimental data.

3.6.2   The Limits of Detectability

It has already become obvious that the sensitivity of vibration spectroscopy to small coverages of adsorbed molecules depends as much on parameters of the spectrometer as on parameters of the molecules and substrate. For impact scattering, we have already estimated count rates between 2 and 10 CPS for an energy width of 5 meV (which corresponds to an overall resolution of $5\sqrt{2} \approx 7$ meV) and a monolayer of adsorbed molecules. Since the dark count rate of detectors is of the order of 1 CPS, something like one-tenth of a monolayer can be detected. This number, however, refers to the present generation of spectrometers, where the lens system has been designed to comply conveniently with standard ultrahigh vacuum sample holders, which require a relatively large distance between the sample and the lens elements, rather than to optimize the acceptance angle. A reduction of M_{opt}^2 by an order of magnitude is possible. This pushes the count rates up by the same factor.

For dipole scattering, the sensitivity limit is of a more fundamental nature: for a metal surface, we have found a continuous background due to scattering from free electrons. If we form the ratio between the intensities of dipole scattering from adsorbate vibrations and this background, we obtain for the ratio of the peak count rate of a loss peak to the count rate of the countinuous background

$$\frac{I_{loss}}{I_{background}} = (2\pi)^3 \frac{m}{k_I} \frac{F_s(\hat{\vartheta}_c)}{F_b(\hat{\vartheta}_c)} \frac{\sigma}{\omega} \frac{n_s e^{*2}}{M_r \varepsilon_\infty^2} \frac{2\ln 2}{\Delta E \sqrt{\pi}}. \qquad (3.102)$$

Here ΔE is the overall energy width of the spectrometer. The factor $2\ln 2/\sqrt{\pi}$ enters in when a Gaussian energy distribution is assumed for the spectrometer transmission versus energy. This equation can be further simplified for small $\hat{\vartheta}_c$,

$$\frac{I_{loss}}{I_{background}} = 16\pi^3 \sin\theta_I \frac{m}{k_I} \frac{\sigma}{\omega} \frac{n_s e^{*2}}{M_r \varepsilon_\infty^2} \frac{2\ln 2}{\Delta E \sqrt{\pi}}. \qquad (3.103)$$

This equation may be used to estimate the detection limit for small quantities of adsorbed molecules. Assuming a nickel surface at room temperature, an energy of 2 eV and a resolution of $\Delta E = 10$ meV, the ratio of the loss count rate to the background becomes one for CO molecules ($e^*/\varepsilon_\infty \sim 0.5e$) at a surface concentration of $n_s \sim 10^{12}$ cm^{-2}, which is roughly 10^{-3} of a monolayer. Of course, one might improve on this limit by going to lower temperatures where σ becomes smaller and the background consequently lower. As we have seen, the expected linear reduction of the background with the temperature T has actually been found experimentally (Fig. 3.5). Therefore the detection limit may be pushed down to $\sim 10^{-4}$ monolayer for strong dipole scatterers, when the experiment can be carried out at low temperatures. For weak dipole scatterers such as hydrogen, the detection limit is of the order of 10^{-1} to 10^{-2} of a monolayer. These numbers refer to a nickel substrate and vary with the conductivity of the substrate.

3.6.3   Intensity Measurements

For submonolayer quantities of adsorbed molecules, the intensity of a dipole loss divided by the elastically reflected peak should be proportional to the coverage, at least for low coverages, where depolarizing effects can be neglected and e^* is a constant. Actual experimental plots of this ratio versus coverage often deviate from this predicted linear relationship, and are notoriously unreproducible. The reason is that the acceptance angle of spectrometers ϑ_c is smaller than the angular distribution of the elastically scattered beam except for well-ordered surfaces. As soon as the surface disorders with increasing coverage, the relative intensity of a loss appears to increase, as if the spectrometer had a large acceptance angle. In this situation, one therefore has no real measure of the intensity since the state of order (or better, disorder) on a surface is unlikely to reproduce very well from one experiment to the next. Furthermore, it becomes almost impossible to compare intensities between different laboratories and in fact all loss spectra published so far have little quantitative meaning. This statement even applies to the relative intensities of the losses when compared among themselves, since the variation in the apparent acceptance angle has a different influence on losses of high and low energy.

Inspection of Eq. (3.102) again shows that rather than the intensity of a loss relative to the elastically scattered beam, the intensity divided by the *background* should be considered, since this ratio is now independent of the state of order on the surface. Even absolute values of e^*/ε_∞ may be obtained this way, provided that instead of the dc conductivity the real infrared loss function for the metal is used. This loss function, again, can be determined by electron energy loss spectroscopy by measuring the background with a lower ΔE. This increases the acceptance angle up to the point where $F_b(\hat{\vartheta}_c)$ approaches the maximum value $\frac{1}{2}\pi^2$ and all the elastic intensity is collected. However, even when the absolute value of e^* is not of interest, the ratio of a loss intensity relative to the background (possibly corrected for the ratio $F_s(\hat{\vartheta}_c)/F_b(\hat{\vartheta}_c)$ if necessary) is a figure that should reproduce in different experiments and between different laboratories.

The background method just described is also useful to determine absolute cross sections for impact scattering. In most spectrometers, the acceptance ϑ_c can be measured only in one direction. Some spectrometers do not allow measurement of the directly transmitted beam at all. There again the bulk loss function of a substrate can be obtained from a measurement at low resolution where ϑ_c is large enough to allow the approximation $F_b(\hat{\vartheta}_c) \cong \pi^2/2$. Then for higher resolution where ϑ_c becomes small, the measured background on a well-ordered surface relative to the elastically scattered beam is a measure of $F_b(\hat{\vartheta}_c)$ and thus of the effective acceptance angle ϑ_c of the spectrometer. Even when the spectrometer is not of circular symmetry, so that the actual acceptance angle in two directions may be different, this provides a reliable measurement of the effective acceptance angle that then can be used to determine the differential cross section for impact scattering.

REFERENCES

1. D. L. Mills, *Surface Sci.* **48**, 59 (1975).
2. E. Evans and D. L. Mills, *Phys. Rev. B* **5**, 4126 (1972); *B* **7**, 853 (1973).
3. E. G. McRae, *Rev. Mod. Phys.* **51**, 541 (1979).
4. J. Daniels, C. V. Festenberg, H. Raether, and K. Zeppenfeld, "Optical Constants of Solids by Electron Spectroscopy" (Springer Tracts in Modern Physics, Vol. 38). Springer-Verlag, Berlin, 1965.
5. H. Froitzheim, H. Ibach, and D. L. Mills *Phys. Rev. B* **11**, 4980 (1975).
6. H. Ibach, *Phys. Rev. Lett.* **24**, 1416 (1970).
7. H. Ibach, *Surface Sci.* **66**, 56 (1977).
8. R. G. Greenler, *J. Chem. Phys.* **44**, 310 (1966); **50**, 1963 (1969).
9. J. D. E. McIntyre and D. E. Aspnes, *Surface Sci.* **24**, 417 (1971).
10. D. W. Berreman, *Phys. Rev.* **130**, 2193 (1963).
11. R. Ruppin and R. Englman, *Rep. Progr. Phys.* **33**, 149 (1970).
12. D. M. Newns, *Phys. Lett.* **60A**, 461 (1977).
13. H. Ibach, *Phys. Rev. Lett.* **27**, 253 (1971).
14. G. D. Mahan and A. A. Lucas, *J. Chem. Phys.* **68**, 1344 (1978).
15. M. Scheffler, *Surface Sci.* **81**, 562 (1979).
16. H. Pfnür, D. Menzel, F. M. Hoffmann, A. Ortega, and A. M. Bradshaw, *Surface Sci.* **93**, 431 (1980).
17. S. Andersson and B. N. J. Persson *Phys. Rev. Lett.* **45**, 1421 (1980).

18.   W. Ho, R. R. Willis, and E. W. Plummer, *Phys. Rev. Lett.* **40**, 1463 (1978).
19.   S. Y. Tong, C. H. Li, and D. L. Mills, *Phys. Rev. Lett.* **44**, 407 (1980); C. H. Li, S. Y. Tong, and D. L. Mills, *Phys. Rev. B* **21**, 3057 (1980); S. Y. Tong, C. H. Li, and D. L. Mills, *ibid.* **24**, 806 (1981).
20.   S. Y. Tong, *Progr. Surf. Sci.* **7**, 1 (1976).
21.   See, e.g., G. J. Schulz, *Rev. Mod. Phys.* **45**, 423 (1973); I. C. Walker, A. Stamatovic, and S. F. Wong, *J. Chem. Phys.* **69**, 5532 (1978).

CHAPTER 4

VIBRATIONAL MOTION OF MOLECULES ADSORBED ON THE SURFACE

4.1 GENERAL REMARKS

4.1.1 Summary of Basic Issues

In Chapter 3, we explored the theory of low-energy electron energy loss spectroscopy, with principal focus on the inelastic scattering from vibrational motions of adsorbed atoms and molecules. The particular examples explored involved geometries sufficiently simple that the atomic motions associated with a given normal mode are easily visualized. For example, in our discussion of scattering from a monolayer of adsorbed CO, our attention was confined to the $c(2 \times 2)$ layer of CO on the Ni(100) surface. Here the molecular axis is perpendicular to the surface, and the site has fourfold symmetry. The fourfold axis ensures that the normal modes involve motions either purely perpendicular to the surface, or motions strictly parallel to it. Application of the dipole and impact scattering selections rules to this case is straightforward.

When more complex molecules, or molecular fragments are bonded to the surface, the analysis of the vibrational normal modes and the application of the selection rules become less straightforward, and the intuition one has from spectroscopy of the gas phase can be misleading. For example, normal modes of the molecule which fail to produce an oscillating electric dipole moment in the gas phase can become dipole active when the entity is bound to the surface. As an example, consider the stretching vibration of a homopolar diatomic molecule (H_2, N_2) in the gas phase. Symmetry requires no oscillating electric dipole moment can be present when the molecular motion is excited. The presence of a reflection plane normal to the bond axis and midway between the two nuclei requires the component of any electric dipole moment parallel to the bond axis to vansih. A pair of mutually orthogonal reflection planes which contain the bond axis forces any component

127

of the electric dipole moment normal to the bond axis to vanish also. (The rotational symmetry about the bond axis present in the gas phase also requires the dipole moment normal to the axis to vanish.) Suppose the molecule is now adsorbed onto the surface, with axis parallel to it. If the stretching vibration is now excited, reflection symmetry in the plane parallel to the surface is no longer present, and we may have an oscillating electric dipole moment *normal* to the surface. This will be so even if the nuclei move parallel to the surface, when the stretching vibration is excited (symmetry no longer requires the stretching mode to involve only motion parallel to the surface). From the physical point of view, if the nuclei of the molecule are pushed closer together, there may be charge transfer between the substrate and the molecule, and this charge transfer alters the electric dipole moment *normal* to the surface.

When the molecule is adsorbed on the surface, we also encounter new normal modes not present in the gas phase. We may also illustrate this by considering the homopolar diatomic molecule examined in the previous paragraph. With two nuclei, we always have six degrees of freedom in total. Considerations of translational invariance require that in the gas phase, there are three degenerate "normal modes" which are rigid-body translations of the molecule as a whole, one in each of the three mutually orthogonal Cartesian directions. There is no restoring force for these motions, so the three rigid-body translations appear as "normal modes" with zero frequency. We also have a second set of degenerate normal modes in the gas phase. These are rigid rotations of the molecule about two mutually perpendicular axes, each normal to the bond axis. There is no restoring force for these two degenerate motions either, though the finite moment of inertia of the molecule renders their excitation energy finite. We are left only with the stretching mode that was the topic of the previous paragraph.

On the surface, the center of mass of the molecule is trapped at the minimum of a three-dimensional potential well with depth given by the binding energy of the molecule to the surface, and with spatial extent controlled by the detailed geometry of the adsorption site. There is thus a finite restoring force for motion of the center of mass of the molecule in any direction, with the consequence that the degrees of freedom associated with rigid body molecular translations in the gas phase now become normal modes of vibration with finite frequency. From the very nature of these modes, one sees that information on these frequencies and eigenvectors provides direct contact with the nature of the bonding between the adsorbed molecule and the surface. One refers to these modes as the "hindered translation modes." In a similar fashion, the molecule can no longer rotate freely, but restoring forces are again encountered as the molecule begins to rotate. We thus have a set of modes referred to as "hindered rotations." When the molecule is adsorbed on the surface, there are finite restoring forces for all molecular motions, including those for which they are absent in the gas phase.

Before we proceed with our discussion, it is worth noting that the terms "hindered rotation" and "hindered translation" are imprecise, and in fact can lead to confusion. We may illustrate this once again by our homopolar molecule. Let the molecule be adsorbed intact, again with its axis parallel to the surface. The z axis is normal to the surface, the molecular axis is parallel to the x axis, and we suppose the adsorption geometry is such that both the xz and yz planes are reflection planes. Now consider translation of the molecular axis is parallel to the x axis, and let $Q^{(1)}$ be the normal coordinate associated with this motion. From the previous paragraph, we have a term in the potential energy of the molecule of the form $\frac{1}{2}k_{11}(Q^{(1)})^2$, where k_{11} is the force constant provided by bonding of the molecule to the substrate. The molecule can also rotate about the y axis; this is "hindered" by bonding to the substrate, and we also have the term $\frac{1}{2}k_{22}(Q^{(2)})^2$, where $Q^{(2)}$ is the normal coordinate that describes the rotational motion about the y axis. At this level of approximation, one has two normal modes, with one clearly a "hindered translation" and one a "hindered rotation." We illustrate the two normal coordinates in Figs. 4.1a and 4.1b. But symmetry also allows terms of the form $k_{12}Q^{(1)}Q^{(2)}$ in the vibrational Hamiltonian, so in fact each normal coordinate is a linear combination of $Q^{(1)}$ and $Q^{(2)}$ in the end. We have two normal modes, each of which involves translation of the center of mass and a rocking motion about the y axis. It makes sense to speak of the two modes as a "hindered rotation" or "hindered translation" only in the

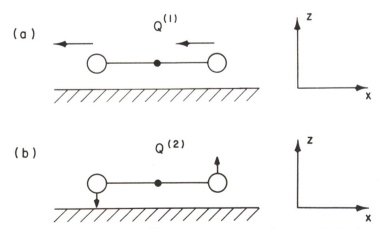

Fig. 4.1 (a) The normal mode $Q^{(1)}$, a "hindered translation," associated with a homopolar diatomic molecule adsorbed with molecular axis parallel to the surface. (b) The normal mode $Q^{(2)}$, a "hindered rotation," for a homopolar diatomic molecule adsorbed with molecular axis parallel to the surface.

special limit where k_{12} has magnitude small compared to both k_{11} and k_{22}. This condition may be satisfied in certain special cases, but certainly not in general. Nevertheless, the terms hindered rotation and hindered translation are useful to distinguish these modes from the genuine eigenmodes of the molecule that involve only relative motions of the constituents.

The foregoing discussion shows that when a molecule, or molecular fragment is adsorbed on the surface, we find new vibrational modes not present in the gas phase. In addition, the eigenvectors and eigenfrequencies associated with the internal vibrations present in the gas phase are altered by the adsorption process. A central issue in surface vibrational spectroscopy is the identification of the new modes, and the frequency shifts and altered character of modes present in the gas phase. The selection rules described in Chapter 3 can play a key role here, since through their use one may place constraints on the character of the atomic motions associated with particular loss features in the vibrational spectrum of an adsorbate. We have seen how this is done in our earlier examples, for simple cases. In the present chapter, we shall undertake a systematic analysis of these issues, with use of the techniques of group theory. Before we turn to this, we restate the selection rules that apply in electron energy loss spectroscopy, and compare them with those that operate in infrared and Raman spectroscopy.

4.1.2   Selection Rules in Surface Vibrational Spectroscopies

We have seen from Chapter 3, and from earlier remarks, that certain selection rules operate in electron energy loss spectroscopy. As we have emphasized, assumptions must be satisfied for these selection rules to be obeyed, i.e., in contrast to many of the selection rules that operate in optical spectroscopy, the selection rules in electron energy loss spectroscopy depend not only on the existence of certain symmetries in the adsorption geometry and scattering configuration, but also on other specific assumptions. For example, in the discussion of selection rules in the impact scattering regime given in Section 3.5.2, we had to ignore differences in the wave function between an outgoing electron with energy $E_S = E_I - \hbar\omega_0$, and the incoming energy E_I, when the wave vector of the scattered electron lies along the specular direction. These assumptions appear satisfied well enough in practice for the selection rules to operate in many important situations. The impact scattering selection rules require the presence of time-reversal symmetry in combination with the presence of reflection or rotation symmetries. If we consider the scattering geometry illustrated in Fig. 3.19, and suppose that there is twofold symmetry around the z axis, this twofold symmetry in combination with time-reversal symmetry requires the impact scattering cross section to vanish when the scattered electron emerges along the specular direction, for scattering from normal modes of vibration with normal coordinate Q_s *odd* under the twofold rotation. Similarly, if the xz plane is a plane of reflection symmetry, then the reflection operation R_{xz} combined with time-reversal symmetry requires the impact scattering amplitude to vanish whenever the scattered electron wave vector lies in the xz plane, for modes with normal coordinate Q_s that is odd under R_{xz}. Suppose for the scattering geometry illustrated in Fig. 3.19, the yz plane is a plane of reflection symmetry. Then the impact scattering amplitude vanishes along the specular direction for modes with normal coordinate odd under the reflection operation R_{yz}. The presence of time-reversal symmetry is also required for this selection rule to hold. The two selection rules involving the combination of time-reversal symmetry and a reflection plane lead to practical consequence only on surfaces where the substrate surface itself has a single mirror plane. Typical low-index surfaces have more than a single mirror plane. As a consequence the scattering cross section might vanish for a particular domain of oriented molecules or atoms in certain sites but will not vanish for the other possible domains.

We then have the dipole selection rule, which states that a vibrational mode with symmetry that generates an oscillating electric dipole moment perpendicular to the surface will produce intense small-angle scattering peaked around the specular direction, confined within the angular range of roughly $\hbar\omega_0/2E_I$, with E_I the impact energy. If the vibrational motion of the molecule generates a dipole moment parallel to the surface, with no component normal to it, then the parallel dipole moment is screened by its image in the (assumed metallic) substrate, and the "dipole lobe" will be absent. As this selection rule is applied, it must be kept in mind that some modes may generate a dipole moment normal to the surface, but its magnitude may be small, or modest in size. If, at the same time, the scattering geometry and mode symmetry is such that the impact scattering amplitude fails to vanish along the specular direction, then the "dipole lobe" may be present, but obscured by the presence of stronger impact scattering.

It is also quite possible to have a situation where dipole scattering is forbidden along the specular direction, but there is no reason for the impact scattering amplitude to vanish there. In this case, we will see no "dipolar lobe," the cross section near the specular direction will be dominated by the impact mechanism, and the cross section remains finite as the scattered electron wave vector is swung through the specular direction. Examples for this case will be discussed in detail later.

We next turn to a discussion of the selection rules which apply to the optical spectroscopy of surface vibrations. There are two principal techniques of concern: infrared and Raman spectroscopy. For the purposes of our discussion, we consider interaction of the incident radiation with a molecule adsorbed on a perfectly smooth crystal surface; as noted in Chapter 1, the Raman spectra reported to date are influenced dramatically by roughness and microscopic inhomogeneities on the surface.

The discussion of the selection rules must begin with an analysis of the electric field in the near vicinity of the surface. We suppose that p polarized radiation is incident on the surface, with θ_I the angle of incidence and E_0 the strength of the electric field in the incident beam. The amplitude of the electric field components in the vacuum just above the crystal surface may then be calculated straightforwardly from the Fresnel formulas of classical electromagnetic theory. For a highly reflecting surface such as a metal, simple approximate expressions prove adequate. If the complex index of refraction of the substrate is written as $n + i\kappa$, and $E_\perp$, $E_\parallel$ are the components of the electric field perpendicular and parallel to the surface, then we have the approximate forms

$$\left| \frac{E_\perp}{E_0} \right| = 2 \sin \theta_I \left[1 - \frac{1}{\cos \theta_I} \left(\frac{n}{n^2 + \kappa^2} \right) + \cdots \right]$$

$$= 2 \sin \theta_I \left[1 - \frac{\alpha_0}{4 \cos \theta_I} + \cdots \right] \tag{4.1}$$

and

$$\left| \frac{E_\parallel}{E_0} \right| = \frac{2}{(n^2 + \kappa^2)^{1/2}} \left[1 - \frac{\alpha_0}{4 \cos \theta_I} + \cdots \right], \tag{4.2}$$

where $\alpha_0 = 4n/(n^2 + \kappa^2)$ is the fraction of the energy in an incident beam absorbed by the substrate at normal incidence.

It is interesting to compare the relative magnitude of the parallel and perpendicular electric field components for typical metals. If we first consider infrared frequencies, where the photon frequency ω is very small compared to the electron plasma frequency ω_p [see Eq. (3.25)], to excellent approximation one has

$$\left| \frac{E_\parallel}{E_0} \right| = \frac{2\omega}{\omega_p} \left[1 + \frac{1}{[\omega\tau(\omega)]^2} \right]^{1/4}. \tag{4.3}$$

In a noble metal such as silver, we have $\hbar\omega_p \cong 9$ eV, while at the frequencies employed in typical infrared studies, $\omega\tau(\omega)$ is comparable to or somewhat larger than unity. With $\hbar\omega_p \cong 10$ eV, for frequencies of interest in infrared spectroscopy, one has $|E_\parallel| \cong 10^{-2}E_0$ at the surface while from Eq. (4.1) with α_0 very small, $|E_\perp| = 2 \sin \theta_I E_0$. Hence, the electric field is very nearly normal to the surface, with only a negligibly small parallel component. This conclusion is not necessarily restricted to flat surfaces. Even for small metal particles, the parallel component of the electric field vector may be small enough to make the selection rule effective enough for practical purposes. The actual calculation of electromagnetic field becomes rather involved for arbitrary geometries. It may be useful, however, to discuss briefly a simple model in order to understand the relevant parameters that enter the problem. Suppose the substrate has the form of a small sphere of radius R. In order to model the interaction with infrared radiation, the sphere may be considered in a uniform electric field ($R \ll \lambda$). The electric field at any point outside the sphere may be calculated from electrostatics in this case,

$$E_r = E_0\left(1 + \tilde{\varepsilon}\,\frac{2R^3}{r^3}\right)\cos\theta,$$

$$E_\theta = -E_0\left(1 - \tilde{\varepsilon}\,\frac{R^3}{r^3}\right)\sin\theta, \quad \text{with} \quad \tilde{\varepsilon} = \frac{\varepsilon - 1}{\varepsilon + 2}. \tag{4.4}$$

Here r is the distance from the center of the sphere, E_0 the field strength at infinite distance, θ the angle between the direction of the uniform field and r, and ε the dielectric constant. The maximum of the ratio between the field components parallel and perpendicular to the surface at a distance d from the surface is then

$$\frac{E_\theta}{E_r}\bigg|_{\text{max}} = \frac{1}{\varepsilon} + \frac{\varepsilon + 2}{\varepsilon}\frac{d}{R} + \cdots. \tag{4.5}$$

Since a typical distance for an adsorbed molecule is 1–2 Å even for a radius $R = 10$–20 Å, the observed intensity ($\sim E^2$) would be suppressed by a factor of 100 for the parallel dipole moments provided, of course, ε is large enough. Qualitatively, the same argument can be made for electron scattering since there (at least in the classical limit) the interaction between the dipole oscillator and the electron can be viewed as originating from the electric field generated by the electron which acts as the driving force on the oscillator.

At visible frequencies, such as those employed in Raman spectroscopy, the situation can be quite different. We consider silver again because many of the recent Raman studies employ this material as a substrate. Since α_0 is still quite small, roughly 0.05 in the visible, we still have $|E_\perp| \cong 2\sin\theta_1\,E_0$ again everywhere but at grazing incidence. But now we have, if a frequency of 2.5 eV is considered, $\kappa = 2.8$ and $n = 0.2$. As a consequence, $|E_{||}| \cong 0.7E_0$, and the parallel component of the electric field is comparable in magnitude to that perpendicular to the surface.

Electromagnetic radiation couples to an adsorbed molecule through interaction with the electric dipole moment $\mathbf{P}$ of the molecule. The dipole moment is modulated by motion of the nucleus, as we have seen earlier, and also has its magnitude influenced by the electric field associated with the radiation. We emphasize this by writing the electric dipole moment as $P_\alpha(Q_s, \mathbf{E})$, where, as in Section 3.5.2, Q_s is the normal coordinate associated with the mode of interest and $\mathbf{E}$ the electric field of the radiation. The interaction energy of the electromagnetic field with the molecule may then be written as

$$V_I = -\sum_\alpha \int_0^{E_\alpha} dE_\alpha\, P_\alpha(Q_s, \mathbf{E}) E_\alpha, \tag{4.6}$$

and we may write

$$P_\alpha(Q_s, \mathbf{E}) = P_\alpha^{(0)} + \sum_\beta \left(\frac{\partial P_\alpha}{\partial E_\beta}\right)_0 E_\beta + \left(\frac{\partial P_\alpha}{\partial Q_s}\right)_0 Q_s$$
$$+ \sum_\beta \left(\frac{\partial^2 P_\alpha}{\partial Q_s E_\beta}\right)_0 E_\beta Q_s + \cdots. \tag{4.7}$$

The third term in Eq. (4.7) gives rise to infrared absorption by the molecular vibrations, and the fourth term leads to Raman scattering of the incident photon. We consider each in turn.

In the infrared regime, we have seen that only the component of the electric field normal to the surface is nonzero. Hence, if we use the subscript $\perp$ to denote the direction perpendicular to the surface, the coupling of infrared radiation to the mode with normal coordinate Q_s is given by

$$V_I^{(3)} = -(\partial P_\perp/\partial Q_s) E_\perp Q_s, \tag{4.8}$$

and the selection rule that operates in infrared spectroscopy may be seen to be the same as the dipole selection rule in electron energy loss spectroscopy. The nuclear motion associated with an infrared active vibrational mode of the substrate must modulate the electric dipole moment component *normal* to the surface.

The Raman terms are more complex in form. One may extract them from Eq. (4.6) by writing the total electric field $\mathbf{E}$ at the surface as the sum of that $\mathbf{E}^{(I)}$ associated with the incoming photon, and the field $\mathbf{E}^{(S)}$ associated with the scattered photon. It is then the cross terms between the incident and scattered fields that lead to the Raman event. So we may write, for the case where the wave vectors of both the scattered photon, the incident photon (both p polarized), and the surface normal lie in one plane, called the scattering plane,

$$V_I^{(4)} = -\alpha_{\perp,\perp} E_\perp^{(I)} E_\perp^{(S)} Q_s - \alpha_{\|,\perp}(E_\|^{(I)} E_\perp^{(S)} + E_\|^{(S)} E_\perp^{(I)}) Q_s$$
$$- \alpha_{\|,\|} E_\|^{(I)} E_\|^{(S)} Q_s. \tag{4.9}$$

Modes of distinctly different symmetry contribute to the various terms in Eq. (4.9). One may illustrate this by a simple example. Consider a CO molecule adsorbed on a site of C_{4v} symmetry, so we have two normal modes polarized normal to the surface, and four polarized parallel to it. Each mode parallel to the surface is twofold degenerate, and we chose eigenvectors of each set so that one mode involves vibrational motion parallel to the scattering plane, and one is perpendicular to it.

Then the modes polarized perpendicular to the surface may lead to Raman scattering either through the first or the last term in Eq. (4.9). The modes with nuclear displacement parallel to the surface and also parallel to the scattering plane are Raman active by virtue of the second term in Eq. (4.9) while those polarized parallel to the surface and normal to the scattering plane are Raman inactive, for the geometry considered in deriving Eq. (4.9). These last modes contribute to the depolarized Raman spectra, i.e., if the incident photon is p polarized, and the scattered photon wave vector lies in the scattering plane, then the modes "silent" in the geometry just considered lead to Raman events in which the scattered photon is s polarized. Also, if the scattered photon is p polarized, with wave vector tipped away from the scattering plane, the modes again may be seen.

So in the Raman spectroscopy of adsorbed molecules, the fact that $E_{\parallel}$ is not suppressed strongly near metal surfaces as it is in the infrared allows access to all of the normal modes of the adsorbed molecule, with appropriate choice of scattering geometry. As remarked earlier, the Raman spectra of adsorbed molecules reported in the literature are influenced strongly by roughness present at the surfaces or interfaces that have been examined. The Raman spectroscopy of well-characterized single-crystal surfaces is in its infancy, and the foregoing arguments suggest that it may prove to be a most useful addition to the surface vibrational spectroscopies presently available.

4.2 THEORY OF THE VIBRATIONAL NORMAL MODES OF ADSORBED MOLECULES

In this section, we examine the theoretical description of the vibrational normal modes of molecules adsorbed on the surface. Here we consider an isolated molecule or adatom adsorbed on an otherwise perfect crystal surface. Our aim is to present the formalism in a general fashion, to assist the group theoretical discussions which form the basis of the remainder of the present chapter. In Chapter 5, where we consider the influence of a surface on the vibrational modes of crystal lattices, we shall turn our attention to the case of an ordered layer of material adsorbed on the crystal surface.

4.2.1 General Theory

We begin by considering a molecule comprised of N atoms in free space, then we examine the effect of bonding to a crystal surface on its vibrational properties. Central to the discussion here, and elsewhere in this volume is the assumption that the adiabatic approximation is valid. The Hamiltonian $\mathcal{H}$ of the molecule, when written out in full detail, depends on both the coordinates and momenta $\{\mathbf{r}_i, \mathbf{p}_i\}$ of the electrons which surround the nuclei, and those $\{\mathbf{R}_i, \mathbf{P}_i\}$ of the nuclei themselves. The adiabatic approximation asserts that the wave function of the molecule may be approximated as the product of two functions. One, the vibrational wave function $\chi(\{\mathbf{R}_i\})$ depends on only the nuclear coordinates, while the second $\Phi(\{\mathbf{r}_i\}, \{\mathbf{R}_i\})$ describes electrons bound to the nuclei. Since Φ depends parametrically on the nuclear positions, the scheme allows the electron distribution to adjust (adiabatically) to nuclear motions.

The scheme provides an explicit prescription for constructing the Schrödinger equation obeyed by the vibrational eigenfunction $\chi(\{\mathbf{R}_i\})$. We have the form

$$\left[\sum_i \frac{P_i^2}{2M_i} + V(\{\mathbf{R}_i\}) \right] \chi(\{\mathbf{R}_i\}) = E\chi(\{\mathbf{R}_i\}), \tag{4.10}$$

with $\mathbf{P}_i$ the momentum operator for the ith nucleus, M_i its mass, E the energy of the eigenfunction $\chi(\{\mathbf{R}_i\})$, and the prescription for constructing the potential function $V(\{\mathbf{R}_i\})$ need not concern us here. A detailed and complete derivation of the adiabatic approximation, with corrections to the basic scheme has been presented by Born and Huang [1]. A more recent analysis of the scheme has been given by Maradudin [2]. When one considers vibrational motion of nuclei in a metal, subtle issues are raised, because the spectrum of electronic excitations is continuous, and extends down to zero energy. Brovman and Kagan have explored the validity of the basic scheme in this case [3], which is of direct interest to us because the molecules and adatoms considered below are commonly adsorbed on metallic substrates.

The equilibrium configuration of the nuclei is found by seeking the set of nuclear positions, $\{\mathbf{R}_i^{(0)}\}$, for which the potential energy is a minimum. We are then concerned with small-amplitude vibrations of the nuclei about their equilibrium position, so we may write $\mathbf{R}_i = \mathbf{R}_i^{(0)} + \mathbf{u}_i$. If $u_{i\alpha}$ is the αth Cartesian component of the displacement of the nucleus away from equilibrium, then on noting that $(\partial V/\partial R_{i\alpha})_0$ vanishes, we may write

$$V(\{\mathbf{R}_i\}) = V(\{\mathbf{R}_i^{(0)}\}) + \frac{1}{2} \sum_{ij} \sum_{\alpha\beta} \left(\frac{\partial^2 V}{\partial R_{i\alpha} \partial R_{j\beta}} \right)_0 u_{i\alpha} u_{j\beta} + \cdots . \tag{4.11}$$

If the molecule is constructed from N atoms, and we suppose it is in free space, then Eq. (4.11) in combination with Eq. (4.9) describes a quantum mechanical set of coupled harmonic oscillators, whose number $3N$ equals the number of degrees of freedom in the molecule. When the expansion of the potential energy function $V(\{\mathbf{R}_i\})$ is terminated after the terms quadratic in the nuclear displacement from equilibrium, as in Eq. (4.11), one has the description of the molecular vibration problem referred to as the harmonic approximation.

The central issue of the theory of molecular vibrations is then to analyze the normal modes of the vibrational Hamiltonian formed from Eq. (4.11) in combination with Eq. (4.10). For any molecule in free space, three of the $3N$ normal modes correspond to rigid-body translations of the entire molecule parallel to the three mutually orthogonal Cartesian axes in space. Since the potential energy $V(\{\mathbf{R}_i\})$ and the special form given in Eq. (4.11), are invariant under rigid translations of the molecule as a whole, there is no restoring force for this translational motion, and the three translational degrees of freedom appear in the normal mode analysis as three modes with frequency identically equal to zero. The eigenvectors then describe rigid translations of the molecule in a particular direction. If we consider eigenvectors associated with any two distinct normal modes of vibration, it is a general theorem that they must be orthogonal. Most particularly, the eigenvector of any mode with nonzero frequency must be orthogonal to the eigenvectors associated with each of the three zero frequency translational modes. From this statement, it is straightforward to prove that the nuclear motions associated with the $(3N - 3)$ normal modes of nonzero frequency must be such that for each of these modes, the center of mass of the molecule remains fixed in space.

The potential energy function $V(\{\mathbf{R}_i\})$ must also be invariant under rigid-body rotations of the molecule. From this it follows that a molecule in free space has rotational degrees of freedom, again with no restoring force provided by the potential energy. The excitation energy of the quantized rotational levels is finite, because the molecule has finite moments of inertia. If we consider a molecule of general form, there are three rotational degrees of freedom, associated ultimately with the three finite principal elements of the moment of inertia tensor. A linear molecule has a finite moment of inertia only about two axes, each of which is orthogonal to the molecular axis. Hence, we have only two rotational degrees of freedom for a linear molecule. For an atom in free space, of course, the three translational degrees of freedom exhaust the possibilities of nuclear motion.

In what follows, it will be convenient to suppose the potential function $V(\{\mathbf{R}_i\})$ may be constructed as a superposition of pairwise interactions between the nuclei, each of which depends on only the internuclear separation $|\mathbf{R}_i - \mathbf{R}_j|$. This central force approximation is used frequently in the description of nuclear vibrations in molecular and solid state physics, and by introducing it here, we can obtain a number of simple and useful results readily generalized to more complex descriptions of the potential function. Thus, we write

$$V(\{\mathbf{R}_i\}) = \tfrac{1}{2} \sum_{ij}' \varphi_{ij}(|\mathbf{R}_i - \mathbf{R}_j|), \qquad (4.12)$$

where the prime on the sum in Eq. (4.12) instructs us to omit the term with $i = j$, and the subscript on φ allows the interaction to assume different forms, for different pairs on nuclei.

As remarked earlier, we must have $(\partial V/\partial R_{i\alpha})_0$ vanish when the derivative is evaluated with the nuclei at their equilibrium position. If $d(ij) = |\mathbf{R}_i^{(0)} - \mathbf{R}_j^{(0)}|$ is the separation between nuclei i and j in the equilibrium configuration, and $\hat{n}(ij) = (\mathbf{R}_j^{(0)} - \mathbf{R}_i^{(0)})/d(ij)$ is a unit vector directed along the line that connects nuclei i and j, then a given nuclear arrangement is in equilibrium only if the condition

$$\sum_j \hat{n}_\alpha(ij)\varphi_{ij}'(d(ij)) = 0 \qquad (4.13)$$

is satisfied for all nuclei i, and for all three choices of α. Here $\varphi_{ij}'(d(ij))$ is the derivative of the interaction function with internuclear separation, evaluated with $|\mathbf{R}_i - \mathbf{R}_j|$ set equal to $d(ij)$. When the condition in Eq. (4.13) is satisfied, one is assured that the nuclei are arranged so that the force on each vanishes identically, when all nuclei are at rest. This does not ensure that the particular arrangement is either stable or is the configuration of lowest energy, but for particular model forms of the functions φ_{ij}, one may derive the associated molecular geometries.

For the central force picture, the terms $V_2(\{\mathbf{u}_i\})$ quadratic in the vibrational amplitude may be written

$$V_2(\{\mathbf{u}\}) = \tfrac{1}{4} \sum_{\alpha\beta} \sum_{ij}' K_{\alpha\beta}(ij)(u_{i\alpha} - u_{j\alpha})(u_{i\beta} - u_{j\beta}), \qquad (4.14)$$

where $K_{\alpha\beta}(ij)$ serves as an effective force constant that couples motion of the pair (ij) in direction α with that in direction β. The factor $\tfrac{1}{4}$ is convenient, because as one sums over all values of i and j, each "bond" is counted twice. We have the explicit form

$$K_{\alpha\beta}(ij) = \frac{\varphi_{ij}'(d(ij))}{d(ij)} \delta_{\alpha\beta} + \left(\varphi_{ij}''(d(ij)) - \frac{\varphi'(d(ij))}{d(ij)} \right) \hat{n}_\alpha(ij)\hat{n}_\beta(ij), \qquad (4.15)$$

where, from inspection, one sees that $K_{\alpha\beta}(ij) = K_{\beta\alpha}(ij)$ and also $K_{\alpha\beta}(ij) = K_{\alpha\beta}(ji)$.

We could now proceed with the discussion of the normal modes of the molecule in the gas phase. However, since it costs us little to add the effect of interaction between the molecule and the substrate atoms at this point, we turn to this question. We shall continue to denote the displacements associated with the nuclei in the adsorbed molecule by $\mathbf{u}_i$ and $\mathbf{u}_j$, as above, and we shall use $\mathbf{u}_n$ and $\mathbf{u}_m$ to denote those associated with substrate atoms. The system Hamiltonian then consists of three distinct pieces,

$$\mathcal{H} = \mathcal{H}_M + \mathcal{H}_S + V, \tag{4.16}$$

where $\mathcal{H}_M$ is the vibrational Hamiltonian of the adsorbed molecule, given above, $\mathcal{H}_S$ that of the substrate, and V the coupling terms between the adsorbed molecule and the substrate. We have, as before,

$$\mathcal{H}_M = \sum_i \frac{P_i^2}{2M_i} + \tfrac{1}{4} \sum_{\alpha\beta} \sum_{ij} K_{\alpha\beta}(ij)(u_{i\alpha} - u_{j\alpha})(u_{i\beta} - u_{j\beta}) \tag{4.17}$$

while the vibrations of the substrate are controlled by

$$\mathcal{H}_S = \sum_n \frac{P_n^2}{2M_n} + \tfrac{1}{4} \sum_{nm} \sum_{\alpha\beta} K_{\alpha\beta}(n, m)(u_{n\alpha} - u_{m\alpha})(u_{n\beta} - u_{m\beta}) \tag{4.18}$$

and the coupling terms will be written in the form

$$V = \tfrac{1}{2} \sum_{in} \sum_{\alpha\beta} K_{\alpha\beta}(i, n)(u_{i\alpha} - u_{n\alpha})(u_{i\beta} - u_{n\beta}). \tag{4.19}$$

It is important to note that when the molecule is adsorbed on a surface, the equilibrium condition is modified, to incorporate the forces exerted on the nuclei of the molecule by the atoms in the substrate. In place of Eq. (4.13), we now have

$$\sum_j \hat{n}_\alpha(ij)\varphi'_{ij}(d(ij)) + \sum_n \hat{n}_\alpha(in)\varphi'_{in}(d(in)) = 0 \tag{4.20a}$$

which, when satisfied, ensures that the nuclei of the adsorbed molecule are located at positions where the net force they experience is zero. Similarly, for the ions of the substrate we have

$$-\sum_i \hat{n}_\alpha(in)\varphi'_{in}(d(in)) + \sum_m \hat{n}_\alpha(nm)\varphi'_{nm}(d(nm)) = 0. \tag{4.20b}$$

By comparing Eqs. (4.20a) and (4.13), we see that when the molecule sits on the surface, it is not possible for the nuclear configuration to remain the same as it is in the gas phase. The bond lengths and bond angles necessarily change as a consequence of the interaction with the substrate, though it is possible for these changes to be small. A principal task of surface vibrational spectroscopy is to learn how the structure of a given molecule absorbed on the surface differs from, or is similar to the gas- or liquid-phase species. Note that if the bond angles and internuclear separations are affected by the adsorption process, the force constants $K_{\alpha\beta}(ij)$ are different for the gas phase and adsorbed forms of the molecule. Also, Eq. (4.20b) shows that the substrate atoms in the near vicinity of the adsorbate are shifted in position, away from locations appropriate to atoms in the surface of a clean crystal.

Since the potential energies of Eq. (4.16) are each quadratic in the nuclear displacement, it is possible, in principle, to diagonalize the Hamiltonian exactly. However, when one faces this procedure in a practical case, one recognizes that the task is now quite formidable, when compared to the analysis of gas phase vibrational spectra. In the latter case, we had $3N$ nuclei, so the Hamiltonian matrix is a $3N \times 3N$ Hermitean forms. For molecules of substantial size, such matrices are readily diagonalized with use of modern computational techniques. If the molecule is adsorbed on a substrate with N_s atoms, we now face a problem for which the dynamical matrix is $3(N + N_s) \times 3(N + N_s)$ in size. If the substrate is regarded as a macroscopic crystal, then clearly direct diagonalization of the vibrational Hamiltonian is out of the question. We shall discuss some numerical procedures that appear to work well, for simple cases explored so far, and we shall also derive approximate procedures that may be applied to modes of certain classes. In Chapter 5, we shall see that additional methods may be used when periodic overlayers are present. For the moment, we confine our attention to the formal procedures associated with the diagonalization process.

It is useful to introduce the mass weighted coordinates $v_{i\alpha}$, defined by means of the relation $v_{i\alpha} = u_{i\alpha}M_i^{1/2}$ and the momenta $\Pi_{i\alpha}$ cannonically conjugate to $v_{i\alpha}$. We have $\Pi_{i\alpha} = M_i^{-1/2}P_{i\alpha}$. The Hamiltonian then becomes

$$\mathcal{H} = \tfrac{1}{2}\sum_{i\alpha}\Pi_{i\alpha}^2 + \tfrac{1}{2}\sum_{\alpha\beta}\sum_{ij}D_{\alpha\beta}(ij)v_{i\alpha}v_{j\beta}$$

$$+ \tfrac{1}{2}\sum_{n\alpha}\Pi_{n\alpha}^2 + \tfrac{1}{2}\sum_{\alpha\beta}\sum_{nm}D_{\alpha\beta}(nm)v_{n\alpha}v_{m\beta}$$

$$+ \tfrac{1}{2}\sum_{\alpha\beta}\sum_{in}D_{\alpha\beta}(in)v_{i\alpha}v_{n\beta} + \tfrac{1}{2}\sum_{\alpha\beta}\sum_{in}D_{\alpha\beta}(ni)v_{n\alpha}v_{i\beta}. \qquad (4.21)$$

We have introduced the elements of the dynamical matrix $D_{\alpha\beta}(ij)$, $D_{\alpha\beta}(in)$, and $D_{\alpha\beta}(nm)$ defined by, with $\tilde{K}_{\alpha\beta}(ij) = K(ij)(M_iM_j)^{-1/2}$,

$$D_{\alpha\beta}(ij) = \delta_{ij}\left[\sum_{j'}\tilde{K}_{\alpha\beta}(ij') + \sum_n\tilde{K}_{\alpha\beta}(in)\right] - (1 - \delta_{ij})\tilde{K}_{\alpha\beta}(ij), \qquad (4.22a)$$

$$D_{\alpha\beta}(in) = D_{\beta\alpha}(ni) = -\tilde{K}_{\alpha\beta}(in), \qquad (4.22b)$$

and

$$D_{\alpha\beta}(nm) = \delta_{nm}\left[\sum_{m'} \tilde{K}_{\alpha\beta}(nm') + \sum_{i} \tilde{K}_{\alpha\beta}(ni)\right]$$
$$- (1 - \delta_{nm})\tilde{K}_{\alpha\beta}(nm). \tag{4.22c}$$

Since the potential energy terms in Eq. (4.21) are a Hermitian quadratic form, they and the kinetic energy may be diagonalized by a unitary transformation. This may be done utilizing the eigenvectors and eigenvalues of the dynamic matrix. If there are N_s atoms in the substrate, we then have $3(N + N_s)$ real eigenvalues ω_s^2, and the eigenvectors $\{e_{i\alpha}^{(s)}, e_{n\alpha}^{(s)}\}$ may be chosen real also, since all elements of the dynamic matrix are real. We have

$$\sum_{\beta j} D_{\alpha\beta}(ij)e_{j\beta}^{(s)} + \sum_{\beta n} D_{\alpha\beta}(in)e_{n\beta}^{(s)} = \omega_s^2 e_{i\alpha}^{(s)} \tag{4.23a}$$

and

$$\sum_{\beta m} D_{\alpha\beta}(nm)e_{m\beta}^{(s)} + \sum_{\beta i} D_{\alpha\beta}(ni)e_{i\beta}^{(s)} = \omega_s^2 e_{n\alpha}^{(s)} \tag{4.23b}$$

as the statement of the basic eigenvalue problem, while the eigenvectors obey the orthonormality relation

$$\sum_{i\alpha} e_{i\alpha}^{(s)}e_{i\alpha}^{(s')} + \sum_{n\alpha} e_{n\alpha}^{(s)}e_{n\alpha}^{(s')} = \delta_{ss'} \tag{4.24a}$$

supplemented by the completeness conditions

$$\sum_{s} e_{i\alpha}^{(s)}e_{j\beta}^{(s)} = \delta_{ij}\delta_{\alpha\beta}, \tag{4.24b}$$

$$\sum_{s} e_{i\alpha}^{(s)}e_{n\beta}^{(s)} = 0, \tag{4.24c}$$

$$\sum_{s} e_{n\alpha}^{(s)}e_{m\beta}^{(s)} = \delta_{nm}\delta_{\alpha\beta}. \tag{4.24d}$$

We may now write the nuclear displacement $v_{i\alpha}$ as a linear superposition of the eigenamplitudes

$$v_{i\alpha} = \sum_{s} e_{i\alpha}^{(s)}Q_s, \tag{4.25a}$$

with a similar relation for the cannonically conjugate momenta:

$$\Pi_{i\alpha} = \sum_{s} e_{i\alpha}^{(s)}\Pi_s. \tag{4.25b}$$

It is a straightforward matter to insert Eqs. (4.25a) and (4.25b) into the Hamiltonian displayed in Eq. (4.27) to verify that it becomes diagonal:

$$\mathcal{H} = \tfrac{1}{2}\sum_{s} \{\Pi_s^2 + \omega_s^2 Q_s^2\}. \tag{4.26}$$

From Eq. (4.26), we see that Q_s is, in fact, equivalent to the normal coordinate of classical mechanics, with Π_s its cannonically conjugate momentum. Note that Eq. (4.25a) and its companions may be inverted to read

$$Q_s = \sum_{i\alpha} e_{i\alpha}^{(s)} v_{i\alpha} + \sum_{n\alpha} e_{n\alpha}^{(s)} v_{n\alpha}, \tag{4.27}$$

and the quantum mechanical operators Q_s and Π_s may be expressed in terms of the boson and annihilation and creation operators a_s, a_s^+ that destroy and create quanta of the normal mode s. We have

$$Q_s = (\hbar/2\omega_s)^{1/2}(a_s + a_s^+) \tag{4.28a}$$

and

$$\Pi_s = i^{-1}(\hbar\omega_s/2)^{1/2}(a_s - a_s^+). \tag{4.28b}$$

This completes our formal discussion of the description of the vibrational motion of a molecule adsorbed on the surface. In the next section, we turn to a discussion of the nature of these vibrational modes.

4.2.2 The Nature of the Vibrational Modes of Adsorbed Molecules

In this section, we pause in the formal development to examine some general features of the normal modes of vibration of a molecule on the surface.

Suppose we begin with the following simple picture, whose regime of validity will be outlined shortly. Consider a molecule adsorbed on the surface, and examine its vibrational normal modes assuming only that the nuclei in the molecule itself may move. Imagine for the moment that all atoms in the substrate are frozen in position so they do not participate in the motion. Clearly this picture is valid in the limit that the mass of the substrate atoms is assumed very large, so it applies to light atoms or molecules adsorbed on heavy metal substrates. This is a case of considerable practical importance.

In this case, in the Hamiltonian of Eq. (4.21) (treated within the framework of classical mechanics for the moment), we may set $\{v_n\}$ and $\{\Pi_n\}$ to zero, and we are left with the vibrational Hamiltonian that describes just the $3N$ degrees of freedom of the molecule. The problem is clearly *not* the same as that encountered in the gas phase, however. As we saw after Eq. (4.20), the bonding of the molecule to the substrate alters the shape of the molecule, so, in general, the entity we are considering is distinctly different in geometrical from and symmetry than in the gas phase. Also, even though the atoms in the substrate are assumed frozen, as the nuclei in the molecule vibrate they "stretch" the bonds between the molecule and the substrate. This enters the mathematics of Section 4.2.1 through the terms in $K_{\alpha\beta}(i, n)$ on the right-hand side of Eq. (4.22a). In the presence of these terms, the vibrational Hamiltonian of the molecule is no longer invariant under rigid body translations and rotations of the molecule. There are now restoring forces present if a rigid-body translation or rotation is attempted. Thus, all $3N$ degrees of freedom become vibrational modes with finite frequency; we have the "hindered translations" and "hindered rotations" encountered in the specific example considered in Section 4.1.

Now suppose we consider one particular normal mode of the molecule, with frequency ω_v when calculated by the approximate method described in the previous paragraph. The next step is to allow the substrate atoms to move in response to the vibrational motion of the nuclei in the molecule. Let ω_M be the maximum phonon frequency of the substrate. A crucial issue is then whether $\omega_v > \omega_M$, or $\omega_v < \omega_M$, since the effect of coupling to the substrate motions is qualitatively different in the two cases. When $\omega_v > \omega_M$, excitation of the molecular vibration induces motion in the nearby substrate nuclei, but the disturbance remains localized to the near vicinity of the adsorbate. In essence, the substrate is driven by a force with frequency too high to excite a propagating disturbance, and the amplitude of the induced displacements must necessarily fall off exponentially with increasing distance from the adsorbate. We shall see below that the frequency of the molecular vibration is shifted *upward* from the value ω_v to $\omega_v + \delta\omega_v$, when the substrate atoms are allowed to move in response to the molecular vibration.

If $\omega_v < \omega_M$, a very different picture applies. Now the molecular frequency is sufficiently low that when the substrate atoms move, in response to motion of the adsorbate nuclei, the disturbance propagates into the substrate. The molecule behaves very much like a point source that launches phonons that propagate into the crystal interior, or along the surface away from the molecule. A consequence is that the molecular vibration acquires a finite lifetime, and no longer remains a true normal mode of the system. This is true even within the harmonic approximation of lattice dynamics. In any case, a detailed calculation of the lifetime is required to assess the width Γ of the molecular vibration. If $\Gamma \ll \omega_v$, the molecular vibrational remains a long-lived feature of the system, and we have a direct analog to the well-known resonance modes in the lattice dynamics defects and impurities in bulk crystal lattices [4]. It is entirely possible to have $\Gamma \approx \omega_v$, in which case the "mode" decays so quickly that it fails to remain a well defined excitation of the system. We shall explore the nature of such resonance modes for a specific example in Chapter 5, when we discuss calculations that explore the $c(2 \times 2)$ and the $p(2 \times 2)$ ordered overlayers of oxygen on the Ni(100) surface.

When a molecule or an adsorbed atom sits on the crystal surface, with the substrate in thermal equilibrium at temperature T, an examination of the *frequency spectrum* of the nuclear motion will always show structure for frequencies below the maximum phonon frequency ω_M of the substrate. This is true even if all $3N$ "molecular" vibrational modes have frequency above ω_M. The adsorbed atom or molecule sits on top of the substrate, the substrate atoms engage in thermal motions with frequency below ω_M, and the adsorbate responds to these motions. Thus, a complete description of the motion of an adsorbed species must include an analysis of the high-frequency modes above ω_M for which, as we have seen, the displacements remain localized around the adsorbate; also it is necessary to explore the frequency regime between 0 and ω_M, where the adsorbate nuclei are "driven" by the thermal motion in the substrate.

The point in the previous paragraph is illustrated very nicely by some recent theoretical studies carried out by Black [5], who has explored the vibrational properties of atoms adsorbed on the Ni(111) surface. We present a brief summary of these calculations.

Suppose we consider the mean square displacement $\langle u_{i\alpha}^2 \rangle_T$, $\langle u_{n\alpha}^2 \rangle$ of the adsorbed atom or an atom in the substrate, with the substrate at temperature T. We may introduce a spectral density function $\rho_{i\alpha}(\Omega)$ which provides a measure of the influence of the frequency range between Ω, and $\Omega + d\Omega$ on the vibrational amplitude of the atom in question. In terms of the spectral density function, one may write [6]

$$\langle u_{i\alpha}^2 \rangle_T = \frac{\hbar}{2M_i} \int_0^\infty \frac{d\omega}{\omega} [1 + 2\bar{n}(\omega)] \rho_{i\alpha}(\omega), \qquad (4.29)$$

where $\bar{n}(\omega) = [\exp(\hbar\omega/k_B T) - 1]^{-1}$, and a relation identical to Eq. (4.29) applies to substrate atoms. If we model our vibrating atom as a simple harmonic oscillator (Einstein oscillator) with frequency ω_0, then $\rho_{i\alpha}(\omega)$ is simply $\delta(\omega - \omega_0)$.

We consider the form of the spectral density function for an oxygen atom adsorbed on the Ni(111) surface. Upton and Goddard [7] have carried out *ab initio* theoretical studies of the interactions of an oxygen atom with the Ni(111) surface, with the substrate modeled as a finite (rigid) cluster of Ni atoms. As the oxygen atom approaches the surface along a line perpendicular to it, and through the hollow site of threefold symmetry to which the oxygen binds, they find a minimum in the potential energy at a distance $R_\perp = 1.20$ Å above the plane which contain the nickel nuclei. They also calculate the curvature of the potential energy curve near the minimum, so knowledge of d_0 and this curvature gives values for $\varphi''_{NiO}(d_0)$ if the oxygen–nickel interaction is assumed to have central force character also confined to nearest neighbors. With this potential energy curve for the frequency of the motion of the oxygen atom normal to the surface, one calculates 514 cm^{-1}, if one assumes the Ni nuclei are held rigidly in place. This is the frequency ω_v of our earlier discussion. If the Ni nuclei are allowed to move as the oxygen vibration normal to the surface is excited, Black finds the frequency ω_v is shifted from 514 cm^{-1} upward to $\omega_v + \delta\omega_v = 578 \text{ cm}^{-1}$, in excellent accord with the value 580 cm^{-1} obtained by Ibach and Bruchmann from electron energy loss studies of ordered overlayers of oxygen on the Ni(111) surface [8]. Thus, for this system, the data are in remarkable agreement with first-principles calculations of the oxygen–nickel force constants.

We have seen that the oxygen is also "driven" by the thermal motions in the underlying Ni substrate, so that the oxygen atom motion normal to the surface occurs not only at the frequency of 580 cm^{-1}, the high-frequency localized mode of the oxygen atom, but also in the frequency range $0\text{--}295$ cm^{-1} where the Ni phonons occur. In Fig. 4.2a, we show the spectral density associated with oxygen motion normal to the surface in this frequency range (solid line), and compare that with motion of a Ni atom in the surface, in the direction normal to the surface (dashed line) and that of a Ni atom

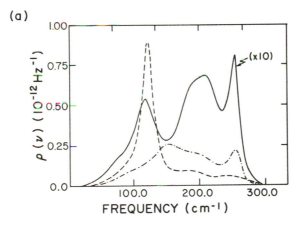

(a)

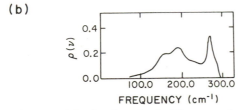

(b)

Fig. 4.2 (a) The spectral density of an adsorbed O atom on the Ni(111) surface, for motion normal to the surface, in the frequency range $0\text{--}295 \text{ cm}^{-1}$ (—). This is compared with that for motion of a surface Ni atom normal to the surface (---) and parallel to the surface ($\cdot - \cdot - \cdot -$). (b) The spectral density for motion of a Ni atom deep in the crystal interior. This is the phonon density of states of the Ni crystal. These calculations have been performed by J. E. Black, [5] used with permission.

parallel to the surface (dot–dash line). We see by comparing the positions of peaks in the various spectral density functions that the oxygen atom is driven by Ni motions both parallel and perpendicular to the surface.

The mean square displacement normal to the surface of the oxygen atom has been calculated to be 8.1×10^{-19} cm^2 for the above model. Roughly 75% of this comes from contributions in the frequency regime *below* the maximum phonon frequency of the substrate, while only 25% is associated with thermal excitation of the high-frequency 580 cm^{-1} mode. Thus, a proper treatment of the frequency spectrum of adsorbate vibrations in the regime from 0 to ω_M is required in addition to an analysis of the high-frequency modes, for a complete description of the vibrational motion of adsorbates. We shall return to this topic in Chapter 5.

While the preceding considerations establish the importance of the low-frequency motions of adsorbates, it is the high-frequency modes which lie well above ω_M that are most commonly studied in electron energy loss experiments. At least for species on transition metal surfaces, ω_M is sufficiently low that experimental study of energy losses between 0 and $\hbar\omega_M$ requires one to look at energy transfers sufficiently small that the signal lies in the tail of the energy distribution of electrons transmitted through the mono-chromator, and is thus difficult to detect. We saw that the high-frequency modes well above ω_M may be described approximately by setting the displacements $\mathbf{v}_n$ of the substrate atoms to zero in Eq. (4.21) and in the sub-sequent expressions. It is possible to obtain a simple and general procedure that provides the first correction to this approximation. We conclude the present subsection with a derivation of this prescription.

We suppose the vibrational modes of the adsorbate–substrate complex may be divided into two distinct sets. One group, with frequencies $\{\omega_v\}$ all have frequency high compared to ω_M. In a typical case, these are modes which are internal modes of vibration of a molecule which (in the limit $\{\mathbf{v}_n\} = 0$) have frequency shifted by bonding to the substrate. Then we have a second set of modes we denote by $\{\omega_p\}$, which have frequencies comparable to ω_M, or smaller. Included in this set are the phonon modes of the substrate, perturbed by the presence of the adsorbate.

The full eigenvalue problem for the high-frequency modes involves the solution of the entire set of equations given in Eq. (4.23). We rewrite these, in notation suitable to the present problem, noting that we wish to consider one of the high-frequency modes:

$$\sum_{\beta j} D_{\alpha\beta}(ij)e_{j\beta}^{(v)} + \sum_{\beta n} D_{\alpha\beta}(in)e_{n\beta}^{(v)} = \omega_v^2 e_{i\alpha}^{(v)} \tag{4.30a}$$

and

$$\sum_{\beta m} D_{\alpha\beta}(n,i)e_{i\beta}^{(v)} + \sum_{\beta i} D_{\alpha\beta}(n,i)e_{i\beta}^{(v)} = \omega_v^2 e_{n\alpha}^{(v)}. \tag{4.30b}$$

We may write this in a more compact form as follows. Suppose we consider an eigenvalue problem in which the adsorbate nuclei are not allowed to vibrate, but remain clamped rigidly in place. We have eigenvectors and eigenvalues associated with this problem, $3N$ in number, and these are generated from

$$\sum_{\beta m} D_{\alpha\beta}(n,m)e_{m\beta}^{(0p)} = \omega_{\mathrm{P}}^{(0)2}e_{n\alpha}^{(0p)}, \qquad (4.31)$$

where it is noted that $D_{\alpha\beta}(n,m)$ is *not* the dynamical matrix of the semi-infinite substrate unperturbed by the adsorbate, but includes both the effect of shifts in nuclear position caused by the adsorbate on the force constants in its vicinity, as well as the restoring force produced by "flexing" the bonds to the adsorbate [the terms in $\sum_i K_{\alpha\beta}(ni)$ in Eq. (4.22c)]. If the eigenvalues and eigenvectors generated from Eq. (4.31) are assumed known, we may construct the function

$$U_{\alpha\beta}^{(0)}(nm;\omega) = \sum_{\mathrm{P}} \frac{e_{n\alpha}^{(0p)}e_{m\beta}^{(0p)}}{\omega^2 - \omega_{\mathrm{P}}^{(0)2}}, \qquad (4.32)$$

and this may be used to solve Eq. (4.30b) for $e_{m\beta}^{(v)}$, with the consequence that the substrate degrees of freedom may be eliminated from Eq. (4.30a). The first step of this procedure gives

$$e_{n\beta}^{(v)} = \sum_{\beta'n'}\sum_{\delta j} U_{\beta\beta'}^{(0)}(nn';\omega_v)D_{\beta'\delta}(n'j)e_{j\delta}^{(v)} \qquad (4.33)$$

and the eigenvalue problem may then be reduced to one defined entirely on the $(3N_s \times 3N_s)$-dimensional subspace spanned by the eigenvectors associated with motion of the adsorbate nuclei. We have

$$\sum_{\beta j} \tilde{D}_{\alpha\beta}(ij;\omega_v)e_{j\beta}^{(v)} = \omega_v^2 e_{\alpha i}^{(v)}, \qquad (4.34)$$

where we must now consider the eigenvalue problem generated by the modified dynamical matrix

$$\tilde{D}_{\alpha\beta}(ij;\omega_v) = D_{\alpha\beta}(ij) + \sum_{n\delta}\sum_{n'\delta'} D_{\alpha\delta}(in)U_{\delta\delta'}^{(0)}(nn';\omega_v)D_{\delta'\beta}(n'j). \qquad (4.35)$$

At first glance, Eq. (4.34) appears to provide an enormous simplification over the original eigenvalue problem defined in Eq. (4.30), since we now can confine our attention to a rather small subspace, of dimension $3N_s \times 3N_s$, in contrast to the original problem, which involved a $[3(N + N_s) \times 3(N + N_s)]$-dimensional space, with N very large. However, the construction of $U_{\delta\delta'}^{(0)}(nn';\omega_v)$ requires one to consider an eigenvalue problem quite comparable in difficulty to the initial one. There is one limit where the procedure simplifies, however. This is the limit $\omega_v^2 \gg \omega_{\mathrm{P}}^{(0)2}$, which will in general apply when we consider high-frequency modes of an adsorbate. We then have

$$U_{\delta\delta'}^{(0)}(nn';\omega_v) \simeq \frac{1}{\omega_v^2}\sum_p e_{n\delta}^{(0p)} e_{n'\delta'}^{(0p)} = \frac{\delta_{nn'}\,\delta_{\delta\delta'}}{\omega_v^2}, \qquad (4.36)$$

so Eq. (4.35) reduces to the much simpler form

$$\tilde{D}_{\alpha\beta}(ij,\omega_v) = D_{\alpha\beta}(ij) + \frac{1}{\omega_v^2}\sum_{n\delta} D_{\alpha\delta}(in)D_{\delta\beta}(nj). \qquad (4.37)$$

The second term in Eq. (4.37) describes the first correction to the effective dynamical matrix of an adsorbed molecule from the fact that the substrate nuclei are "dragged" into motion by vibration of the entity on the surface. From the physical point of view, this term describes an indirect interaction between nucleus i and nucleus j of the adsorbate, through the substrate as an intermediary. In practice, one can generate the first correction term to the infinite substrate mass approximation by replacing ω_v^2 on the right-hand side of Eq. (4.37) by the zero-order value $\omega_v^{(0)2}$ obtained by diagonalizing $D_{\alpha\beta}(ij)$ alone, then calculating the new frequency by finding the eigenvalue of $\tilde{D}_{\alpha\beta}(ij,\omega_v)$ which reduces to $\omega_v^{(0)2}$ in the limit the substrate mass becomes infinite.

We conclude this subsection by applying Eq. (4.37) to a simple case illustrated in Fig. 4.3a. We consider an adsorbed atom with mass M_A, which sits on a hollow site of fourfold symmetry. The substrate atoms have mass M_s, and we suppose the adsorbed atom couples to the adsorbate atoms by means of a nearest-neighbor central potential $\varphi(r)$. For this site there are three normal modes of vibration of the adsorbed atom. The first is a motion normal to the surface, and there are two degenerate modes, each one of which involves motion of the adatom parallel to the surface.

If d_0 is the distance between the adsorbate and each of the nearest-neighbor substrate atoms, the adatom is in an equilibrium position only if [Eq. (4.20a)] $\varphi'(d_0) = 0$, and in the limit of infinite substrate mass, the

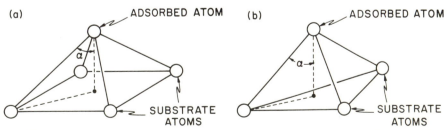

Fig. 4.3 An adsorbed atom that sits on a hollow site of (a) fourfold symmetry and (b) threefold symmetry.

frequencies $\omega_\perp^{(0)2}$ and $\omega_{||}^{(0)2}$ for motion perpendicular to or parallel to the surface are given by

$$\omega_\perp^{(0)} = 2 \cos \alpha [\varphi''(d_0)/M_A]^{1/2}, \tag{4.38a}$$

$$\omega_{||}^{(0)} = \sqrt{2} \sin \alpha [\varphi''(d_0)/M_A]^{1/2}, \tag{4.38b}$$

where α is the bond angle illustrated in Fig. 4.3a. The procedure outlined above then generates the following expressions, which include the first correction to Eqs. (4.38) from the finite substrate mass:

$$\omega_\perp^2 = \omega_\perp^{(0)2}\left(1 + \frac{M_A}{4M_s \cos^2 \alpha}\right) \tag{4.39a}$$

and also

$$\omega_{||}^2 = \omega_{||}^{(0)2}\left(1 + \frac{M_A}{2M_s \sin^2 \alpha}\right). \tag{4.39b}$$

Similar relations can be readily worked out for an adsorbed atom which sits above a hollow site of threefold symmetry. Here we have $\varphi'(d_0) = 0$ again as the equilibration condition, and

$$\omega_\perp^{(0)} = \sqrt{3} \cos \alpha [\varphi''(d_0)/M_A]^{1/2}, \tag{4.40a}$$

$$\omega_{||}^{(0)} = \sqrt{\tfrac{3}{2}} \sin \alpha [\varphi''(d_0)/M_A]^{1/2}, \tag{4.40b}$$

with the first correction from the finite substrate mass given by

$$\omega_\perp^2 = \omega_\perp^{(0)2}\left(1 + \frac{M_A}{3M_s \cos^2 \alpha}\right) \tag{4.41a}$$

and

$$\omega_{||}^2 = \omega_{||}^{(0)2}\left(1 + \frac{2M_A}{3M_s \sin^2 \alpha}\right). \tag{4.41b}$$

The equations for the corrected frequencies for a number of geometries are summarized in Table 4.1, where $\varphi''(d_0)$ has been replaced by f.

Black has explored the frequencies of vibration of hydrogen, oxygen, and sulfur adatoms on the Ni(100) and the Ni(111) surface, after using the first-principles calculations of Upton and Goddard [7] to obtain the parameters α and $\varphi''(d_0)$, as discussed earlier when we examined the frequency spectrum of the vibrational motion of oxygen on the Ni(111) surface. In Table 4.2, we compare Black's exact result with the predictions of Eq. (4.41). For hydrogen, where M_A/M_s is very small, Eq. (4.41) gives results in excellent accord with the complete analysis. Even for sulfur, where $M_A/M_s = 0.54$

TABLE 4.1

Nearest-Neighbor Force Constant Model for
Adatom Stretching Modes

Site	Mode	Character	Frequency
On top	Σ^+	T_z	$\omega^2 = \dfrac{f}{M_s}\left(1 + \dfrac{M_s}{M_A}\right)$
Twofold bridge	A_1	T_z	$\omega^2 = \dfrac{f}{M_s}\left(1 + 2\,\dfrac{M_s}{M_A}\cos^2\alpha\right)$
	B_1	T_x	$\omega^2 = \dfrac{f}{M_s}\left(1 + 2\,\dfrac{M_s}{M_A}\sin^2\alpha\right)$
Threefold bridge	A_1	T_z	$\omega^2 = \dfrac{f}{M_s}\left(1 + 3\,\dfrac{M_s}{M_A}\cos^2\alpha\right)$
	E	T_x, T_y	$\omega^2 = \dfrac{f}{M_s}\left(1 + \dfrac{3}{2}\,\dfrac{M_s}{M_A}\sin^2\alpha\right)$
Fourfold bridge	A_1	T_z	$\omega^2 = \dfrac{f}{M_s}\left(1 + 4\,\dfrac{M_s}{M_A}\cos^2\alpha\right)$
	E	T_x, T_y	$\omega^2 = \dfrac{f}{M_s}\left(1 + 2\,\dfrac{M_s}{M_A}\sin^2\alpha\right)$

[a] M_s and M_A, masses of the substrate and adatom, respectively; α, angle between bond direction and the surface normal; f is the same as $\varphi''(d_0)$ in the text.

TABLE 4.2

Frequencies for Adatoms on the Ni (111) Surface

Adatom	Distance above surface plane (Å)	Mode	Frequency (cm^{-1})		
			$M_s = \infty$	Exact	Eq. (4.41)
H	0.79	$\perp$	1217	1232	1232
		$\parallel$	1567	1579	1578
O	1.20	$\perp$	514	578	568
		$\parallel$	443	514	505
S	1.50	$\perp$	393	471	457
		$\parallel$	270	376	357

and one is unsure that the first correction provided by Eq. (4.41) is accurate, the approximate expression gives frequencies for the perpendicular and parallel modes that differ from the exact results by only 3 and 5%, respectively.

In Table 4.3, we summarize the theoretical calculations of the frequencies of the perpendicular vibrations of the three adatoms on the Ni(100) and

Ni(111) surfaces, and compare them with the frequencies obtained from electron energy loss spectroscopy. The theory and experiment agree very well. We regard this as most impressive, since the theoretical values of the frequencies have been calculated from first principles, with no adjustable parameters. The only assumption that has been employed is that the oxygen interacts with only the nearest-neighbor Ni atoms via an interaction of central force character. This assumption is required in order to extract from the calculations of Upton and Goddard information sufficient to describe the influence of motions of the Ni atoms parallel to the surface on the motion of the adatom. For simple adsorbate–substrate combinations such as those considered in Tables 4.2 and 4.3, we have reached the point where one can make quantitative contact between theory and measurements of adsorbate vibration frequencies. Since the vibrational frequencies are very sensitive to details of the adsorption site geometry, one can have very high confidence in the site assignments in these cases.

TABLE 4.3

Frequencies of Adatom Vibrations Normal to the Surface;
Comparison between Theory and Experiment
for Low-Index Ni Surfaces[a]

Surface	Adatom	Distance above surface plane (Å)	Frequency (cm^{-1})	
			Theory	Experimental
Ni(100)	H	0.30	631	605 [9]
	O	0.88	430	435 [8]
	S	1.24	355	355–371 [9]
Ni(111)	H	0.79	1232	1120 [10]
	O	1.20	578	580 [8]
	S	1.50	471	—

[a] From Black [5], used with permission.

We now turn to a discussion of the influence of the symmetry of the adsorption site geometry on the frequency spectrum and eigenvectors associated with adsorbate normal modes. We shall place emphasis on how the modes of the adsorbed species differ from those in the gas phase. While we have discussed general aspects of this topic in the present section, through use of group theory one may classify the normal modes of vibration of specific adsorbate–substrate combinations in a systematic manner. We turn next to a brief summary of the methods of group theory, then to applications of these to surface vibrational spectroscopy.

4.3 ELEMENTS OF GROUP THEORY

4.3.1 Introductory Remarks

In Section 4.2, we discussed the theory of molecular vibrations, with
emphasis on the relationship between the vibrations of the molecule in the
gas phase to those when the molecule is adsorbed on a surface. We saw that
in general that the bond lengths and bond angles of the adsorbed entity
differ from those in the gas phase, i.e., the basic symmetry of the molecule
is altered by adsorption. This means that modes degenerate in the gas phase
may be split after adsorption, while we encounter new finite frequency modes,
such as the hindered rotations and hindered translations. Those features of
the vibrational spectrum specific to the adsorbed state provide direct
contact with the geometry of the surface.

While the methods outlined in Section 4.2, expanded to take account of
the noncentral nature of the two-body interaction when necessary, can
provide us with a detailed description of these phenomena, in fact most of
the qualitative features may be deduced by noting that adsorption of the
molecule lowers the symmetry of its environment. The consequences of this
may be analyzed by the methods of group theory. Much of the remainder of
the present chapter will be devoted to a systematic study of the relationship
between molecular vibrations in the gas phase, and those in the adsorbed
state, for symmetry groups commonly realized in molecular physics, and
for molecules bound to adsorption sites of specified symmetry. Before we
turn to this discussion, in the present section we review the concepts of
group theory. There are by now a number of excellent textbooks that cover
this topic in detail, with eplicit applications to the theory of molecular
vibrations [11], so we shall summarize the principal features of the method
without proof.

4.3.2 Group Theory and Harmonic Vibrations

If we consider a molecule with N nuclei, or any other array of vibrating
masses of interest, there will exist certain symmetries in all but the most
complex systems. These are the rotations and reflections that leave the
geometry unchanged. The ensemble of all such symmetry operations for a
particular system is referred to as the symmetry group of the system. The
symmetry group includes among its elements the identity operation I,
which when applied leaves all nuclei in place and coordinate axes unchanged.
Given any two symmetry elements R_1 and R_2, the product of R_1 and R_2 is
defined as the action of R_2 followed by that of R_1. The fundamental mathematical property that allows one to refer to any particular collection of

mathematical operators $\{I, R_1, \ldots, R_N\}$ as a group is that the product $R_i R_j$ of any pair in the collection is also a member of the set. That is, we have $R_i R_j = R_k$, where R_k is also a member of the original set $\{I, R_1, \ldots, R_N\}$. One may easily verify that the collection of all possible symmetry operations that leave a molecule unchanged in shape or orientation is closed under the operation of multiplication as just defined, so this set of symmetry operations forms a group in the mathematical sense of the word.

Suppose we consider the action of a given symmetry operation R_i on the $3N$-dimensional space spanned by the set of vectors $\{\mathbf{u}_i\} = \{u_{1x}, u_{1y}, \ldots, u_{Nz}\}$ that describe the displacement of the nuclei in our molecule from their equilibrium position. The action of R_i on this space may be represented by a $3N \times 3N$ matrix $M_{i\alpha;j\beta}(R_i)$ as follows. We label the position of each nucleus by the index $1, \ldots, N$, and let $\{\mathbf{u}_i'\}$ denote the arrangement of the coordinates after the action of R_i. For example, if R_i exchanges the nucleus at site 3 with that on site 1, while the orientation of the coordinate axis is left unaltered, then after the symmetry operation we have $\{\mathbf{u}_i'\} = \{u_{3x}, u_{3y}, u_{3z}, \ldots, u_{Nx}, u_{Ny}, u_{Nz}\}$. Since each element in the set $\{\mathbf{u}_i'\}$ may be expressed as a linear combination of the original set $\{\mathbf{u}_i\}$, we may write

$$u_{i\alpha}' = \sum_{j\beta} M_{i\alpha,j\beta}(R_i) u_{j\beta}. \tag{4.42}$$

The momenta $\{\mathbf{P}_i\}$ transform under the influence of R_i in exactly the same way as the coordinates $\{\mathbf{u}_i\}$, since momentum and coordinate operators have the same transformation properties under rotation and reflection.

The set of matrices $\{M(I), M(R_1), \ldots, M(R_N)\}$ formed in this fashion obeys the same multiplication laws as the collection of abstract symmetry operators $\{E, R_1, \ldots, R_N\}$. Such a set of matrices is said to form a *representation* of the group $\{I, R_1, \ldots, R_N\}$. There are two distinctly different kinds of representations of a group, *reducible representations* and *irreducible representations*.

We may appreciate the nature of these two kinds of representations by explicitly constructing a reducible representation. Consider a $6N$-dimensional vector space spanned by the coordinates and momenta of the nuclei in our molecule. The basis vector is thus $(u_{1x}, u_{2x}, u_{3x}, \ldots, p_{Nx}, p_{Ny}, p_{Nz})$. Quite clearly, a $6N$-dimensional representation of our symmetry group G is provided by the set of matrices

$$N(R_i) = \begin{pmatrix} M(R_i) & 0 \\ 0 & M(R_i) \end{pmatrix}. \tag{4.43}$$

Now to each of the matrices in Eq. (4.43) apply a unitary transformation $\mathbf{U}$ to form a new set $N'(R_i) = \mathbf{U}N(R_i)\mathbf{U}^+$.

The matrices $N'(R_i)$, which in general may have all elements nonzero, also form a representation of G *equivalent* to the original set $N(R_i)$ (i.e., all are related by the same unitary transformation U). Also, by construction, we have a prescription for breaking down the set $N'(R_i)$ into the block diagonal form given in Eq. (4.43). We do this by applying the *same* unitary transformation U^+ to each of the matrices $N'(R_i)$.

Any representation $N(R_i)$ of the group G which may be cast into block diagonal form

$$N(R_i) = \begin{pmatrix} M_1(R_i) & & 0 \\ & M_2(R_i) & \\ 0 & & M_n(R_i) \end{pmatrix} \tag{4.44}$$

by application of the *same* unitary transformation to each of the matrices $N(R_i)$ is said to be a *reducible representation* of the group. An *irreducible representation* is one for which no such unitary transformation exists.

Let $(\xi_1, \ldots, \xi_n)$ be the basis vector for an n-dimensional representation $N(R_i)$. Action of $N(R_i)$ on the basis vector produces a new entity $(\xi_1', \ldots, \xi_n')$ in which, in general, each ξ_i' is a linear combination of all the $\{\xi_i\}$ of the original basis vector. Let the representation $\{N(R_i)\}$ be reducible, and let it be decomposed into block diagonal form by a unitary transformation that yields a new basis vector $(\eta_1, \ldots, \eta_{n_1}, \eta_{n_1+1}, \ldots, \eta_{n_1+n_2}, \ldots)$, where n_1, $n_2, \ldots$ are the dimensions of the blocks that appear on the diagonal. Clearly we have $n_1 + n_2 + \cdots = n$. Now that basis vector has been broken up into subsets, and the elements of each subset, $(\eta_1, \ldots, \eta_{n_1}), (\eta_{n_1+1}, \ldots, \eta_{n_1+n_2}), \ldots$ transform among themselves only and are unmixed with members of other subsets upon application of the new representation matrices.

Each symmetry group with a finite number of elements (only such groups are considered here) has a finite number of irreducible representations. The mathematical apparatus of group theory provides explicit prescriptions for decomposing a general reducible representation into its basic components, which are all irreducible representations.

We are now ready to make contact with the application of group theory to the theory of molecular vibrations. Consider a set of nuclei, with dynamical matrix $D_{\alpha\beta}(i, j)$. In Section 4.2, we employed a notation which distinguished between the nuclei associated with adsorbed molecules and those in the substrate. Here we do not do this, since it would make the following equations cumbersome, with little gain in clarity.

Consider an eigenvector $e_{i\alpha}^{(s,r)}$ which is an eigenvector of the dynamical matrix $D_{\alpha\beta}(ij)$ with frequency ω_s. Here we suppose the vibrational mode has degeneracy n, so $r = 1, 2, \ldots, n$ is an additional index appended to the eigenvector to label which particular eigenvector we have selected. The eigenvalue equation then reads, in the present notation,

$$\sum_{\beta j} D_{\alpha\beta}(i,j)e_{j\beta}^{(s,r)} = \omega_s^2 e_{i\alpha}^{(s,r)}, \qquad (4.45)$$

and any linear combination $\sum_r \lambda_r e_{j\beta}^{(s,r)}$ of the n degenerate eigenvectors will also satisfy Eq. (4.45).

Now let us rotate or reflect the molecule with respect to the coordinate system by application of one of the elements R_i of the symmetry group G. The set of mass weighted coordinates $(v'_{1x}, \dots, v'_{Nz})$ that describe the molecule after application of the symmetry operation are related to the original set $(v_{1x}, \dots, v_{Nx})$ by a $3N$-dimensional matrix $M(R_i)$ that is a member of a $3N$-dimensional (generally reducible) representation of G.

We can find the form of the dynamical matrix $D'_{\alpha\beta}(i,j)$ that describes the molecule after application of the symmetry operation R_i. The vibrational potential energy must be, in dyadic notation,

$$V = \tfrac{1}{2}v' \cdot D' \cdot v' = \tfrac{1}{2}v \cdot (M^{-1}(R_i) \cdot D' \cdot M(R_i)) \cdot v, \qquad (4.46)$$

and the last form must be identical to $\tfrac{1}{2}v \cdot D \cdot v$. Hence

$$D' = M(R_i) \cdot D \cdot M^{-1}(R_i). \qquad (4.47)$$

Since $M(R_i)$ is unitary, D' has the same eigenvalue spectrum as D; application of R_i to the molecule cannot alter the eigenvalue spectrum nor the degeneracy of any particular mode. Let $e'^{(sr)}$ be one of the n degenerate eigenvectors associated with D', with eigenvalues ω_s^2. The combination $M^{-1}(R_i) \cdot e'^{(sr)}$ is easily seen to be an eigenvector of the *original* dynamical matrix D. Hence, this object must be expressible as a linear combination of the eigenvectors $e^{(sr)}$ of D. If we let $e''^{(sr)} = M^{-1}(R_i) \cdot e'(sr)$, then the fact that $e''^{(sr)}$ and $e'^{(sr)}$ are both eigenvectors of the same matrix with the same eigenvalues means that we write

$$e_{i\alpha}^{(sr)} = \sum_{r'=1}^{n} \lambda^{(rr')}(R_i)e_{i\alpha}^{''(sr')}. \qquad (4.48)$$

One may show by applying two symmetry operations in succession that the $n \times n$ matricies $\lambda(R_i)$ obey the group multiplication rules. Hence they form an n-dimensional representation of G, and the eigenvectors associated with the n-fold degenerate mode of frequency ω_s form the basis for a representation of G.

The fundamental assertion of group theory, as it applies to the present problem, is that the eigenvectors $e^{(sr)}$ form the basis for an *irreducible representation* of G. Thus, if we know the dimensionality of all the irreducible representations of a symmetry group, we have information on the degeneracies possible for the vibrational modes of the molecule. The apparatus of group theory provides a prescription for the following procedure, which also tells us how many normal modes are present associated with each particular irreducible representation. If we begin with the $3N \times 3N$ matrices $M(R_i)$, generally reducible as already noted, with the methods of group theory it is quite a simple matter to see how many times each irreducible representation appears on the diagonal, once the matrices $\mathbf{M}(R_i)$ have been reduced to block diagonal form

$$\mathbf{M}(R_i) = \begin{pmatrix} \mathbf{M}_1(R_i) & & 0 \\ & \mathbf{M}_2(R_i) & \\ 0 & & \ddots \end{pmatrix}. \tag{4.49}$$

Before we proceed to state these rules, we require one more concept. The concept is best illustrated if we turn to a specific example of symmetry group. We consider the group explored extensively in the text by Tinkham [11], that of a triangle. In Fig. 4.4a, we show a triangle, with the various operations that leave it invariant also illustrated. There are six elements of its symmetry group. We always have the identity operation I; there are then three reflections σ_v through the lines indicated, and two $120°$ rotations C_3 about the axis normal to the plane. One is a $120°$ rotation in the clockwise sense, and one is counterclockwise. This collection of operators forms the symmetry group C_{3v}.

The three σ_v operations are very similar in nature. One says that the collection of these three form a *class*. The operators of a symmetry group may always be arranged in such subunits. Formally, two operations R_1 and R_2 are said to belong to the same class if there exists a third operation R_3 in

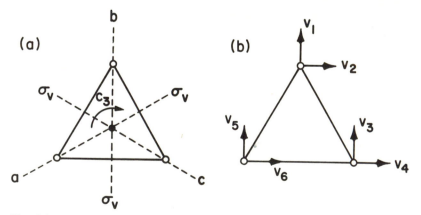

Fig. 4.4 (a) The symmetry operations of three masses, each placed at the vertex of an equilateral triangle. (b) The displacements $(u_1, \ldots, u_6)$ that describe the vibrational motion of the masses in the plane of the triangle.

the group that links R_1 and R_2 via the relation

$$R_2 = R_3^{-1} R_1 R_3. \tag{4.50}$$

We may use Eq. (4.50) to see that $\sigma_v^{(a)}$ and $\sigma_v^{(b)}$, the reflections through axes a and b, belong to the same class. One may see this by choosing $R_1 = \sigma_v^{(a)}$, $R_2 = \sigma_v^{(b)}$, and R_3 the threefold clockwise rotation C_3. The identity I always forms a class by itself; the three σ_v operations constitute a second class, as do the pair of threefold rotations.

We now have two very simple rules which we may easily apply to all the symmetry groups of interest in our theory of molecular vibrations. The first states that the number of irreducible representations of a group is finite, and equal to the number of *classes*. Hence our symmetry group of the triangle has three and only three irreducible representations. The second rule states that if l_i is the dimensionality of one of the irreducible representations and h the number of symmetry operators in the group, including the identity, we have

$$\sum_i l_i^2 = h, \tag{4.51}$$

where the sum is overall irreducible representations. For our group of the triangle, h, often called the order of the group, is 6, and we know we have three irreducible representations. We can see at once that we must have two irreducible representations of dimension one, and one of dimension two, since this is the only combination of three integers whose squares sum to 6.

If we think of the triangle in Fig. 4.4a as a planar molecule, and consider the vibrational normal modes with displacements parallel to the plane of the figure, we have $3 \times 2 = 6$ normal modes in total. From the previous paragraph, we see at once that the modes must either be nondegenerate and transform under the symmetry operations R_i as one of the two one-dimensional irreducible representations, or we may possibly have twofold degenerate modes. But it is not possible to have a degeneracy larger than two. Application of the two rules of the previous paragraph is very simple, and the conclusions drawn from it are powerful.

A compact and useful source of information on the nature of the ir-reducible representations, and the properties of the eigenvectors associated with the normal modes associated with them, is contained in the *character table* of the group. Given a matrix $\mathbf{M}(R_i)$ associated with a particular representation of G, the character $\chi(R_i)$ is defined simply as the *trace* of $\mathbf{M}(R_i)$,

$$\chi(R_i) = \sum_j M_{jj}(R_i). \tag{4.52}$$

Since the representation matrices obey the same multiplication rules as the group elements themselves, from Eq. (4.50) combined with the invariance of the trace under unitary transformation, the matrices associated with all operators of a given class have the same character.

In the character table, the characters of the representation matrices associated with the irreducible representations are arranged in a particular format. The columns are labeled by the various classes, and the rows by symbols which label the irreducible representations. The character table is thus a square array, and for the simple groups of interest here, a set of rules give normalization and orthogonality properties of both rows and columns. These rules enable one to easily construct the character table. Let $\chi^{(i)}(C_k)$ be the character of the matrices associated with operators of class C_k for the irreducible representation i. If N_k is the number of operators in class C_k, the orthonormality relation for the rows reads

$$\sum_k N_k \chi^{(i)}(C_k)\chi^{(j)}(C_k)^* = h\delta_{ij}, \qquad (4.53)$$

while for the columns

$$\sum_i \chi^{(i)}(C_k)^*\chi^{(i)}(Cl) = (h/N_k)\delta_{kl}. \qquad (4.54)$$

One begins with the first row, devoted always to the *symmetric* representation A_1 (sometimes called the identity representation). This is a one-dimensional irreducible representation, present for all groups, whose eigenvector is invariant under all group operations. Thus, one has $+1$ with each entry in the row. The first column is obvious also, since the trace of the identity operation is equal to the dimensionality of the irreducible representation. The remaining elements are uniquely determined by the requirement that Eqs. (4.53) and (4.54) be satisfied. We give the character table of a number of common groups in Table 4.4; the symmetry group of the triangle is C_{3v}. We shall comment later on the various functions that appear in the character table.

We are frequently given information on the nature of a representation of G known to be reducible. We wish to know, once the unitary transformation is found that brings each representation into block diagonal form [Eq. (4.49)], how many times a given irreducible representation appears on the diagonal. Without the need to have explicit representation matrices in hand, this question may be answered with use of the character table, and the following. Let $\chi(C_k)$ be the character associated with operators of class C_k, for the representation assumed reducible. Then if a_i is the number of times irreducible representation i appears on the diagonal after the representation has been cast into block diagonal form, we have

$$a_i = h^{-1}\sum_k N_k \chi^{(i)}(C_k)^*\chi(C_k). \qquad (4.55)$$

TABLE 4.4

Character Tables and Number of Modes in Each Representation of Surface Point Groups with at Least One Symmetry plane[a]

C_s	I	σ_{xz}		$N = 2m + m_{xz}$
A'	+1	+1	z, x, R_y	$3m + 2m_{xz}$
A''	+1	-1	y, R_x, R_z	$3m + m_{xz}$

C_{2v}	I	C_2	σ_{xz}	σ_{yz}		$N = 4m + 2m_{xz} + 2m_{yz} + m_0$
A_1	+1	+1	+1	+1	z	$3m + 2m_{xz} + 2m_{yz} + m_0$
A_2	+1	+1	-1	-1	R_z	$3m + m_{xz} + m_{yz}$
B_1	+1	-1	+1	-1	x, R_y	$3m + 2m_{xz} + m_{yz} + m_0$
B_2	+1	-1	-1	+1	y, R_x	$3m + m_{xz} + 2m_{yz} + m_0$

C_{3v}	I	C	σ		$N = 6m + 3m_v + m$
A_1	+1	+1	+1	z	$3m + 2m_v + m_0$
A_2	+1	+1	-1	R_z	$3m + m_v$
E	+2	-1	0	x, y, R_x, R_y	$6m + 3m_v + m_0$

C_{4v}	I	C_4	C_4^2	σ_v	σ_d		$N = 8m + 4m_v + 4m_d + m_0$
A_1	+1	+1	+1	+1	+1	z	$3m + 2m_v + 2m_d + m_0$
A_2	+1	+1	+1	-1	-1	R_z	$3m + m_v + m_d$
B_1	+1	-1	+1	+1	-1		$3m + 2m_v + m_d$
B_2	+1	-1	+1	-1	+1		$3m + m_v + 2m_d$
E	+2	0	-2	0	0	x, y, R_x, R_y	$6m + 3m_v + 3m_d + m_0$

C_{6v}	I	C_6	C_6^2	C_6^3	σ_v	σ_d		$N = 12m + 6m_v + 6m_d + m_0$
A_1	+1	+1	+1	+1	+1	+1	z	$3m + 2m_v + 2m_d + m_0$
A_2	+1	+1	+1	+1	-1	-1	R_z	$3m + m_v + m_d$
B_1	+1	-1	+1	-1	+1	-1		$3m + 2m_v + m_d$
B_2	+1	-1	+1	-1	-1	+1		$3m + m_v + 2m_d$
E_1	+2	+1	-1	-2	0	0	x, y, R_x, R_y	$6m + 3m_v + 3m_d + m_0$
E_2	+2	-1	-1	+2	0	0		$6m + 3m_v + 3m_d$

[a] The symbol m stands for the number of nonequivalent atoms which are not on any element of symmetry; m_{xy}, m_{yz}, m_v, m_d stand for the number of nonequivalent atoms on a symmetry plane, but not on rotation axis; m_0 is the number of atoms on the rotation axis.

We do need the characters $\chi(C_k)$ of the reducible representation but, as we see in the next paragraph these can often be found from simple considerations.

Let us turn to the question of the normal modes of the triangular arrangement of masses given in Fig. 4.4. If we consider the modes where the motion of the masses is confined to the xy plane, we have six degrees of freedom in total. We show in Fig. 4.4b the displacements associated with the mass weighted coordinates $(v_1, v_2, \ldots, v_6)$. These six quantities form the basis for a representation of the symmetry group of the triangle, and from the character table we know it is necessarily reducible. The problem of enumerating the normal modes is equivalent to finding the unitary transformation that block diagonalizes the six-dimensional matrices $M(R_i)$ associated with this representation, or equivalently we require the eigenvectors associated with the matrices when they are cast in block diagonal form.

It is not difficult to find the characters $\chi(R)$ associated with the six-dimensional reducible representation. That associated with the identity is clearly 6. Consider the σ_v operation through the plane labeled b in Fig. 4.4a. Under this operation, $v_1 \rightarrow +v_1$, $v_2 \rightarrow -v_2$, while the remaining transformations are described by off-diagonal entries in the transformation matrix. Hence $\chi(\sigma_v) = 0$, and one sees also that $\chi(C_3) = 0$ since under C_3 we have nonzero entries in the representation matrices only off the diagonal. Thus, for our example $a_i \equiv \chi^{(i)}(I)$. Hence, among the six normal modes, we must have one that transforms according to A_1, one as A_2, and two modes each of which is twofold degenerate, and each set of eigenvectors forms a basis for the two-dimensional irreducible representation E of C_{3v}.

A number of the eigenvectors may be written down quickly, and the remainder can frequently be deduced from orthonormality requirements. For example, from classical mechanics, we know for our problem of the planar molecule, we must find two modes with frequency $\omega_s^2 \equiv 0$. These correspond to rigid-body translations of the molecule, and there is no restoring force for such motions (if the molecule is in free space). The eigenvectors associated with the two center-of-mass translations are illustrated in Fig. 4.5a. The eigenvectors associated with these motions must be basis vectors for an irreducible representation of C_{3v}, and this can only be E. By constructing the characters, or by working out the representation matrices [noting the eigenvectors transform under group operations like the pair of coordinates (x, y) of a particle in the plane], one may readily verify that this is so.

From classical mechanics, we know also that one normal mode corresponds to a rigid rotation of the triangle about an axis through its centroid and normal to the plane. The eigenvector associated with this motion is illustrated in Fig. 4.5b. The set of characters associated with this mode is

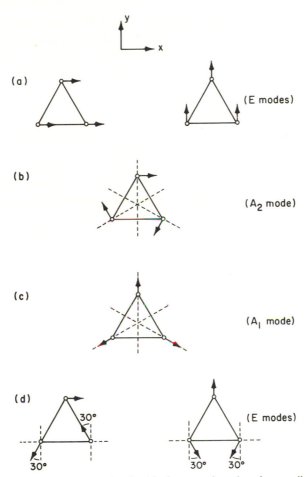

Fig. 4.5   The eigenvectors associated with the normal modes of an equilateral triangle, with motion confined to the xy plane.

worked out readily, and we find it forms the basis for the one-dimensional representation A_2.

We are left with one mode that transforms like A_1, and a twofold degenerate mode with eigenvector that transforms like E. The A_1 mode can also be sketched at once: it is the symmetric breathing mode illustrated in Fig. 4.5c.

The remaining set of E modes are a bit trickier to find. One notes that each must involve a motion that leaves the center of mass of the molecule at rest, since only the two translation modes in Fig. 4.5a can lead to center of mass motion. Also, each eigenvector must be orthogonal to the four eigenvectors shown in Figs. 4.5a–4.5c. We begin by seeking an eigenvector which has the top-most mass moving horizontally to the right. Orthogonality to the translation mode in the x direction along with the A_1 mode requires this mode to have the x components of the motion of the two lower masses moving to the left, each with equal amplitude equal to that of $\frac{1}{2}$ the amplitude of the top-most mass. This cannot be all there is to the eigenvector, since our construct so far is not orthogonal to the A_2 mode. We provide the lower two masses with equal and opposite vertical displacements (this leaves the center of mass invariant) with magnitude (relative to that of the horizontal motion of the topmost mass) so the eigenvector is properly orthogonal to that of the A_2 mode. This uniquely fixes the magnitude of the vertical displacements, and we have an eigenvector as illustrated in the left hand portion of Fig. 4.5d. The second member of our degenerate mode must be orthogonal to the first, and also to the four given in Figs. 4.5a–4.5c. The unique eigenvector with these properties is readily worked out, and is given on the left-hand side of Fig. 4.5d. It remains to verify explicitly that the two modes represented in Fig. 4.5b do indeed form the basis for the irreducible representation E of the group C_{3v} of the molecule. It is a simple matter to do this, and verify that we have indeed found a proper set of eigenvectors.

The preceding example illustrates the great power of group theory. We started with a problem characterized by six degrees of freedom, and from the methods described in Section 4.2 we would be led to a 6×6 matrix to diagonalize to find the eigenfrequencies and eigenvectors of the various normal modes. This is a formidable task, by algebraic methods. From group theory combined with other elementary principles, we have been able to deduce the form of each of the six eigenvectors. Given any dynamical matrix deduced from any model of the internuclear forces (we had made no assumption as to the nature of this interaction, notice), we may readily find simple algebraic expressions for the normal mode frequencies.

It is useful to include in the character table, one piece of information other than the characters associated with the various irreducible representation matrices. One typically provides certain elementary functions which prove to be basic functions for the irreducible representations. These give one an intuitive feeling for the basic "shape," as required by symmetry, of the various eigenvectors. In Table 4.4, a set of such functions is provided. The symbol R_z stands for the pattern of displacements associated with an infinitesimal rotation about the z axis, while R_x and R_y for that produced by an infinitesimal rotation about the x and y axes, respectively.

We have one task assisted by group theory to discuss. We shall frequently wish to discuss the normal modes of a molecule distorted very slightly from a high-symmetry form. In Section 4.2, we saw that when a molecule is adsorbed on a surface, it is distorted in form, with the consequence that the configuration of nuclei has symmetry lower in the adsorbed state than the gas-phase configuration. If the distortion is small, the eigenfrequencies and eigenvectors differ only slightly from those in the gas phase, though these differences are crucial to study since, as remarked earlier, they provide direct access to the nature of the adsorption geometry.

A crucial issue easily addressed by group theory when the symmetry-breaking perturbation is small is the splitting of an initially degenerate mode into a multiplet. In first-order perbation theory, such splittings are found by diagonalizing the Hamiltonian matrix of the *full* problem, with lowered symmetry, on the subspace spanned by the eigenvectors of the mode degenerate in the absence of the symmetry breaking terms in the Hamiltonian.

Let the molecule in the gas phase have symmetry group G. Then in the adsorbed state, the system is described by a smaller group G'. In general, G' will be a *subgroup* of G formed by discarding those operators which no longer remain good symmetry operations. A set of eigenvectors which forms the basis of an *irreducible* representation of G also necessarily form a basis for a representation of G', but in general this representation will be reducible. By applying the rule in Eq. (4.55), with the sum on C_k ranging over the classes of G' and $\chi(C_k)$ the character associated with the irreducible representation of G (now a reducible representation of G'), we may see how the reducible representation of G may be decomposed into irreducible representations of G' through suitable application of a unitary transformation. The Hamiltonian matrix formed as described in the previous paragraph must then break up into block diagonal form under action of this transformation, and we can see how the initially degenerate mode breaks up into a multiplet when the symmetry is lowered.

We return to the example of the equilateral triangle examined above. Suppose we lower the symmetry by reducing the angle between the topmost vertex by a small amount, so we have an isoceles triangle rather than an equilateral triangle. The symmetry group of the configuration now consists of only the two operators $(I, \sigma_v^{(b)})$, with $\sigma_v^{(b)}$ the reflection through line b of Fig. 4.4a. This group has only two irreducible representations, each of dimension one. There is the symmetric representation A_1 with characters $(1, 1)$ and A_2 with $(1, -1)$. There can thus be no degeneracy after the apex angle is reduced below $60°$; the twofold degenerate mode illustrated in Fig. 4.5d must then split into a pair of nondegenerate modes. We easily see from Eq. (4.55) that we obtain one A_1 mode of G', and one A_2 mode after the symmetry is lowered. In fact, the eigenmode on the left-hand side of Fig. 4.5d is the A_2 mode of G', while the right-hand mode is the A_1 mode.

Once the symmetry is lowered from G to G', the displacements of the lower two masses need not be canted from the y and x directions by the

30° angles illustrated in Fig. 4.5. The eigenvectors depicted there are approximate eigenvectors for the new problem only in the limit that the symmetry-breaking perturbation is small.

Through reasoning similar to this, it is possible to construct a series of tables that describe how the degeneracy of gas-phase vibrational modes is lifted, when the molecule is adsorbed on the surface and is influenced by the lower symmetry. Such tables are called correlation tables. In the next section, we present and discuss such tables for a sequence of molecular symmetry groups and adsorption site geometries encountered in practice.

More generally, when one wishes to know the eigenvectors associated with the new low-symmetry geometry, to see how they are modified by the lowered symmetry, one may proceed as follows. We first write the vibrational Hamiltonian out using the eigenvectors of the high-symmetry form as a basis. The Hamiltonian, of course, will not be diagonal in this basis. Now from this set of eigenvectors, we can form (or find) sets which are eigenvectors of the irreducible representations of G'. No terms in the vibrational Hamiltonian can couple eigenvectors which are associated with *different* irreducible representations of G'. Hence, we have partially diagonalized the Hamiltonian. The labor required to complete the calculation is reduced, and one can gain considerable insight into the nature of the full solution by breaking the problem down into two steps.

Consider again the case of the triangular molecule, with one vertex reduced below 60°. If the molecule is in the gas phase, the two translational modes in Fig. 4.5a remain exact eigenmodes of the Hamiltonian, from elementary considerations.[1] The A_2 mode in Fig. 4.5b, with normal coordinate Q_{A_2}, is also an A_2 mode of G', as is the eigenvector of the left-hand member of Fig. 4.5d. We call this eigenvector Q_{E_a}, and its partner in the right-hand side of Fig. 4.5d is called Q_{E_b}. The A_1 mode in Fig. 4.5c is also an A_1 mode of G', as is Q_{E_b}.

If we write the vibrational Hamiltonian of the isoceles triangle in terms of the eigenvectors of the equilateral form, the most general structure consistent with symmetry G' is then

$$V = \tfrac{1}{2}[k_{11}Q_{E_a}^2 + 2k_{12}Q_{E_a}Q_{A_2} + k_{22}Q_{A_2}^2]$$
$$+ \tfrac{1}{2}[k_{33}Q_{E_b}^2 + 2k_{34}Q_{E_b}Q_{A_1} + k_{44}Q_{A_1}^2]. \qquad (4.56)$$

---

[1] These two zero-frequency modes remain degenerate, despite the fact that G' has only one-dimensional irreducible representations. This is an example of *accidental degeneracy*, and has its origin in the fact that the Hamiltonian has symmetry higher than that required by the operators in G' alone. If they were bound to a substrate by a spring attached to its center of mass, these modes would indeed split.

We now have two 2×2 submatrices to diagonalize, and each of the new nondegenerate vibration eigenvectors is a linear combination of Q_{E_a} and Q_{A_2} for one pair and Q_{E_b} and Q_{A_1} on the other. As Q_{A_2} is mixed into Q_{E_a} by the coupling term proportional to k_{12}, we obtain an eigenvector with the property that the displacement of the lower two masses no longer makes a $30°$ angle with the y axis. The admixture between Q_{E_a} and Q_{A_2} is determined by geometry; one of the two modes must describe the rigid-body rotation of the molecule, and there is one linear combination of Q_{E_a} and Q_{A_2} that describes such a motion. The second eigenvector that emerges from the 2×2 matrix must be orthogonal to the one that describes the uniform rotation. In essence, the constants k_{11}, k_{12}, and k_{22} that enter Eq. (4.56) cannot assume arbitrary values. If they are derived from any dynamical matrix invariant under rigid-body rotations of the molecule, their ratio must be such that one of the two eigenvectors describes the rigid-body rotational mode. There is no general principle that determines the admixtures of Q_{E_b} and Q_{A_1} that form the eigenvectors of the second 2×2 matrix. Here we must have a microscopic model that leads to specific forms for k_{33}, k_{34}, and k_{44}.

We see that when the strength of the symmetry-breaking perturbation is weak, the eigenvectors of the low-symmetry species are well approximated by those of the high symmetry form, though splittings frequencies first order in the perturbation occur. As the strength of the perturbation is increased, the eigenvectors are altered to assume a form with features specific to the low-symmetry state. If the strength of the symmetry-breaking perturbation is increased from very weak to moderate or strong, the no-crossing rule tells us that while the eigenvector of a given mode may be altered in form through admixture with other eigenmotions of the high-symmetry structure, as in Eq. (4.56), it will be part of a basic set for the same irreducible representation of G' as one deduces from the weakly perturbed limit. Thus, by formally regarding the symmetry-breaking perturbation as weak, one obtains a correct description of the symmetry properties and degeneracies present in the low-symmetry configuration. In practice, one needs additional information to decide how strongly the eigenvectors are perturbed in nature from those appropriate to the gas phase. One may expect that for physisorbed systems, where bonding to the substrate is weak, the influence of the substrate is modest.

While the information derived from group theory in combination with physical reasoning is not sufficiently complete that we can uniquely determine all the eigenvectors associated with the in-plane vibrations of a triangular array of masses with symmetry equivalent to that of an isoceles triangle, we can say quite a bit about their general form, and determine some of the forms. The remaining sections of this chapter will place primary attention on the systematic application of group theory to the analysis of the relationship between vibrational modes of adsorbed molecules, and their relation to those of the gas-phase species.

4.4 MOLECULAR VIBRATIONS AND SURFACE POINT GROUPS

After having reviewed some elements of groups theory, we now turn to a more detailed discussion of the point groups relevant to molecules that are adsorbed on surfaces. Adsorption of a molecule on a surface in general lowers the symmetry. For the ethylene C_2H_4 molecule, e.g., which in the gas phase belongs to the point group $D_{2h} \equiv V_h$, at least one symmetry plane is removed when the bonding to the surface is established, since for an adsorbed molecular complex no symmetry plane parallel to the surface plane can exist. The number of point groups of surface complexes is therefore relatively small. These point groups are C_1, C_s, C_{2v}, C_{3v}, C_{4v}, C_{6v}, C_2, C_3, C_4, C_6. Of these 10 point groups the last three are of little importance. Therefore only a very limited set of groups needs to be discussed in detail here.

4.4.1 Character Tables and the Number of Modes in Each Representation

The potential symmetry elements of adsorbate complexes are rotation axes oriented perpendicular to the surface plane and also symmetry planes perpendicular to the surface. In our discussion we shall henceforth refer to the direction normal to the surface as the z direction. This is in accord with the standard notation in group theory with, however, the exception of the point group C_s where the z axis is typically assumed to be perpendicular to the single symmetry plane of this point group. Since on a surface the breakdown of the symmetry to C_s is frequently encountered, this notation would be not conveniently handled and we therefore assume the y axis to be perpendicular to the single symmetry plane of C_s rather than the z axis.

The character tables of the point groups are compiled in Tables 4.4 and 4.5. The tables also allow one to determine the number of modes in each representation as well as to determine which modes are observable in specular reflection for impact and dipole scattering. We begin the explanation of how to use the tables with the simple point group C_s. The only element of symmetry here is the symmetry plane σ_{xz} which, as mentioned before, is oriented perpendicularly to the surface. Vibrational modes may behave either symmetric (even) or antisymmetric (odd) under mirror re-

TABLE 4.5

Character Tables and Number of Modes in Each Representation of
Surface Point Groups with a Rotational Axis Only[a]

C_2	I	C_2		$N = 2m + m_o$
A	+1	+1	z, R_z	$3m + m_o$
B	+1	-1	x, y, R_x, R_y	$3m + 2m_o$

C_4	I	C_4	C_4^2		$N = 4m + m_o$
A	+1	+1	+1	z, R_z	$3m + m$
B	+1	-1	+1		$3m$
E	+2	0	-2	x, y, R_x, R_y	$3m + m_o$

C_3	I	C_3		$N = 3m + m_o$
A	+1	+1	z, R_z	$3m + m_o$
E	+2	-1	x, y, R_x, R_y	$3m + m_o$

C_6	I	C_6	C_6^2	C_6^3		$N = 6m + m_o$
A	+1	+1	+1	+1	z, R_z	$3m + m_o$
B	+1	-1	+1	-1		$3m$
E_1	+2	+1	-1	-2	x, y, R_x, R_y	$3m + m_o$
E_1	+2	-1	-1	+2		$3m$

[a] With the exception of C_2 these point groups are of little practical importance.

flection at σ_{xz}. By symmetric we mean that the polarization vectors of the motion of atoms which relate to each other through the operation of mirror reflection are transformed into each other without a change of sign, (character $+1$) while the sign is reversed for an antisymmetric vibration (character -1).

In Section 4.2, we saw how group theory may be used to ennumerate the number of normal modes that are associated with each irreducible representation of the point group of the molecule. We considered in detail a planar molecule with identical mass at each vertex of a triangle. For more complex molecules, it is convenient to proceed as follows. The molecule may be viewed as composed of a finite number of subsets of nuclei, each subset with the property that under group operations, nuclei within the subset are interchanged but there is no interchange of nuclei between different subsets. Clearly, all nuclei within a given subset must have the same mass. From the point of view of group theory, each subset may be treated as a molecule with fewer nuclei than that in the real molecule. With the methods of Section 4.2, we may determine the number of normal modes of each subset associated with the various irreducible representations of the point group, and the nature of their eigenvectors. The number of modes of the *entire molecule* associated with a given irreducible representation is then found by summing up the number of modes of the subsets associated with the representation in question.

In essence, what is done is to break the normal mode analysis of the full molecule into two steps. If the molecule can be decomposed into n subsets, we find the number of modes and the normal coordinates $Q_{A_i}^{(1)}, \ldots, Q_{A_i}^{(a)}$ associated with irreducible representation A_i for each subset. The potential energy of the entire molecule then will contain a term of the form $\frac{1}{2}\sum_{ij} k_{ij} Q_{A_i}^{(i)} Q_{A_i}^{(j)}$ and there will be no other terms that describe motions associated with the irreducible representation A_i. Diagonalization of this part of the Hamiltonian yields a number of normal modes of A_i symmetry of the whole molecule equal in number to the number of normal coordinates $Q_{A_i}^{(1)}, \ldots, Q_{A_i}^{(n)}$ associated with the normal modes of A_i symmetry of the various subsets. Each eigenvector of the whole molecule is a linear combination of the eigenvectors $Q_{A_i}^{(1)}, \ldots, Q_{A_i}^{(n)}$ of the subsets.

We now turn to the example illustrated in Fig. 4.6 where we have depicted an adsorbed molecular complex resembling a water molecule. The symmetry plane is assumed to be perpendicular to the plane of the molecule.

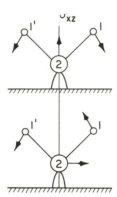

Fig. 4.6 Illustration of symmetric and antisymmetric eigenmodes of a molecule adsorbed on a hard-wall surface.

The atoms designated as 1 and 1′ are transformed into each other by the symmetry operation, while atom 2 lies on the plane and therefore belongs to a second, different subset. Since atom 1 has three degrees of freedom we can build a total of three symmetric and antisymmetric combinations. Therefore, if we denote the number of atoms which do not lie on the mirror plane by m and count atoms that transform into each other by the symmetry elements of the point group (here only σ_{xy}) only once ("equivalent atoms"), the atoms off the symmetry plane contribute $3m$ vibrational modes to both representations of the point group. For an atom on the mirror plane the situation is different. Here the antisymmetric vibration must be polarized perpendicularly to the mirror plane. Each atom on the mirror plane therefore contributes to one antisymmetric vibration, i.e., one vibration to the representation A''. The symmetric vibrations of the atom on the mirror plane are polarized within the plane. According to the two degrees of freedom within the plane, each atom on the mirror plane contributes two vibrations to the A' representation. We denote the number of atoms on the mirror plane by m_{xz}. The number of modes of A' and A'' character are then $3m + 2m_{xz}$ and $3m + m_{xz}$, respectively. The two expressions are given in the last column of the character table. The total number of atoms of the molecular complex is $N = 2m + m_{xz}$. The factor of 2 arises since m counts atoms off the plane only once. The sum of the number of modes in all representations is equal to the total number of degrees of freedom $3N$. Mode counting in the remaining point groups follows essentially the same scheme. There we have to make an additional distinction between atoms lying on one symmetry plane only and atoms lying on the rotation axis where symmetry planes cross each other. We denote the number of inequivalent atoms lying on one symmetry plane only by m_{xz}, m_{yz}, m_v (v, vertical), and m_d (d, diagonal), respectively. As in C_s, each inequivalent atom contributes two vibrations to the representation which is symmetric (character $+1$) under the mirror operation through the corresponding σ plane, and one vibration to the antisymmetric (character -1) representation. For an atom on the twofold axis of C_{2v}, one vibration is polarized parallel to the C_2 axis. This makes the vibration symmetric with respect to C_2, σ_{xz}, and σ_{yz}. It therefore belongs to the A_1 representations. The two other vibrations of that atom must be oriented perpendicular to the C_2 axis and perpendicular or parallel to either symmetry plane. The atoms on the C_2 axis therefore contribute one mode to each of the B representations.

Point groups such as C_{3v}, C_{4v}, C_{6v}, ... with more than a twofold rotation axis have two-dimensional irreducible E representations, since the description of such rotations requires the simultaneous transformation of at least two coordinates. Vibrations which belong to the E representations therefore have a twofold degeneracy. The number of modes in the E representations are most conveniently accounted for by building the possible symmetric and antisymmetric combinations with respect to the symmetry planes first. The difference between the number of modes in these nondegenerate representations and the total number of the degrees of freedom of the set of equivalent atoms in question is then the contribution to the number of modes in the E representation. Let us take C_{3v} as an example: If only one atom lies on a symmetry plane, the set of equivalent atoms contains three with a total of nine degrees of freedom. Two modes are associated with representations symmetric with respect to σ_v, and one with an antisymmetric representation. The remaining six degrees of freedom contribute three doubly degenerate vibrations in E. If this simple procedure proves difficult to apply in a particular case, then in Section 4.2 we have presented formal methods that may be used to find the number of distinct sets of E modes. We shall also look into more details of the degenerate vibrations when we discuss specific examples.

Before proceeding further to the issue of selection rules, a comment on the procedure of mode counting may be useful. In Fig. 4.6 we have replaced the surface by a hard wall of proper symmetry. The phonon modes of the substrate are therefore neglected. As discussed before we may also replace the hard wall by a set of nearest and, if necessary, next-nearest neighbor surface atoms for a better model of the symmetry and bonding within an actual surface site. The number $3N$, with N atoms in the adsorbed molecule, is then the additional number of eigenmodes introduced into the system through the adsorption of a molecule. The number of modes actually discernable as sharp features in a spectrum of a surface-sensitive technique is, however, a different matter. First some of the additionally introduced eigenmodes may be of low frequency, comparable to the phonon frequencies of the substrate. Whether such a mode develops an appreciable localization on the surface depends on the strength of the coupling and the mass ratio of substrate atoms and the atoms within the molecule. We have discussed this point earlier in the present chapter, and we shall turn to it once again in Chapter 5. In addition, vibrations originally belonging to the phonon bands of the substrate may be pulled up in frequency and develop a high degree of localization. Therefore the modes counted in the character tables are those that would be observable provided that their frequency is well above the phonon range. For the frequently encountered case of adsorbed CO, NO, and hydrocarbons or more complex organic molecules on a metal substrate, the internal vibrations of the molecule itself are in general observed since these frequencies are high compared to the substrate phonon range.

The adsorption of the molecule on the surface also inhibits the translation and rotations of the molecule. For the eigenmodes originating from free translations or rotations of the gas phase molecule, the term hindered translations and rotations has proved useful since this terminology immediately reveals that a certain mode has no counterpart in the spectrum of genuine modes of the molecule. Because of hindered translations and rotations, the total number of eigenmodes with finite frequency for the adsorbed molecule is $3N$ and not $3N - 6$ as for multiatom molecules in the gas phase. The mode corresponding to the translation in the direction normal to the surface typically gives rise to a mode whose frequency lies above the phonon band, at least for strongly chemisorbed molecules. The same can be said for the hindered rotations around an axis parallel to the surface. Vibrations arising from hindered translations parallel to the surface or a hindered rotation around the surface normal can, however, be rather low in frequency and mix with the phonon band. The representations which contain hindered translations and rotations are marked by the symbols x, y, z and R_x, R_y, R_z in the next-to-last column of Tables 4.4 and 4.5.

In conclusion, the procedure of mode counting must be exercised with prudence and it is useful to envisage as well which atom and which bond makes the most significant contribution to an eigenmode. This typically allows a rough estimate of the frequency range in which the mode is to be expected.

4.4.2 Eigenmodes and Selection Rules

We now address the question as to which modes are observed in electron energy loss spectroscopy, in view of the selection rules. For dipole scattering the relevant matrix element was [Eq. (3.64)] $\langle \psi_i | \mu_z' | \psi_f \rangle$ with μ_z the z component of the dipole moment operator and φ_i and φ_f the initial and final vibrational states. For a nonvanishing cross section the matrix element $\langle \varphi_i | \mu_z | \varphi_f \rangle$ must be totally symmetric with respect to all symmetry operations of a point group (see also Section 4.3). This is achieved when the product $\varphi_i \varphi_f^*$ transforms as the dipole moment μ_z. The z component of the dipole moment has the same transformation properties as a translation in the z direction. Inspection of Tables 4.4 and 4.5 shows that the z translation and thus μ_z transform totally symmetrically. Therefore the product $\varphi_i \varphi_f^*$ must also transform totally symmetrically. Let us for the moment consider the excitations of fundamentals, e.g., excitations from the vibrational ground state of the molecule, which is totally symmetric, into the first excited state of any eigenmode. It then follows that the first excited state must be totally symmetric. For dipole scattering we are therefore led to state the selection rule in a language more precise than used: *Only those vibrations which belong to the totally symmetric representations A_1, A', and A, respectively, are observed as fundamentals in dipole scattering.* This is the so-called "surface selection rule" applicable to dipole scattering of electrons and infrared spectroscopy of molecules adsorbed on any material which in its electronic properties, size, and shape effectively screens the dipole moment parallel to the surface to an extent that coupling to the parallel dipole moment can be safely neglected. The selection rule is not limited to two-dimensional flat surfaces we have discussed in Section 4.1.2.

The selection rule states which modes are dipole-allowed in terms of the representation to which they belong. It does *not* refer to the *polarization* of the motion of individual atoms within a molecule; the transformation properties of the *dipole moment operator* and not the atomic displacements enter the selection rule. In the introduction we have already discussed the example of a diatomic molecule adsorbed on the surface in such a way that the molecular axis is oriented parallel to the surface. A C—C species is a possible candidate for a strongly bonded surface molecule with parallel orientation.

The carbon–carbon stretching vibration is polarized almost parallel to the surface. Nevertheless, this motion belongs to an A_1 representation and is therefore *dipole allowed*. There is therefore nothing peculiar about systems where this mode is observed to contribute to dipole scattering. The physical picture one might associate with the occurrence of a dynamical dipole moment of this vibration is that the motion of the atom might alter the amount of charge donated from the orbitals of the C—C species into the substrate. Of course, from group theory nothing can be said about the magnitude of the dipole moment, and we shall in fact see that the dipole moment depends on the substrate material.

While the motion of atoms *parallel* to the surface can give rise to a *perpendicular* dipole moment, *vice versa* a perpendicular dipole moment is not always generated although the motions of atoms have a perpendicular component. This is the case for the E modes, where the dipole moment is strictly parallel to the surface. The motions of atoms which are not situated on the rotation axis have a perpendicular component of motion, however, such that the *sum* of the perpendicular components over the set of equivalent atoms *vanishes*.

In Chapter 3 we have shown that for impact scattering and specular reflection the cross section should vanish when the eigenmode is odd under a C_2 operation. *Vice versa*, this implies that all other modes may contribute to a spectrum obtained under the condition of specular reflection. While the intensity of impact scattering is frequently lower than that of dipole scattering, one has also observed impact scattering and dipole scattering to be of the same order of magnitude, particularly for hydrocarbons where the dynamic dipole moment for hydrogen vibrations is low. Therefore when a spectrum is observed in specular reflection, one must not infer that all observed losses are modes of A_1 type. This information is only obtained when the intensity measured as a function of the scattering angle assures that the spectrum does contain dipole scattering by the sharp drop off in intensity characteristic of dipole scattering. Further inspection of the character tables shows that the impact scattering selection rule becomes significant only in a limited number of systems which are best discussed in connection with our case studies in Section 4.4.4.

4.4.3 Correlation Tables

The character tables and the procedure of mode counting has allowed us to determine the number of eigenmodes in each representation. We have not addressed the questions of how the modes of the surface complex relate to the modes of the gas-phase molecule and their representation. In the case of a diatomic molecule such as CO and NO the only eigenmode of the molecule in the gas phase is the stretching vibration (i.e., we exclude rigid-body rotations and translations). The corresponding mode for the adsorbed molecule is then, by its frequency, easily discriminated from hindered translations or rotations. For polyatomic molecules, however, the analysis of which mode of the adsorbate complex derives from which mode of the free molecule becomes more involved. The benzene molecule, for example, has a total of 20 eigenmodes in the gas phase of which 10 are degenerate (see Appendix B). In the gas phase a CH bending (A_{2u}) mode has a dynamic dipole moment perpendicular to the plane of the molecule. In contrast to this, a CH stretch (E_{1u}), a ring stretch (E_{1u}), and a CH bending (E_{1u}) have a dipole moment parallel to the plane of the molecule. In the case of a very weak adsorption therefore, either the A_{2u} mode or the E_{1u} modes would become dipole active depending on the orientation of the molecule. Stronger adsorption may additionally break the symmetry, other modes could become dipole active, and the degeneracy of the eigenmodes of gas-phase benzene could be lifted. With a set of character tables at hand, one can, in fact, determine into which representation of an assumed surface complex the modes slip, with certain assumed symmetry elements. It is therefore possible to determine the orientation and symmetry of the molecule in the adsorbed state from the number, frequency, and intensity of the dipole active modes observed for the adsorbed molecule. To facilitate this procedure we have worked out correlation tables for the most important point groups of molecules. (See Table 4.6.)

The procedure by which such tables are generated is relatively straightforward for nondegenerate modes. There the characters of all symmetry elements within a representation of an assumed surface point group must match the characters for the symmetry elements which are common with the corresponding representation of the point group of the free molecules. The method is somewhat more involved for degenerate representations, and we have outlined the method that may be used here in Section 4.2. The sum of the characters of the irreducible representations into which the degenerate representation splits is equal to the character of the degenerate representation for any common symmetry element:

$$\chi_E = \chi_1 + \chi_2 + \cdots. \tag{4.57}$$

As an example, we may consider the splitting of the E representation of C_{4v} when going to the subgroup C_{2v}. A common symmetry element is the rotation axis $C_2 = C_4^2$. The character of the E representation of C_{4v} for this symmetry element is -2. The only way to fulfill Eq. (4.57) is to assume splitting into B_1 and B_2 which both have the character -1 for the same symmetry element in C_{2v}. Equation (4.57) is then also fulfilled for the σ planes. The same conclusion follows from the procedures in Section 4.2.

TABLE 4.6

Correlation Tables for the Surface Subgroups of D_{6h}, D_{4h}, D_{3h}, D_{2h}, D_{2d}, T_d, and O_h.

D_{6h}	C_{6v} (σ_v,σ_d)	C_{3v} (σ_v)	C_{3v} (σ_d)	C_s (σ_d)	C_s (σ_v)	C_{2v} (σ_v,σ_d)	C_{2v} (σ_v,σ_h)	C_s (σ_h)	C_s (σ_v)	C_{2v} (σ_h)	C_s (σ_h)	C_s (σ_d)	C_6	C_3	C_2
A_{1g}	A_1	A_1	A_1	A'	A'	A_1	A_1	A'	A'	A_1	A'	A'	A	A	A
A_{1u}	A_2	A_2	A_2	A''	A''	A_2	A_2	A''	A''	A_2	A''	A''	A	A	A
A_{2g}	A_2	A_2	A_2	A''	A''	A_2	B_1	A'	A''	B_1	A'	A''	A	A	A
A_{2u}	A_1	A_1	A_1	A'	A'	A_1	B_2	A''	A'	B_2	A''	A'	A	A	A
B_{1g}	B_2	A_2	A_1	A'	A''	B_2	A_2	A'	A''	B_2	A'	A'	B	A	B
B_{1u}	B_1	A_1	A_2	A''	A'	B_1	A_1	A''	A'	B_1	A''	A''	B	A	B
B_{2g}	B_1	A_1	A_2	A''	A''	B_1	B_2	A'	A''	A_2	A'	A''	B	A	B
B_{2u}	B_2	A_2	A_1	A'	A'	B_2	B_1	A''	A'	A_1	A''	A'	B	A	B
E_{1g}	E_1	E	E	$A''+A'$	$A''+A'$	B_1+B_2	A_2+B_2	A''	$A''+A'$	A_2+B_2	A''	$A'+A'$	E_1	E	B
E_{1u}	E_1	E	E	$A'+A''$	$A'+A''$	B_2+B_1	A_1+B_1	A'	$A'+A''$	A_1+B_1	A'	$A''+A'$	E_1	E	B
E_{2g}	E_2	E	E	$A'+A''$	$A''+A'$	A_2+A_1	A_1+B_1	A'	$A''+A'$	A_1+B_1	A'	$A''+A'$	E_2	E	A
E_{2u}	E_2	E	E	$A''+A'$	$A''+A'$	A_2+A_1	A_2+B_2	A''	$A''+A'$	A_2+B_2	A''	$A''+A'$	E_2	E	A

(continued)

Table 4.6 (continued)

D_{3h}	C_{3v} (σ_v)	C_s (σ_v)	C_{2v} (σ_h σ_v)	C_s (σ_h)	C_s (σ_v)
A_1'	A_1	A'	A_1	A'	A'
A_1''	A_2	A''	A_2	A''	A''
A_2'	A_2	A'	B_1	A'	A''
A_2''	A_1	A''	B_2	A''	A'
E'	E	$A'+A''$	A_1+B_1	A'	$A'+A''$
E''	E	$A'+A''$	A_2+B_2	A''	$A'+A''$

D_{3d}	C_{3v} (σ_d)	C_s	C_3
A_{1g}	A_1	A'	A
A_{1u}	A_2	A''	A
A_{2g}	A_2	A''	A
A_{2u}	A_1	A'	A
E_g	E	$A'+A''$	E
E_u	E	$A'+A''$	E

D_{4h}	C_{4v} ($\sigma_v\sigma_d$)	C_{2v} (σ_v)	C_{2v} (σ_d)	C_s (σ_v)	C_s (σ_d)	C_4	C_2
A_{1g}	A_1	A_1	A_1	A'	A'	A	A
A_{1u}	A_2	A_2	A_2	A''	A''	A	A
A_{2g}	A_2	A_2	A_2	A''	A''	A	A
A_{2u}	A_1	A_1	A_1	A'	A'	A	A
B_{1g}	B_1	A_1	A_2	A'	A''	B	A
B_{1u}	B_2	A_2	A_1	A''	A'	B	A
B_{2g}	B_2	A_2	A_1	A''	A'	B	A
B_{2u}	B_1	A_1	A_2	A'	A''	B	A
E_g	E	B_1+B_2	B_1+B_2	$A'+A''$	$A'+A''$	E	B
E_u	E	B_1+B_2	B_1+B_2	$A'+A''$	$A'+A''$	E	B

(continued)

Table 4.6 (continued)

D2d	Cs	C2v
A₁	A'	A₁
A₂	A''	A₂
B₁	A'	A₂
B₂	A'	A₁
E	A'+A''	B₁+B₂

D2h (σyz, σxz)	C2v (σyz, σxz)	Cs (σyz)	Cs (σxz)	C2v (σyz, σxy)	Cs (σyz)	Cs (σxy)	C2v (σxz, σxy)	Cs (σxz)	Cs (σxy)
A_g	A₁	A'	A'	A₁	A'	A'	A₁	A'	A'
A_u	A₂	A''	A''	A₂	A''	A''	A₂	A''	A''
B_1g	A₂	A''	A''	B₁	A''	A'	B₁	A'	A'
B_1u	A₁	A'	A'	B₂	A'	A''	B₂	A'	A''
B_2g	B₁	A''	A'	A₂	A''	A''	B₂	A'	A''
B_2u	B₂	A'	A''	A₁	A'	A'	B₁	A''	A'
B_3g	B₁	A'	A''	B₂	A'	A''	A₂	A''	A'
B_3u	B₁	A''	A'	B₁	A''	A'	A₁	A'	A'

(continued)

Table 4.6 (continued)

T_d	C_{2v}	C_s	C_{3v}	C_s		O_h	C_{4v}	C_{3v}
A_1	A_1	A'	A_1	A'		A_{1g}	A_1	A_1
A_2	A_2	A''	A_2	A''		A_{1u}	A_2	A_2
E	$A_1 + A_2$	$A' + A''$	E	$A' + A''$		A_{2g}	B_1	A_2
F_1	$A_2 + B_1 + B_2$	$A' + 2A''$	$A_2 + E$	$A + 2A''$		A_{2u}	B_2	A_1
F_2	$A_1 + B_1 + B_2$	$2A' + A''$	$A_1 + E$	$2A' + A''$		E_g	$A_1 + B_1$	E
						E_u	$A_2 + B_2$	E
						F_{1g}	$E + A_2$	$E + A_2$
						F_{1u}	$E + A_1$	$E + A_1$
						F_{2g}	$E + B_2$	$E + A_1$
						F_{2u}	$E + B_1$	$E + A_2$

We demonstrate the use of the correlation tables with the specific example of benzene C_6H_6 chemisorbed on the (111) surfaces of platinum and nickel. The sample spectra are shown in Fig. 4.7. The most prominent feature of the loss spectra is a doublet at 830/910 and 730/820 for the platinum and nickel surfaces, respectively. Comparison to C_6D_6 has shown that both losses are associated with a hydrogen motion. Further investigation has also shown [12] that the relative intensity of the two losses within the doublet depends on temperature and coverage. This suggests that benzene may adsorb in two different sites each giving rise to a different frequency for the same type of vibration. With the help of the frequency table for gas-phase benzene in Appendix B and the compatibility table we can now determine the orientation of the benzene molecule when it is adsorbed on the surface. Let us assume that the benzene stands upright with its ring perpendicular to the surface. The frequency table in Appendix B tells us, as already discussed that then, even if the surface would not break the symmetry, the E_{1u} mode at 1038, 1486, and 3063 cm^{-1} should give rise to strong dipole losses. If we further consider the actual symmetry of upright standing benzene, the correlation table shows that at least the A_{1g}, B_{1u}, and E_{2g} modes would become dipole active as well. Additional losses therefore ought to be observed near 992, 1010, 606, 1178, and 1596 cm^{-1}, which is clearly at variance with the spectrum. We

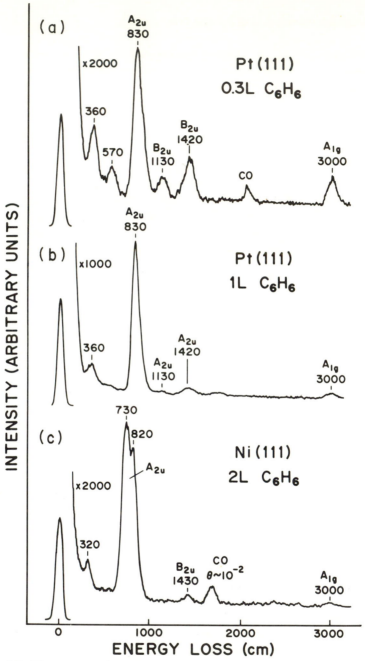

Fig. 4.7 Electron energy loss spectrum of Benzene (C_6H_6) adsorbed on a Pt(111) surface at (a) 0.3 langmuir and (b) 1 langmuir. The modes are assigned according to the representations of the point group of gaseous benzene. Their symmetry on the surface is A_1 (dipole losses). The A_{2u} mode is a doublet resulting from two different binding sites. (c) For benzene adsorbed on Ni(111) at 2 langmuirs the intensity of the A_{1g} and B_{2u} modes is much lower indicating that the adsorption breaks the symmetry only weakly.

therefore now test the proposition that benzene lies with its ring parallel to the surface. Now the A_{2u} CH bending mode (673 cm^{-1}) becomes dipole active just by the orientation of the molecule on the surface and the observed doublet could be associated with this mode. Chemical interaction with the surface could break down the symmetry to C_{6v} or C_{3v} depending on the adsorption site and strength of the interaction. Under the symmetry C_{6v} the A_{1g} modes become active. For C_{3v} we have two different choices of the symmetry planes. One would make the B_{1g} and B_{2u} modes additionally active and in the second case the B_{1u} and B_{2g} modes became active. We turn again to the frequency table and obtain the set of dipole active frequencies for either proposition. These are summarized in Table 4.7. Clearly the number of observed frequencies favors the point group $C_{3v}(\sigma_d)$, although there is some remaining uncertainty about the assignment of the 570 cm^{-1} loss.

TABLE 4.7

Predicted Set of Frequencies of
adsorbed Benzene Assuming Various Symmetries

Oriented	D_{6h}	C_{6v}	$C_{3v}(\sigma_d)$	$C_{3v}(\sigma_v)$	Observed[a]
A_{2u}	673	673	673	673	830/910
A_{1g}		3062	3062	3062	3000
		992	992	992	
B_{2u}			1310		1420
			1150		1130
B_{1u}				3068	
				1010	
B_{2g}				995	
				703	

[a] Observed frequencies for benzene (C_6H_6) on Pt(111).

It is also worthwhile to consider the relative intensities of the losses. On the nickel surface benzene is relatively weakly adsorbed. Therefore the A_{1g} and B_{2u} losses which arise through symmetry breaking are very weak. For platinum, the relative intensity of the losses depends on the coverage. For small coverage the A_{1g} and B_{2u} losses are relatively more intense than for higher coverages. In Fig. 4.8 we have plotted the intensity of the A_{1g} and B_{2u} losses relative to the A_{2u} loss as a function of coverage. The lowering of the sensitivity of the spectrometer with higher loss energies (Section 3.6) was taken into account. From this behavior of the intensities it may be inferred that the distortion of the benzene molecule through the interaction with the surface becomes less for high coverages.

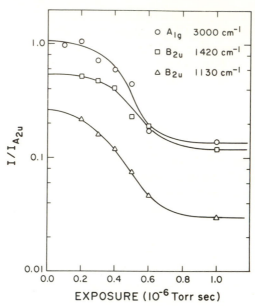

Fig. 4.8 Ratio of intensities of the A_{1g} and B_{2u} modes relative to the A_{2u} mode as a function of exposure. With higher surface coverage the interaction between the molecule and the surface becomes weaker and thus the symmetry-breaking effect of the surface which has made the A_{1g} and the B_{2u} modes dipole active.

Another example for the relation between adsorption strength and the intensity of losses related to the breaking of the molecular symmetry is provided by the comparison of the spectrum of ethylene C_2H_4 on Pt(111) and Ag(110) [13]. On silver, ethylene is weakly adsorbed with the molecular plane parallel to the surface. Consequently the spectrum (Fig. 4.9) shows only the symmetric CH wagging mode which is the only mode of the gas-phase molecule that has a dipole moment perpendicular to the plane of the molecule (Appendix B). On platinum, however, ethylene bonds strongly and modes that become dipole active through symmetry-breaking exhibit a strong dipole moment.

We may cast these observations into the following rule: *For weakly chemisorbed molecules, the strongest dipole losses are those infrared active modes of the molecule in the gas phase which produce a perpendicular dipole moment when the molecule is adsorbed. The relative strength of dipole losses resulting from the breakdown of the symmetry is a measure of the adsorption strength.*

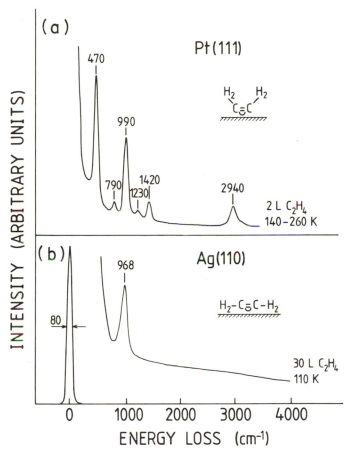

Fig. 4.9   Electron energy loss spectrum in specular reflection for ethylene on (a) Pt(111) and (b) Ag(110) (from Ibach and Lehwald [28], and Backx, de Groot, and Biloen [13], used with permission). For ethylene on silver, the surface hardly breaks the symmetry of the molecule. Therefore only the B_{1u} CH wagging mode with its dipole moment perpendicular to the plane of the molecule becomes observable as a dipole loss. The orientation of the molecule follows automatically from these considerations since the strongly IR-active B_{2u} and B_{3u} modes (see also frequency tables, Appendix B) which have their dipole moment oriented within the plane of the molecule, are not observed for the adsorbed molecule. On the platinum surface ethylene is strongly bonded. This is indicated by the CH stretching vibration which is typical for an sp^3 complex. Strong bonding to the surface breaks the symmetry and new modes become dipole active.

4.4.4 The Eigenmodes of Simple Molecules Adsorbed on Surface Sites of High Symmetry

In this section we shall back up the more abstract discussion of the surface point groups with a few examples. We also intend to illustrate the use of the character and correlation tables in connection with the frequency tables of Appendix B. We have already looked into the relatively complex situation of adsorbed benzene in the preceding section. Let us now turn to the comparatively simple and well-studied CO molecule and investigate the eigenmodes and their symmetry in different bonding sites.

We begin with CO on a site where the CO molecule stands upright on the surface bridging between two substrate atoms, as in Fig. 4.10. We assume that the surface complex has two symmetry planes, one being the plane of the surface molecular complex and the other being perpendicular to this plane. The surface complex then also has a twofold axis C_2 oriented along the CO band with the carbon and the oxygen nucleus on the C_2 axis. The point group of this complex is therefore C_{2v}. Examples for C_{2v} sites are the bridging sites on a (100) and a (110) surface of an fcc lattice. As long as the interaction with the second layer underneath is neglected, the bridging sites on the (111) surface of an fcc lattice also have C_{2v} symmetry. Since both atoms of the molecule lie on the C_2 axis the number m_0 (Table 4.4) is 2 and $m_{xy} = m_{xz} = 0$. The surface complex, therefore, has two eigenmodes of A_1, B_1 and B_2 character each, which account for the total of six degrees of freedom. No mode of A_2 character exists for this surface complex. Furthermore, the orientation of the eigenvectors is completely determined by symmetry in this case. As long as the surface atoms are considered to be of infinite mass, the eigenvalue equation for the system breaks up into three quadratic equations.

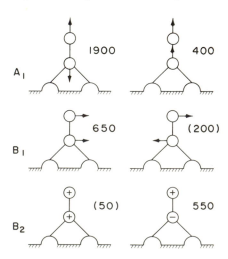

Fig. 4.10 The six eigenmodes of CO on a two fold bridging site with an assumed C_{2v} symmetry. Only the CO stretching vibration is a genuine mode of the gas-phase molecule. The rest of the surface modes are hindered translations and rotations. The frequencies (in cm^{-1}) are approximate values taken from a cluster calculation (from Richardson and Bradshaw [14], used with permission) and should be used as a guideline only. The frequencies of two modes fall into the phonon band and may therefore not be localized on the surface.

As already discussed in Section 4.4.2, the dipole active modes of this system are the two A_1 modes, the CO stretching, and the surface-CO mode. A typical spectrum is shown in Fig. 4.11. The CO coverage in this case is such that a $c(4 \times 2)$ diffraction pattern is observed and only bridging sites of the Ni(111) surface are occupied. The remaining B_1 and B_2 vibrations are odd under C_2 rotation. The selection rule for impact scattering therefore forbids the observation of these modes in specular reflection, while they should become

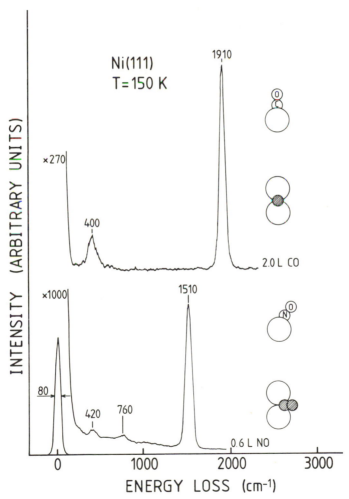

Fig. 4.11 Spectra of CO and NO on a Ni(111) surface at $T = 150°$K (from Erley, Wagner, and Ibach [15], and Lehwald, Yates, and Ibach [16], used with permission). The two losses for CO correspond to the CO stretching and the Ni–C stretching vibration. The canted orientation of NO makes the Ni–N–O bending mode dipole active as well.

observable in the off-specular direction. So far, however, no experimental evidence for impact scattering has been reported for this system. If the CO molecule is placed on the surface such that the symmetry becomes C_{3v}, C_{4v}, or C_{6v}, then the number and nature of the A_1 modes does not change. Therefore the dipole loss spectrum of the CO in various sites of high symmetry is distinguishable only on the basis of characteristic frequency shifts (to be discussed later) but not by the number of observed dipole losses. However, as soon as the molecule tilts and the symmetry is reduced to C_s, additional modes become active. Let us assume that the molecule tilts out of the plane of Fig. 4.10. Inspection of Fig. 4.10 shows that the tilt makes the two B_2 modes dipole active. We also may consult the correlation tables to obtain this result: tilting the molecule out of the plane removes the σ_{xz} plane. The appropriate correlation table is found in a subsection of the correlation table of D_{2h}. According to this table the A_1 and B_2 modes of C_{2v} become A' modes of C_s and are therefore dipole active which is the result already obtained.

We have already discussed that not all dipole active modes may be observed in a spectrum. An additional side condition is that this frequency must be well above the phonon bands of the substrate. Consulting Fig. 4.10 again tells us that one of the B_2 (now A') mode corresponds to a translation parallel to the surface in such a way that the nearest-neighbor bonds are, to first order, not strained. The frequency of this mode is therefore low for the CO molecule in the C_{2v} site. When the molecule is canted, the hindered translation mixes with the hindered rotation since the eigenvectors of the mode are no longer required to be parallel to the surface. Therefore the two A' modes derived from the two B_2 modes may not be quite as far apart in frequency as suggested by the numbers in Fig. 4.10.

An example where one additional mode is observed in an energy loss spectrum is also shown in Fig. 4.11. In NO the antibonding π^* orbitals are partially filled. Therefore NO has a higher tendency toward canted positions compared to CO in order to have the extra electron in the π^* orbital participating in the bonding. Depending on whether the bonding to the surface involves further donation of electrons into the π^* orbital or removal, the NO stretching vibration can shift down or up over a wide range. Figure 4.11 is a demonstration that the different electronic structure of CO and NO gives rise to a different local symmetry after bonding to a nickel surface.

We now turn to the discussion of the eigenmodes of a triatomic molecule like H_2O in the principal surface point groups. We assume that the water molecule adsorbs through the lone pair orbital of the oxygen atom to produce a surface complex as depicted in Fig. 4.6. There is spectroscopic evidence that this type of bonding exists. However, in the adsorption systems studied so far [17, 18] the water molecule tended to cluster via O–H–O hydrogen bridges. Therefore, until now spectra of isolated, adsorbed water molecules are not reported probably because the temperature was not low enough. Another example of the same configuration would be a methylene (CH_2) group.

We again begin with the C_{2v} point group. We now have $m_0 = 1$ and $m_{xz} = 1$. According to Table 4.4 the molecule has $3A_1$, $1A_2$, $3B_1$, and $2B_2$ modes. The hydrogen atoms contribute two modes to the A_1 representation one mode to A_2, two modes to B_1 and one mode to B_2. The oxygen (carbon) atom contributes each one mode to A_1, B_1, and B_2. The nature of those nine modes is easily understood with the help of the characters. The two totally symmetric H modes are the symmetric stretching and symmetric bending (scissor) mode. The A_2 is a frustrated rotation (twisting mode) of the H_2O molecule around the C_2 axis. The two H modes of the B_1 representation are the antisymmetric stretching and the bending (rocking) mode and the final H mode in B_2 is a tilt-out of the plane of the molecule, frequently addressed as the "wagging" mode. The center oxygen or carbon atom vibrations are typically lower in frequency due to the larger mass. As in the case of the CO molecule the eigenvectors of the center atom are completely determined by the symmetry elements and polarized along the x, y, and z axis, respectively. Of all nine modes, three modes (two H modes and one oxygen or carbon mode) are dipole active. In specular reflection, where the dipole losses are observed, the observation of the A_2 mode (O–H twisting) is not forbidden by the selection rule applicable to impact scattering. For a CH_2 group, where the dipole moments are comparatively small, the CH_2 twisting mode might appear in the spectrum on the same scale of magnification. One therefore really has to investigate the nature of the scattering mechanism by measuring the intensity as a function of the scattering angle before the local symmetry can be determined from the number of observed modes.

Just as in the case of the CO molecule, we may break the symmetry down by removing the σ_{xz} or the σ_{yz} plane and the C_2 axis which makes either the B_1 or the B_2 modes active. The A_2 mode becomes active when both symmetry planes are removed (point group C_2 or C_1).

A final example in this section is a NH_3 or CH_3 group on a C_{3v} site. The C_{3v} point group has a doubly degenerate representation E. Three hydrogen vibrations, the "degenerate stretching" the "degenerate deformation" and the rocking mode (also referred to as the ν_d, δ_d, ρ-modes) belong to the E representation. Another H mode is the A_2 torsion mode. The two H modes of A_1 are the symmetric stretching and deformation (ν_s mode and δ_s mode). These two modes plus one surface–nitrogen and surface–carbon mode, respectively, are dipole active. However, none of the modes has to have a vanishing cross section for impact scattering even in specular reflection! Again the local symmetry can be inferred only from the spectrum once the scattering mechanism is understood.

After giving so much consideration to the surface point groups, the reader might be under the impression that the formation of high-symmetry complexes is the most natural way of establishing a molecule–surface bond. This is not so. Even on low-index surfaces with sites of high symmetry, molecules may build adsorbate complexes of low symmetry or even with no element of symmetry at all. If we also take high-index surfaces into account, highly symmetric surface groups are the exception rather than the rule. Symmetry considerations gain this importance not because symmetric surface species are abundant but because, once they are prepared, they are excellent candidates for further studies that can and will improve our understanding of surface bonding.

4.4.5 Overtones, Combination Bands, and Multiple Energy Losses

A perfectly harmonic dipole oscillator with energy levels

$$E_n = \hbar\omega_0(n + \tfrac{1}{2})$$

absorbs light when the frequency of light ω_L is equal to ω_0. Another way of saying this is that photon absorption causes a transition from E_n to E_{n+1}, i.e., $\Delta n = 1$. Although it is possible to excite an harmonic oscillator up to $n = 2, 3, 4, \ldots$, as long as the excitation is accomplished by light, it must be done stepwise and no absorption bands occur at multiples of ω_0. The force field for the motion of nuclei in a molecule or solid is, however, not simply quadratic in the displacements from the equilibrium positions of the nuclei but contains higher-order "anharmonic" terms. It is via these terms that the motion of nuclei in an eigenmode has a slight deviation from a perfect harmonic motion, and as a result the light may couple to the oscillator at multiples of ω_0. These excitations are called "overtones." The deviation from the perfect quadratic force field also shifts the energy levels of the oscillator. Overtones therefore are typically observed at a somewhat lower frequency than a multiple of the fundamental frequency. Also, the "fundamental" frequency which is the energy difference between the ground state and the first excited state divided by $\hbar$ is not equal to the ground state frequency for an anharmonic oscillator. A further effect of anharmonicity is that the eigenmodes obtained in the harmonic approximation are no longer perfect eigenstates. Therefore, in addition to multiples of eigenmode frequencies, combinations of different eigenmodes lead to absorption. Within the framework of a simple but explicit model, we shall address a number of these issues once again at the end of Chapter 5.

Absorption due to overtones and combination bands is relatively weak. With the limited signal-to-noise ratio in electron energy loss spectroscopy and surface infrared reflection spectroscopy, they are not always detected. An example for a combination band for a surface species is shown in Fig. 4.12. Nevertheless, since the sensitivity of spectrometers is improving (Section 6.4), a brief discussion of the selection rules applicable to dipole excitation of overtones and combination bands may be useful. It must also be noted that an overtone or combination band may indeed become rather strong when the frequency of the combination is close to a fundamental of the same representation. Such a situation is referred to as Fermi resonance. For a further discussion of this point see Ref. [19]. More examples will also be discussed in Section 6.4. In order to apply the selection rule to overtones and combination bands, the representation of these have to be determined. If all contributing modes are nondegenerate, the character of the combination is a product of the characters of all contributing eigenmodes for each symmetry element G:

$$\chi^G(n_i v_i + m_j v_j + \cdots) = [\chi^G(v_i)]^{n_i} [\chi^G(v_j)]^{m_j} \cdots. \qquad (4.58)$$

Application of this rule to all symmetry elements allows us to determine the complete set of characters and thus the representation of the combination band $n_i v_i + m_j v_j + \cdots$. For a dipole-allowed combination all characters must be $+1$. Therefore any overtone of a totally symmetric vibration and any even harmonic of all other vibrations is dipole active. In particular this latter statement may offer an interesting potential for the observation of dipole-forbidden frequencies via dipole coupling at least for strong dipole scatterers

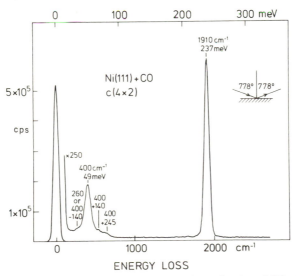

Fig. 4.12   Phonon sidebands around the metal–carbon vibration of CO adsorbed on the Ni(111) surface in a $c(4 \times 2)$ overlayer.

When degenerate eigenmodes are participating, the determination of the resulting representations is more complex. For the surface point groups, all even harmonics of degenerate vibrations also have a perpendicular dipole moment. This is in fact easy to understand even without resorting to group theory. Let us consider the motion of an adatom adsorbed on top of a substrate atom. The degenerate mode (Table 4.4) is a motion parallel to the surface. Anharmonicity, however may cause the atom to move back and forth on a path which is slightly curved rather than strictly parallel to the surface. Therefore a perpendicular dipole moment may be created which is symmetrical on either side of the equilibrium position. This dipole moment oscillates with twice the frequency of the eigenmode (Fig. 4.13) and therefore gives rise to an absorption band at $2\omega_0$.

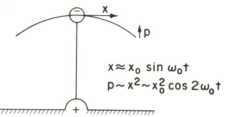

$$x \approx x_0 \sin \omega_0 t$$
$$p \sim x^2 \sim x_0^2 \cos 2\omega_0 t$$

Fig. 4.13 Illustration for the excitation of a second harmonic of a dipole-forbidden vibration. In the harmonic approximation x and z modes are strictly decoupled. Anharmonicity, however, makes the parallel (E) motion of the atom slightly curved. This produces a dynamic dipole moment oscillating at twice the frequency of vibration. Note that the term "anharmonic" here refers to a force field expressed in Cartesian coordinates. A valence force field with bond stretching and bending force constants, even though perfectly harmonic in the coordinates describing stretching and bending, is anharmonic in Cartesian coordinates.

While in infrared absorption spectroscopy a harmonic oscillator absorbs only at the fundamental frequency ω_0, in electron energy loss spectroscopy energy losses at multiples of $\hbar\omega_0$ occur even for a perfectly harmonic oscillator, as we appreciate from Section (3.3.5.). We consider this issue once again, from a different viewpoint. This is related to the nonperiodic nature of the force that drives the oscillator. A simple Hamiltonian that models the situation considers an harmonic oscillator perturbed by a time-dependent force which is independent of the oscillator amplitude,

$$\mathcal{H} = \hbar\omega_0(a^+ a + \tfrac{1}{2}) + F(t)(\hbar/2m\omega_0)^{1/2}(a^+ + a). \qquad (4.59)$$

Here $F(t)$ is the time-dependent force exerted on the oscillator by the electron, which is here regarded as a classical point particle moving along a well-defined trajectory. The Schrödinger equation for this model Hamiltonian has an exact solution for arbitrary $F(t)$. The model applies to dipole excitation in the limit of sufficiently high-electron energy so that the electron trajectory is little perturbed by the inelastic interaction. For impact scattering in the adiabatic approximation with rigid muffin-tin potentials as discussed in Chapter 3, the model is suitable as well. In the literature on molecular excitations of free molecules, the limit in which the model Hamiltonian applies is called the "impulse limit," with the understanding that in this limit the lifetime of any compound state formed from the electron and the free molecule is short compared to $1/\omega_0$. More typical experiments in gas-phase electron–molecule scattering are, however, concerned with resonances of longer lifetime.

The probability for the excitation from the ground state to the nth excited state after the time t for the model Hamiltonian is

$$P_{0 \to n} = \frac{1}{n!} |G(t)|^{2n} e^{-|G(t)|^2} \tag{4.60}$$

with

$$G(t) = \frac{i}{(2m\hbar\omega_0)^{1/2}} \int_0^t e^{i\omega_0\tau} F(\tau) \, d\tau.$$

For a periodic excitation by a classical field, $F(\tau) \sim e^{-i\omega\tau}$, $G(t)$ becomes a δ function as $t \to \infty$. Therefore a classical field loses energy by driving the oscillator only when $\omega = \omega_0$. The Poisson distribution of Eq. (4.60) then just corresponds to the classical amplitude of the oscillator. For a quantum particle like an electron, the final probability distribution for the oscillator levels after the interaction is completed must correspond to the probability distribution in the energy loss spectrum of the electron. Therefore as long as only a single harmonic oscillator is involved, and to the extent that one may neglect the finite momentum transfer, the loss spectrum of multiple losses is a Poisson distribution (see Section 3.3.5 for a more complete discussion of multiple losses and the conditions necessary for the Poisson distribution to be realized in practice). This Poisson distribution was indeed observed in electron energy loss spectra on ZnO (Fig. 4.14) and later also for a layer of NiO grown on a Ni(100) substrate [21]. For dipole scattering from a monolayer, however, the one-phonon excitation probability is typically of the order of 10^{-2} to 10^{-3}. The probability for double losses is therefore already very small for the strongest dipole scatters.

For impact scattering, multiple losses as well as overtones plus combination bands are easily observed. Both typically come within the same order of magnitude as the single loss. Again, the multiple losses are excited in a sequence of scattering events on different molecules. The energy losses due to multiples therefore occur exactly at multiples of the fundamental losses. In comparison to the picture just discussed where the Poisson distribution results from a sequence of uncorrelated single scatterings, excitation

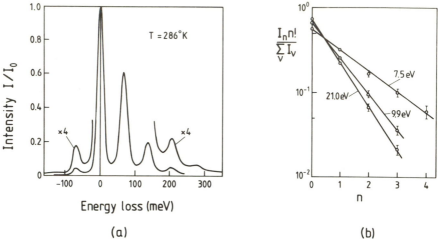

(a)   (b)

Fig. 4.14 (a) Electron energy loss spectrum of surface phonons in ZnO (from Ibach [20], used with permission). The surface phonons here are dielectric surface waves as described in Chapter 3. The dipole interaction is strong because the dipole moment of atoms within a layer of about 100 Å thickness contribute (Chapter 3) to the cross section. Similar strong excitation of surface phonons will occur on all ionic materials with strong IR absorption, oxides in particular. (b) The Poisson distribution of intensities.

of overtone- and combination-band losses in a single scattering event becomes possible when the anharmonicity of the internuclear potential is allowed for. An example for the latter case is shown in Fig. 4.15. In surface spectroscopy so far, little attention has been paid to overtone losses, although they could provide extremely important information on the shape of the potential curve and on the dissociation limit for individual bonds within an adsorbed species. We shall discuss this in more detail in Section 6.4.

4.4.6 Subsurface Atoms

Segregation of impurities to the surface and dissolution of adsorbate atoms into the bulk are common phenomena in surface science. It therefore seems important to understand whether and to what extent vibrations of dissolved atoms are accessible to electron energy loss spectroscopy. Unfortunately there is no easy answer. Again, we have to distinguish between dipole and inpact scattering. In impact scattering the probing depth of electrons into the bulk is determined by the mean free path λ. The inelastic intensity from a vibrating atom localized at a depth d below the surface is reduced by a factor $\exp(-2d\lambda)$. The problem here is the mean free path. For 50- to 100-eV electrons, it is of the order of 5 to 10 Å, largely independent of the material. For lower electron energies the mean free path can become rather large for

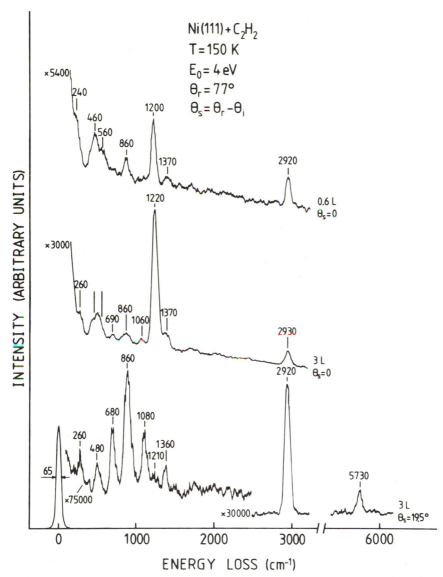

Fig. 4.15 In-and-out off-specular spectrum of C_2H_2 on Ni(111) at $T = 150°K$; $E_0 = 4$ eV, $\theta_r = 77°$, and $\theta_s = \theta_r - \theta_i$. Note that for impact scattering the intensity of the overtone of the CH stretching vibration is of the same order of magnitude as the single loss.

wide band gap semiconductors and metals whose electronic properties resemble a free electron gas. For silver, gold, and aluminum, values between 20 and 50 Å have been measured for electron energies between 5 and 10 V [22], and there is reason to believe that the mean free path increases further with decreasing electron energy. Since for electron optical reasons which we have discussed in Chapter 2, typical electron energies are between 1 and 10 eV, we must conclude that subsurface atoms are accessible to electron energy loss spectroscopy unless they are deep in the bulk.

In dipole scattering the screening of a dipole moment potentially produced by a dissolved atom is the key factor. In a classical screening model, Rahman *et al.* [23] have calculated the screening of the field of a dipole imbedded in a conducting substrate. They find that the potential of a dipole oriented either parallel or perpendicular to the surface is essentially the same as if the dipole were placed on the surface, however, with an additional screening factor $\exp(-2k_D d)$, where d is again the depth of the dipole below the surface and k_D the inverse screening length. From this model we conclude that atoms dissolved in a semiconductor are again accessible to electron energy loss spectroscopy because the screening length is large. For a metal surface where the typical screening length is less than 1 Å, one may infer that the screening is very effective. Probably dipole losses cannot be observed from atoms dissolved in metals. A possible exception are atoms between the first and second layers because then $k_D d$ is of the order of 1. Another reason may also lead us to assume that dipole scattering of atoms between the first and second layers may be observable. Except for the lightest atoms, the vibrations of these atoms will not be all that localized, with the consequence that motion of the subsurface species will excite motion in the surface atoms. A localized motion of a surface metal atom will produce a dipole moment by itself, most particularly when the electronic environment of the metal atom is altered by the chemical bonding to the subsurface dissolved atom.

The change of the chemical nature of the substrate becomes even more important when strong bonds are established between the impurity atoms and the substrate atoms. The growth of an oxide may serve as an example. As long as few oxygen atoms are dissolved in the metal substrate, dipole scattering may be small. Once clusters of oxides are formed, however, the scattering intensity should increase sharply. Thus electron energy loss spectroscopy should be very sensitive to cluster growth or a metal–insulator transition on a surface.

At the time of writing this book, no definite evidence for any vibrational losses originating from subsurface impurities is available (with the exception of energy losses from a complete oxide layer). Nevertheless, there seems to be interesting physics as well as practical applicability involved in this issue which would warrant further studies.

4.5   SOME PRINCIPLES IN MODE ASSIGNMENT

In the preceding sections we have been concerned primarily with several general principles of molecular dynamics on surfaces and how they relate to the technique of electron energy loss spectroscopy. Relatively little attention was paid to the values of the eigenfrequencies themselves and to the relations between frequencies and the structure and bonding of the molecular surface groups. In fact, one of the major applications of vibration spectroscopy is to use the observed vibration spectrum not only for the analysis of symmetry and structure but for the chemical analysis of the adsorbed species. The key to this kind of chemical analysis is the fact that certain molecular groups may be discriminated from others by their characteristic frequency spectrum. Characteristic frequencies of free molecules and molecular subgroups are well documented [19, 24–27) and can therefore be used for the analysis of surface species. Frequencies of a few selected molecules are summarized in Appendix B.

4.5.1   Group Frequencies

As discussed in detail in Section 4.2, eigenfrequencies of a molecule or a surface complex are obtained as solutions of a secular equation. The order of the secular equation is equal to the number of eigenmodes in each representation of the molecular point group. By varying the coupling between any two atoms or the mass of any atom, all eigenmodes of the molecule are affected. Therefore a vibrational frequency is in principle a property of the molecule as a whole and no subgroup of the molecule is independent of the rest of the molecule. We have shown in Eq. (4.37), however, that under certain provisions the dynamical matrix describing the entire molecule plus the substrate can be written as an expansion where the first term is a dynamical matrix of the molecule coupled to a rigid surface and a remaining term describing the coupling to the phonon spectrum. This coupling is small when the masses of substrate atoms is large. The same concept can be applied to describe the frequencies of a subgroup of light atoms within a molecule, as we have remarked earlier. If we take the CH, CH_2, CH_3 groups as examples, the vibrational amplitude of the carbon atom in a CH stretching or bending vibration is relatively small, since its mass is 12 times larger than the hydrogen mass. In a first-order approximation one may describe the frequency spectrum of a CH_x group as if the mass of the carbon atom were infinitely large. In this approximation the CH_x groups have a frequency spectrum independent of the rest of the molecule as long as no direct coupling of the H atom to other atoms occurs.

In the spirit of the approximation described above, one may also define a local symmetry for a CH_x group and classify the characteristic eigenmodes

TABLE 4.8

Nomenclature of Eigenmodes

Local point group	Example	Stretching modes		Bending modes			
				A', δ, in-plane bend		A'', π, out of plane bend	
C_s	$=\!C\!<^H_H$	A', ν, stretch		A', δ, in-plane bend		A'', π, out of plane bend	
C_{2v}	$=\!C\!<^H_H$	A_1, ν_s s stretch	B_1, ν_a a stretch	A_1, δ_s scissor	B_1, ρ_r rocking	B_2, ρ_w wagging	A_2, ρ_t twisting
C_{3v}	$=\!C\!-\!H\,(H,H)$	A_1, ν_s s stretch	E, ν_d d stretch	A_1, ρ_s s deform	E, ρ_d rocking	E, δ_d d deform	A_2, ρ_t torsion

^{a} s, symmetric; a, antisymmetric; d, degenerate.

according to the representations of the local point groups. This scheme then leads to the standard nomenclature of CH modes, which is summarized in Table 4.8. In addition to the point-group notation symbols, v, π, δ, ρ are in use and also more descriptive notation such as "stretching," "deformation," "scissor" modes.

A picture of the orientation of the eigenvectors is easily obtained with the group representation symbol and the character tables (Table 4.4). To elucidate the method with one example, we consider the B_1 rocking mode of the CH_2 group. According to the character table of C_{2v} (Table 4.4), a B_1 mode is antisymmetric with respect to the C_2 axis and the symmetry plane perpendicular to the CH_2 plane, however, symmetric to the CH_2 plane itself. The eigenvectors therefore lie within the CH_2 plane (σ_{xz} plane) and have already been depicted in Fig. 4.6. The notation of Table 4.8 is not only applicable to CH_x groups but to all subgroups of molecules that have the appropriate local symmetry. Examples are NH_x, SH_x, SiH_x, PH_x, but also all subgroups where the hydrogen is replaced by a halide as in 1,2-dibromoethylene ($CH_2=CBr_2$). Each of the molecular subgroups has a set of eigenfrequencies which, though somewhat dependent on the environment, is characteristic enough in most cases to be used to identify the existence of a certain subgroup within a molecule. In Tables 4.9 and 4.10 we quote the frequencies for the CH_x and CD_x groups. The values are obtained by taking

TABLE 4.9

Characteristic Frequencies of Molecular Subgroups CH_x
(frequencies from [24])

CH	stretch	bend
$>C<^H$	3028 ± 17	1230 ± 68
$=C<^H$	3084 ± 41	1045 ± 230
$\equiv C-H$	3330 ± 18	641 ± 65

CH_2	s-stretch	a-stretch	scis	wag	twist	rock
$>C<^H_H$	2932 ± 58	2990 ± 56	1447 ± 21	1267 ± 90	1213 ± 77	875 ±127
$=C<^H_H$	3003 ± 14	3092 ± 17	1414 ± 28	896 ± 37	728 ±197	968 ±110

CH_3	s-stretch	d-stretch	s-deform	d-deform	rock
$-C<^H_H$ (H)	2908 ± 41	3003 ±137	1352 ± 88	1445 ± 28	1029 ±140

TABLE 4.10

Characteristic Frequencies of Molecular Subgroups CD_x
(frequencies from [24])

CD	stretch	bend
$>C<^D_{\ }$	2244 ± 23	823 ± 74
$=C<^D_{\ }$	2309 ± 26	788 ±155
$\equiv C-D$	2597 ± 87	541 ± 70

CD_2						
	s-stretch	a-stretch	scis	wag	twist	rock
$>C<^D_D$	2145 ± 53	2241 ± 61	1103 ± 77	996 ±132	877 ± 86	668 ±155
$=C<^D_D$	2224 ± 21	2316 ± 53	1042 ± 31	723 ± 25	526 ±141	802 ±141

CD_3					
	s-stretch	d-stretch	s-deform	d-deform	rock
$-C\overset{D}{\underset{D}{<}}D$	2105 ± 24	2233 ± 45	1052 ± 80	1049 ± 25	809 ±146

the average of the appropriate frequencies for a large number of molecules as listed in Shimanouchi's tables [24]. The standard deviation from the average is also listed in Tables 4.9 and 4.10. Deviations are generally larger for the deuterated groups. This is basically a consequence of the smaller m_C/m_D mass ratio which is less favorable for the concept of local subgroup frequencies. We can use the standard deviation as a measure of how characteristic a certain frequency is. We realize from the tables that the stretching vibrations, the scissor vibrations, and the symmetric deformation vibration typically fall within a relatively narrow frequency range. Also, this frequency range does not overlap with other modes of the same subgroup. These frequencies are therefore easily assigned in a spectrum of a known molecule and may also serve as a clue for the existence of a molecular subgroup within an unknown chemical. We also notice that the CH stretching frequencies have characteristic downward shifts in frequency when the carbon bond goes from triple-bonded carbon to single-bonded carbon. This frequency reduction results from a somewhat smaller CH stretching force constant for higher coordinated carbon atoms. The stretching frequencies therefore also serve as an indication for the type of bonding in which the carbon atom is engaged.

Characteristic frequency shifts also occur when the carbon atom bonds to atoms with a large electronegativity difference. An example for a large

TABLE 4.11

Frequencies of the Symmetric
Deformation Modes (in cm^{-1}) of Methyl Halides
and Electronegativity Difference
between Carbon and the Halides

	CH$_3$I	CH$_3$B$_r$	CH$_3$C	CH$_3$F
Δ	0	0.3	0.5	1.5
δ_s (cm^{-1})	1252	1306	1355	1464

shift is the frequency shift of the symmetric deformation mode of methyl-halides (Table 4.11).

With some precautions, characteristic frequencies may not only be assigned to clearly separated molecular subgroups but even to individual bonds. We illustrate this concept again with hydrocarbons. Table 4.12 lists the CC stretching vibrations of a number of hydrocarbons [24]. The frequencies can be clearly grouped into three separate classes averaging around

TABLE 4.12

CC Frequencies of Hydrocarbons

Molecule	ν(C—C)	ν(C=C)	ν(C≡C)
Ethane CH$_3$—CH$_3$	995		
Ethylene CH$_2$=CH$_2$		1623	
Acetylene CH≡CH			1974
Propane CH$_3$—CH$_2$—CH$_3$	869⎱ 1054⎰ 962		
Propylene CH$_2$—CH=CH$_2$	919	1647	
Allene CH$_2$=C=CH$_2$		1071⎱ 1956⎰ 1513	
Methylacetylene CH$_2$—C≡CH	931		2142
Butane CH$_3$—CH$_2$—CH$_2$—CH$_3$ (*trans*)	837⎱ 1009⎰ 968 1059		
1,3-Butadiene CH$_3$=CH—CH=CH$_2$	1196	1596⎱ 1630⎰ 1613	
2-Butyne CH$_3$—C≡C—CH$_3$	725⎱ 1152⎰ 1024		
Butadiene CH≡C—C≡CH	874		2020⎱ 2184⎰ 2102
Average frequency	~950	~1600	~2100

950, 1600, and 2100 cm^{-1} which are the characteristic frequencies of a CC single, double, and triple bonds, respectively. The fact that even for a long-chain molecule the frequencies fall into such ranges may be surprising. It relates however, to a well-known result in solid state physics: if we consider the eigenmodes of an infinite linear chain with two alternating force constants f_1 and f_2 and equal masses, the frequencies at the zone boundary are $M\omega^2 = 2f_1$ and $M\omega^2 = 2f_2$, which is the same result as obtained for two atoms coupling to each other by the force constants f_1 or f_2. Since the dispersion curve at the zone boundary is flat, the portion of the Brillouin zone around the zone boundary provides the largest contribution to the spectral distribution of frequencies. The model, however, applies only when different force constants (or different masses) are involved. A molecule such as allene (Table 4.12) with two double bonds and equal masses has two largely separated CC frequencies falling into the single-and triple-bond frequency range. Only the average of the symmetric and antisymmetric CC frequencies falls into the characteristic double-bond range. Symmetric and antisymmetric splitting also occurs with propane, however, not quite as dramatically.

4.5.2   Isotope Shifts

To a high order of approximation the internuclear force constants within a molecule or a surface molecular group remains unchanged when nuclei are replaced by isotopes. Therefore frequency spectra experimentally obtained for a variety of different isotopes of the same molecule allow one to determine the parameters of the internuclear force field in great detail. But even without a detailed quantitative normal coordinate analysis, substantial information is obtained from isotope shifts. For example, isotopically labeled molecules can be used to analyze individual steps in a surface chemical reaction. Here the precise amount of the frequency shift is often less important that the observation of which mode is shifted. A simple example is the decomposition of NO on a surface. By letting $N^{14}O^{16}$, $N^{15}O^{16}$, and $N^{14}O^{18}$ decompose in separate experiments, the surface vibrations of oxygen and nitrogen are easily distinguished. The spectra may then be compared to vibrational spectra of adsorbed oxygen, nitrogen and coadsorbed oxygen and nitrogen. This comparison can eventually lead to a determination of the binding sites involved in the dissociation process.

Other important chemical information is obtained when spectra of deuterated hydrocarbons are compared to the spectra of the protonic species. On surfaces, unsaturated hydrocarbons either form π complexes with the surface (see Fig. 4.7, benzene) or "rehybridize" to strongly bonded sp^2- or sp^3-type bonding. Deuteration of the molecule allows one to identify the modes associated with the CC bond, and from this frequency the order of the intramolecular carbon bonds (Table 4.12) may be inferred. Deuteration also discriminates the hydrogen modes. The number and symmetry of hydrogen modes allows us to determine CH, CH_2, or CH_3 subgroups within a molecule, as we have discussed in the preceding section.

After these qualitative considerations we now turn to the quantitative aspects of isotope shifts. When a hydrogen atom is replaced by deuterium one might expect the frequency of all H modes to be reduced by a factor given by the square root of the mass ratio $\sqrt{2}$. Actually, observed shifts tend to be smaller for two reasons. The first reason is the anharmonicity of the potential. While the energy levels of a harmonic oscillator are equally spaced, the energy differences tend to become smaller for higher levels in realistic potentials. We shall study this effect in greater detail in Section 5.4. The reduction of the energy differences due to the anharmonicity is noticeable even for the fundamental. The effect is stronger for hydrogen than for deuterium, since deuterium has a smaller lever spacing and the fundamental frequency for deuterium is less affected by the anharmonicity. Therefore the ratio of the fundamentals of the H_2 and D_2 molecule is 1.390 and not $\sqrt{m_0/m_H} = 1.414$.

In polyatomic molecules the ratio can be even significantly smaller than 1.39. This is a consequence of the coupling of the hydrogen-associated eigenmodes to other eigenmodes of the molecule. For the same reason the CC frequencies in hydrocarbons do not remain constant when hydrogen is replaced by deuterium. Typically a small downshift is observed also. Therefore the assignment of modes to CC or CH bond becomes a nontrivial procedure. Of course any proposed assignment can be checked by a full, normal coordinate analysis. But there are other, less involved means, that are helpful as well. One of them is the Teller–Redlich rule. Another is associated with the unique feature of electron energy loss spectroscopy of employing two different scattering mechanisms which we shall discuss in the next section.

The Teller–Redlich rule [19] states that within the harmonic approximation for two isotopic species labeled 1 and 2, we have

$$\prod_i \left[\omega_i(1)/\omega_i(2)\right] = \prod_j \left[m_j(2)/m_j(1)\right]^{v_j/2}. \qquad (4.61)$$

The index i denotes all vibrations within a certain representation of the point groups to which the surface species belong, the index j denotes all inequivalent atoms, and v_j is the number of vibrations this atom contributes to a certain representation. These numbers are the numbers $m, m_0, m_{xy}, \ldots$ in the character Tables 4.4 and 4.5. Any proposed assignment which includes the assignment of a mode to a certain representation should be consistent with this rule except for a small deviation of the order of 2% resulting from anharmonic effects as already discussed. When the Teller–Redlich rule is applied to molecules on surfaces, a problem arises with the low-lying eigenmodes (hindered translations and rotations) of the adsorbate complex which are frequently inaccessible to the experiment. This frequency ratio can be treated approximately as if the molecular complex would rotate or translate freely,

$$\prod_i \frac{\omega_i(1)}{\omega_i(2)} = \prod_j \left(\frac{m_j(2)}{m_j(1)}\right)^{v_i/2} \left(\frac{M(2)}{M(1)}\right)^{-t/2} \prod_k \left(\frac{I_k(2)}{I_k(1)}\right)^{-1/2}. \tag{4.62}$$

Here M is the total mass of the molecule, t the number of hindered translations within the representation, I_k the moment of inertia, and k the number of hindered rotations, again of course within the representation which is considered. The index i now enumerates all but the hindered motions. The proposition of using the ratios for free translations can easily be tested with those hindered translations which are observable: To provide an example, for ethylene of Pt(111) and Ni(111), the frequency ratio for the metal–carbon vibration is experimentally [28, 29] found to be 1.056 ± 0.015 while the square root of the mass ratio of C_2D_4 and C_2H_4 is 1.069.

We illustrate the application of the Teller–Redlich rule with two examples, acetylene on platinum and nickel. Frequencies and their proposed assignment [28, 29] are listed in Table 4.13. For acetylene on platinum, the proposed symmetry is C_s with the symmetry plane perpendicular to the C–C axis. In Table 4.13 the A' modes are listed. The two metal–carbon modes of the same representation are treated as hindered translations. The frequency product, 2.54, then compares rather favorably with the value calculated from the mass ratios. The deviation may be even within a range that is caused by anharmonic effects since one also must allow for an error margin of 1 to 2% in the experimentally determined product. For C_2H_2 on Ni(111) with no symmetry element, three translations and two rotations must be considered to account for the carbon–metal vibrations. (Actually, on of the carbon–metal vibrations has also been measured and this frequency ratio for C_2H_2 and C_2D_2 is equal to the expected value.) To account for the rotations, the moments of inertia have to be calculated from a model for the structure. For simplicity, we assume the linear structure of gaseous C_2H_2 for which the ratio of $I(D)/I(H)$ is 1.39. The theoretical value of the frequency product is calculated with this value. The deviation between the experimental and theoretical values is now larger, probably because of the oversimplification of the treatment of the hindered rotations. Nevertheless, even with such simplification, the Teller–Redlich rule is still a useful test

TABLE 4.13

Modified Teller–Redlich Test on Proposed
Surface Symmetries and Mode
Assignment for Acetylene

C_2H_2	Pt(111)	Ni(111)	Ni(111)
Symmetry	C_s	C_1	Test: C_s
ν_{CH}	3010/2245	2 × 2920/2190	2920/2190
ν_{CC}	1310/1260	1220/1190	1220/1190
δ_{CH}	985/730	1080/890	1080/890
		1370/1090	
ρ_{CH}	770/570	690/540	690/540
		860/640	
T	2	3	2
R	—	$R_x R_z$	
$\prod_i \dfrac{\omega_i(1)}{\omega_i(2)}$	2.54	4.77	2.12
Theor.	2.626	5.15	2.626
Deviation	3.5%	8%	24%

on a proposed symmetry and assignment: In an earlier proposition for the assignment of modes, the symmetry for C_2H_2 on Ni(111) was assumed to be the same as for C_2H_2 on Pt(111) [31]. Then the symmetric A' subset of the frequencies would have to satisfy the Teller–Redlich rule. The last column in Table 4.13 shows that the deviation is substantial. In fact, there seems to be no argument which could justify a deviation as large as 24% in this case since the only unspecified frequency ratios are of translational character for which, as we have shown, the square root of the mass ratio is a good number. Therefore the earlier proposition of C_s symmetry of C_2H_2 on nickel can be ruled out on the basis of the Teller–Redlich rule argument. We shall see in the upcoming section that further evidence for C_1 symmetry (no symmetry element) is available for this system. Therefore we conclude this section with the comment that the Teller–Redlich test with the modifications proposed here is—though not always significant—a useful test on a proposed assignment and one should not abstain from applying it.

4.5.3   Angular Profiles

A unique feature of electron energy loss spectroscopy is the availability of scattering processes with distinctively different sensitivities to modes of different symmetries. Dipole-active modes are observed when the analyzer collects inelastic electrons emitted at an angle close to specular reflection. Energy losses resulting from impact scattering typically have a broad angular distribution. This angular distribution, though apparently without sharp structures, still exhibits significant features which depend on the details of the multiple scattering and the normal coordinates of the vibrational modes. At this point the theory of impact scattering is still in an early stage of development. Likewise, the experimental material on the angular distribution of intensities is scarce and only a few systems have been studied. But even without a detailed understanding, studies of angular distributions of intensities are useful in several ways which we shall briefly discuss here.

The angular profile of energy losses depends on the normal coordinates of the vibrational modes. The intensity of the inelastic scattering from one particular mode may be larger at a certain scattering angle and weaker at another. Consequently, when spectra are recorded at different scattering angles, the appearance of the spectra may differ substantially, since at different scattering angles, they may be dominated by different modes. Figure 4.15 shows an example of two spectra of C_2H_2 on Ni(111), one spectrum obtained in specular reflection, the other with the monochromized beam rotated by 19.5°. We see that modes which are barely detectable in specular reflection are easily distinguished off specular and vice versa. Therefore angular profiles allow us to detect the complete set of eigenmodes even with the moderate resolution of electron energy loss spectroscopy. This is a substantial advantage over surface infrared spectroscopy.

Angular profiles also provide helpful information for the assignment of modes. In Fig. 4.16 the angular profiles of the modes of C_2H_2 and C_2D_2 are plotted. The profiles are already grouped into carbon- and hydrogen-associated modes. Clearly the profiles exhibit remarkable similarities when one compares the profiles of the same normal mode for C_2D_2 and C_2H_2. Using key features of the profiles such as intensity of dipole scattering, intensity of impact scattering, and the existence of a minimum in the transition region, one is able to pair most of the C_2H_2 modes to the C_2D_2 modes even without considering whether the frequency ratio between the two modes would be reasonable. Outstanding examples are the CC mode which has the largest dipole moment in either case and the CH mode which appears to have no di ole contribution to the scattering. Nevertheless, since all bending modes show the dipole enhancement and are therefore dipole active, the CH stretching mode cannot be a dipole forbidden mode. Furthermore, as all bending modes of hydrogen are dipole active, the surface point groups of C_2H_2 cannot have an element of symmetry, as we have discussed already in the previous section.

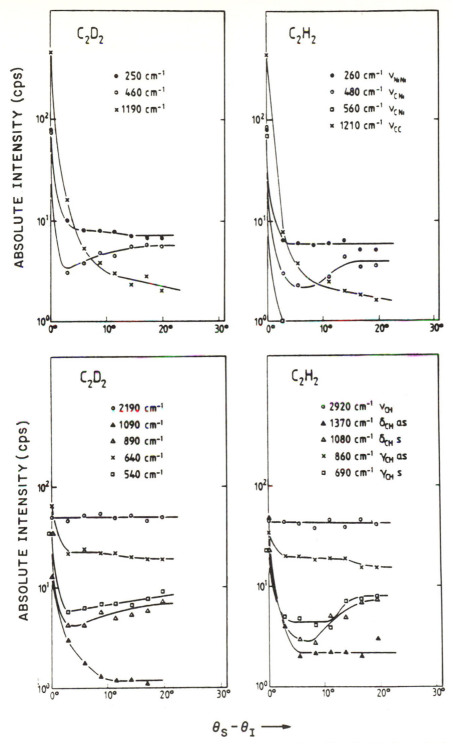

Fig. 4.16 Absolute intensities of energy losses as a function of $\theta_S - \theta_I$ with the angle of the scattered electrons kept fixed at $\theta_S = 77°$. The primary energy is 4 eV. The angular profiles for C_2D_2 and C_2H_2 are grouped by using the same symbol for the data points in corresponding eigenmodes (from Ibach and Lehwald [30], used with permission).

Despite the remarkable similarities in the profiles of corresponding modes in Fig. 4.16, there are also differences. After all, the normal coordinates of the modes in C_2D_2 and C_2H_2 are not identical. In hydrocarbons with CH_2 and CH_3 groups, mode coupling between the CH bending modes and the carbon modes can become more important. The amplitude of vibration for the carbon atoms in then substantially different in the deuterium and the proton species. More experimental and theoretical exploration is needed to determine whether comparison of angular profiles in such cases can still serve as a means for mode assignment.

REFERENCES

1. M. Born and K. Huang, "Dynamical Theory of Crystal Lattices," p. 166 ff. Oxford Univ. Press, London and New York, 1954.
2. A. A. Maradudin, *in* "Dynamical Properties of Solids (G. K. Horton and A. A. Maradudin, eds.), Vol. 1. North-Holland Publ., Amsterdam, 1973.
3. E. Brovman and Yu. Kagan, *in* "Dynamical Properties of Solids" (G. K. Horton and A. A. Maradudin, eds.), Vol. 1. North-Holland Publ., Amsterdam, 1973.
4. A. S. Barker and A. J. Sievers, *Rev. Mod. Phys.* **47**, *Suppl.* 2 (1975).
5. J. E. Black, *Surface Sci.* **100**, 555 (1980); *"Vibrations at Surfaces"* (R. Caudano, R. Gilles, A. A. Lucas, eds.). Plenum, New York, 1981.
6. J. E. Black, B. Laks, and D. L. Mills, *Phys. Rev. B* **22**, 1818 (1980).
7. T. H. Upton and W. A. Goddard III, "ISISS 1979, Surface Science: Recent Progress and Perspectives." Chem. Rubber Publ. Co., Boca Raton, Florida, 1980; *Phys. Rev. Lett.* **46**, 1635 (1981).
8. H. Ibach and D. Bruchmann, *Phys. Rev. Lett.* **44**, 36 (1980).
9. S. Andersson, *Surface Sci.* **79**, 389 (1979); *Chem. Phys. Lett.* **55**, 185(1978).
10. W. Ho, N. Di Nardo, and E. W. Plummer, *J. Vac. Sci. Technol.* **17**, 134 (1980).
11. M. Tinkham, "Group Theory and Quantum Mechanics." McGraw-Hill, New York, 1964. In Chapter 7, group theory is applied to the normal modes of polyatomic molecules.
12. S. Lehwald, H. Ibach, and J. E. Demuth, *Surface Sci.* **78**, 577 (1978).
13. L. Backx, C. P. M. de Groot, and P. Biloen, *Surface Sci.* **6**, 256 (1980).
14. N. V. Richardson and A. M. Bradshaw, *Surface Sci.* **88**, 255 (1979).
15. W. Erley, H. Wagner, and H. Ibach, *Surface Sci.* **80**, 612 (1979).
16. S. Lehwald, F. T. Yates, and H. Ibach, *Proc. IVC8-ICSS4-ECOSS3, Cannes, Suppl. Rev. Le Vide, Les Couches Minces*, **201**, 221 (1980).
17. B. A. Sexton, *Surface Sci.* **94**, 435 (1980).
18. H. Ibach and S. Lewald, *Surface Sci.* **91**, 187 (1980).
19. G. Herzberg, "Molecular Spectra and Molecular Structure," Vol. II, Infrared and Raman Spectra of Polyatomic Molecules. Van Nostrand–Reinhold, New York, 1945.
20. H. Ibach, *Phys. Rev. Lett.* **27**, 253 (1971).
21. G. Dalmai-Imelik, F. C. Bertolini, and F. Rousseau, *Surface Sci.* **63**, 67 (1977).
22. H. Ibach, ed, *in* "Electron Spectroscopy for Surface Analysis" (Topics in Current Physics 4), p. 4. Springer-Verlag, Berlin, Heidelberg, New York, 1977.
23. T. S. Rahman, J. E. Black, and D. L. Mills, *Phys. Rev. B* **25** (1982).
24. T. Shimanouchi, "Tables of Vibrational Frequencies," Consolidated Vol. I, NSRDS-NBS 39; Volume II, *J. Phys. Chem. Rev. Data* Vol. 6, No. 3, 993 (1977).

25. K. Nakamoto, "Infrared Spectra of Inorganic and Organic Coordination Compounds." Wiley, New York, 1963.

26. L. M. Sverdlov, M. A. Korner, and E. P. Krainov, "Vibrational Spectra of Polyatomic Molecules." Wiley, New York, 1974.

27. E. Maslowski, "Vibrational Spectra of Organometallic Compounds." Wiley, New York, 1977.

28. H. Ibach and S. Lehwald, *J. Vac. Sci. Technol.* **15**, 407 (1978).

29. S. Lehwald and H. Ibach, *Surface Sci.* **89**, 425 (1979).

30. H. Ibach and S. Lehwald, *J. Vac. Sci. Technol.* **18**, 625 (1981).

31. J. E. Demuth, and H. Ibach, *Surface Sci.* **85**, 365 (1979).

CHAPTER 5

VIBRATIONS AT CRYSTAL SURFACES; ORDERED ADSORBATE LAYERS AND THE CLEAN SURFACE

5.1 GENERAL REMARKS

In Chapter 4, we examined the vibrational motion of an atom or molecule adsorbed on the surface of a crystal, with emphasis on the relationship of the motions of the adsorbate to those of the same species in the gas phase. We saw that in a variety of ways, the surface profoundly modifies the gas-phase vibrational spectrum; the study of these new features provides information on the geometry of the adsorption site. While the theoretical discussions in Chapter 4 supposed the adsorbate to be a single entity, isolated from its neighbors, this is clearly an idealization. In any of the surface-sensitive experimental probes of surface vibrations, the surface must be covered with a finite density of adsorbates before a signal may be observed.

There are, then, two possibilities. One may probe the vibrational properties of an array of adsorbates that form an ordered overlayer, commensurate with the underlying substrate if one is in the limit where the adsorbate is chemisorbed. Or, if the adsorbate coverage is either low or such that an ordered configuration does not occur, the adsorbate layer will be disordered. In the low-coverage limit, the mean distance between adsorbates is large, and clustering effects will be unimportant, if island formation is not encountered. Here the analysis of the previous chapter applies directly, but only a selected set of systems may be probed in this low-coverage limit by the vibrational spectroscopies we have discussed, simply because the signal becomes weak. At high coverages, or if island formation is encountered, short-range order within the adsorbate layer influences the measured spectrum and the interpretation of data taken on such systems is difficult, if one is interested in the simple and basic questions that have attracted our attention so far. It is much simpler to interpret data taken under circumstances where the adsorbate layer is ordered, and the surface thus exhibits a well-defined low-energy electron diffraction pattern. We then have clear information in hand on some basic features of the surface geometry, and the nature of the short-range order within the adsorbate layer is understood. In many circumstances, particularly when one probes vibrational modes of an adsorbate with frequencies high compared to those of the phonon modes of the substrate, the isolated adsorbate picture of Chapter 4 continues to apply to very good approximation. One finds the frequencies shift only very little with coverage and thus are affected only slightly by interaction with nearby absorbates. This is true in a large and important range of circumstances. At the same time, we are led to inquire, in principle, into the description of the normal modes of an ordered layer of adsorbates, coupled also to the substrate atoms. The question becomes crucial when the characteristic vibrational frequencies of the adsorbate lie close to, or possibly within, the phonon bands of the host. This chapter is devoted to this and related topics; before we are through, we shall appreciate that these are unique aspects of this topic directly accessible to experimental study by electron energy loss spectroscopy.

While our interest until now has been directed primarily toward the description of the vibrational modes of molecules and atoms adsorbed on the crystal surface, the vibrational motion of atoms in and near the surface layer of clean crystal surfaces is also of great intrinsic interest. By electron energy loss spectroscopy, a number of clean crystal surfaces have been explored, and recent work examines features in the electron energy loss spectrum which lie well below the maximum phonon frequency of the substrate, for adsorbate covered surfaces. This leads us to explore not only the discussion of high-frequency vibrations in ordered adsorbate layers, but the general description of the coupled motion of adsorbate–substrate combinations at frequencies well below the maximum phonon frequency of the substrate. A special case is, then, the clean crystal surface.

Before we begin, we must introduce some notation and coventions, since the geometry is more complex than that considered in Chapter 4.

5.2 FORMAL DESCRIPTION OF THE VIBRATIONS OF ORDERED ADSORBATE LAYERS ON CRYSTALS

5.2.1 Description of the Surface Geometry and General Theory

We shall require a notation more explicit than that used in Chapter 4. In the course of setting up the notation to be employed here, we shall encounter several features of surface geometry that enter a number of discussions of the physics of the crystal surface.

We begin by considering an ordered layer of adsorbate molecules, and assume for simplicity that all molecules sit on equivalent sites. It then follows that our attention is confined to adsorbate layers commensurate with the underlying substrate. For chemisorbed systems, ordered overlayers are mostly commensurate. For weakly physisorbed systems, lateral interactions between adsorbates are of crucial importance, with the consequence that the overlayers need not be commensurate. The physics of such intriguing systems is complex, and lies beyond the scope of the present volume.

The adsorbate layer may be regarded as a finite number n of planes of nuclei, with each plane parallel to the crystal surface. For example, if we consider an ordered array of CO molecules adsorbed with the carbon atom bound to the substrate, we have first one plane of carbon nuclei and then one plane of oxygen nuclei as the outermost planes of the surface complex. We may always define a two-dimensional unit cell for any ordered array of adsorbates. The location of each nucleus in the adsorbate layer is then specified as follows. Let $l_\parallel$ be a vector which lies parallel to the surface, directed to a reference point in the unit cell within which the nucleus of interest is located. Then the particular plane that contains the nucleus is labeled l_z. There may possibly be several nuclei within this unit cell and in this plane, so we need one more label κ that specifies which of the sites within the plane l_z is of concern. Finally, the position of the nucleus is given by the vector $\mathbf{R}_0(l_\parallel l_z \kappa)$. This notation is complex, but necessary for what follows.

If the overlayer is commensurate with the substrate, as we assume here, then the two-dimensional unit cell appropriate to the adsorbate layer is necessarily a unit cell of the substrate layers parallel to the surface as well. However, it need not be the smallest unit cell that may be chosen for the substrate layers. We illustrate this in Fig. 5.1 for a particular case. Figure 5.1 shows the geometrical arrangement of atoms in the (100) surface of an fcc crystal, under the assumption that the atoms in the surface sit in positions

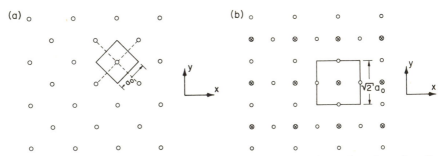

Fig. 5.1 (a) The two-dimensional unit cell associated with the (100) surface of an fcc crystal. (b) The encircled crosses show the location of nuclei associated with a $c(2 \times 2)$ overlayer formed on the (100) surface of an fcc crystal. This unit cell has twice the area of that shown in (a).

identical to those occupied by bulk atoms within a (100) plane. We also show one possible choice of unit cell for this plane of atoms. The unit cell contains one substrate atom, and has area a_0^2, where a_0 is the separation of two nearest-neighbor atoms. In Fig. 5.1b, the same layer of substrate atoms is sketched (open circles), and we also show superimposed on the substrate an adsorbate layer with the $c(2 \times 2)$ structure (encircled crosses). This structure is commonly encountered in studies of atoms and small molecules adsorbed on the (100) surface of the fcc transition metals. Each adsorbate atom sits on a site of fourfold symmetry, and we again indicate a choice of unit cell for the structure. The area of the new unit cell is now $2a_0^2$, and each unit cell contains two adsorbate atoms; there are twice as many substrate atoms as adsorbate atoms. The basic two-dimensional unit cell of the problem is clearly that of the adsorbate layer, under circumstances such as this where the unit mesh in the outermost layer is larger, though commensurate with that appropriate to the atomic planes deep in the bulk.

If one examines the atomic arrangement in the surface layer of a single crystal with perfectly clean surface, it is not at all necessary that the atomic arrangement in the outermost layers be identical to that in the bulk. The atoms may shift off the lattice positions appropriate to the bulk crystal, with the consequence that the symmetry of the surface layer is lower than the bulk. This phenomenon is referred to as surface reconstruction; low-energy electron diffraction studies of reconstructed surfaces show additional Bragg spots between those appropriate to an atomic plane in the bulk [1]. The unit cell of the surface layer is once again larger than that associated with a bulk atomic plane, and once again it is this surface unit cell that is basic to the semi-infinite crystal. For example, the displacement of the atoms occurs not just in the surface alone, but necessarily extends down into the crystal. Even if the shift off the bulk position is very small, several atomic planes below the surface, it is in principle nonzero and the surface unit cell must be applied to each layer of the crystal as a consequence.

We discuss next the formal description of the vibrational normal modes of the adsorbate–substrate complex just described. For the purposes of this section, it will not be necessary to distinguish between a layer of adsorbate nuclei and a layer of nuclei associated with the substrate. After the formal description of the problem is completed, we shall then turn to a sequence of specific examples that illustrate the nature of the vibrational spectrum in detail.

The vibrational Hamiltonian of the array is now written, with attention to the discussion of the previous section,

$$\mathscr{H} = \sum_{l_{\parallel}l_z\kappa} \frac{P^2(l_{\parallel}l_z\kappa)}{2M(l_z\kappa)} + \frac{1}{2}\sum_{\alpha\beta}\sum_{\substack{l_{\parallel}l_z\kappa \\ l_{\parallel}'l_z'\kappa'}} \Phi_{\alpha\beta}(l_{\parallel}l_z\kappa; l_{\parallel}'l_z'\kappa')u_{\alpha}(l_{\parallel}l_z\kappa)u_{\beta}(l_{\parallel}'l_z'\kappa'). \quad (5.1)$$

This is simply the vibrational Hamiltonian that formed the basis of the discussion of Chapter 4 rewritten in notation appropriate to the present problem. Given a model of the interatomic interactions, it is necessary to find the equilibrium position of the various atomic planes first. Near the surface, the separation between adjacent atomic planes of substrate nuclei need not be the same as in the bulk, for example. Dobrzynski and Maradudin have presented a detailed analysis of this question, including the temperature variation of the spacing between adjacent atomic planes near the surface, for a model of a bcc crystal [2]. Also, the position of the nuclei within a given plane must be determined. After this task has been completed, following the methods outlined in Chapter 4, an appropriate two-dimensional unit cell may be chosen, and the vibrational potential energy is, to lowest order, a quadratic form in the displacement $u_\alpha(l_{||}l_z\kappa)$ away from the equilibrium position.

There is one principal difference between the vibrational Hamiltonian in Eq. (5.1), and that used in Chapter 4. This is that $\Phi_{\alpha\beta}(l_{||}l_z\kappa, l'_{||}l'_z\kappa)$ necessarily depends on only the *difference* $l_{||} - l'_{||}$, for the ordered structures of interest here. Quite clearly, for any array with order of a periodic character in the two dimensions parallel to the surface, the interaction between the nucleus $(l_{||}l_z\kappa)$, and that at $(l'_{||}l'_z\kappa')$ depends on their *relative* positions, along a plane parallel to the surface. We shall exploit this fact shortly, to simplify the vibrational analysis.

In Chapter 4, we viewed the basic theoretical problem as one of diagonalizing the vibrational Hamiltonian directly, with the Hamiltonian a quadratic form in both the momenta and the displacements of the atoms. Here we shall consider the problem from the point of view of classical mechanics; the two approaches yield precisely the same eigenvalue spectrum and the same eigenvectors, and the formulas obtained from one are readily translated into the second by simple reinterpretation of the quantities involved. We use the classical approach, and state the prescription for transforming the expressions over to their quantum-mechanical form later.

It is straightforward to find the equation of motion of the displacement $u_\alpha(l_{||}l_z\kappa)$ from Eq. (5.1). We have

$$M(l_z\kappa)\ddot{u}_\alpha(l_{||}l_z\kappa) = - \sum_{\beta l_{||}'l_z'\kappa'} \Phi_{\alpha\beta}(l_{||}l_z\kappa; l'_{||}l'_z\kappa')u_\beta(l'_{||}l'_z\kappa'). \qquad (5.2)$$

We shall seek eigensolutions of Eq. (5.2) in which $u_\alpha(l_{||}l_z\kappa)$ varies with time like $\exp(-i\omega t)$, and we also transform to the mass weighted coordinates $\xi_\alpha(l_{||}l_z\kappa)$ defined by the relation

$$u_\alpha(l_{||}l_z\kappa) = [M(l_z\kappa)]^{-1/2}\xi_\alpha(l_{||}l_z\kappa). \qquad (5.3)$$

Then Eq. (5.2) assumes the form

$$\omega^2\xi_\alpha(l_{||}l_z\kappa) - \sum_{\beta l_{||}'l_z'\kappa'} D_{\alpha\beta}(l_{||}l_z\kappa; l'_{||}l'_z\kappa')\xi_\beta(l'_{||}l'_z\kappa') = 0, \qquad (5.4)$$

where we have introduced the dynamical matrix

$$D_{\alpha\beta}(l_{||}l_z\kappa;l'_{||}l'_z\kappa') = \frac{\Phi_{\alpha\beta}(l_{||}l_z\kappa;l'_{||}l'_z\kappa')}{[M(l_z\kappa)M(l'_z\kappa')]^{1/2}}, \tag{5.5}$$

which has built in the symmetry property

$$D_{\alpha\beta}(l_{||}l_z\kappa;l'_{||}l'_z\kappa') = D_{\alpha\beta}(l'_{||}l'_z\kappa';l_{||}l_z\kappa). \tag{5.6}$$

Until now, there is no difference from the physical point of view between the present discussion and that presented in Chapter 4, although the notation has been changed. We now exploit the fact that the dynamical matrix depends only on the difference between $l_{||}$ and $l'_{||}$. We do this by noting that when the dynamical matrix is endowed with this property, Eq. (5.4) admits solutions with the form

$$\xi_\alpha(l_{||}l_z\kappa) = \exp[i\mathbf{Q}_{||} \cdot \mathbf{R}_0(l_{||}l_z\kappa)]e_\alpha(\mathbf{Q}_{||};l_z\kappa), \tag{5.7}$$

where $\mathbf{Q}_{||}$ is a two-dimensional wave vector that is parallel to the crystal surface. If this form is inserted into Eq. (5.4), then one finds

$$\omega^2 e_\alpha(\mathbf{Q}_{||};l_z\kappa) - \sum_{\beta l'_z\kappa'} \left\{ \sum_{l_{||}'} D_{\alpha\beta}(l_{||}l_z\kappa;l'_{||}l'_z\ \kappa') \right.$$

$$\times \left. \exp[-i\mathbf{Q}_{||} \cdot (\mathbf{R}_0(l_{||}l_z\kappa) - \mathbf{R}_0(l'_{||}l'_z\kappa'))] \right\} e_\beta(\mathbf{Q}_{||};l'_z\kappa') = 0. \tag{5.8}$$

Because the dynamical matrix depends on only the combination $l_{||} - l'_{||}$, the quantity in braces in Eq. (5.8) is necessarily independent of $l_{||}$. We define

$$d_{\alpha\beta}(\mathbf{Q}_{||};l_z\kappa,l'_z\kappa') = \sum_{l_{||}'} D_{\alpha\beta}(l_{||}l_z\kappa;l'_{||}l'_z\kappa')$$

$$\times \exp[-i\mathbf{Q}_{||} \cdot (\mathbf{R}_0(l_{||}l_z\kappa) - \mathbf{R}_0(l'_{||}l'_z\kappa'))]. \tag{5.9}$$

Then noting that for each choice of $\mathbf{Q}_{||}$ we obtain a whole spectrum of eigenvalues, we replace ω^2 by $\omega_s^2(\mathbf{Q}_{||})$, where s is a subscript which labels a particular eigenfrequency associated with the wave vector $\mathbf{Q}_{||}$. As the chapter proceeds, we shall discuss a number of general features of this eigenvalue spectrum. Equation (5.8) then becomes

$$\omega_s^2(\mathbf{Q}_{||})e_\alpha^{(s)}(\mathbf{Q}_{||};l_z\kappa) - \sum_{\beta l_z'\kappa'} d_{\alpha\beta}(\mathbf{Q}_{||};l_z\kappa,l'_z\kappa')e_\beta^{(s)}(\mathbf{Q}_{||};l'_z\kappa') = 0. \tag{5.10}$$

The matrix $d_{\alpha\beta}(\mathbf{Q}_{||};l_z\kappa,l'_z\kappa')$ is Hermitean, in the sense that if we regard it as a matrix with each row labeled by the combination $(\alpha l_z\kappa)$ and each column by $(\beta l'_z\kappa')$, then Eq. (5.6) leads to the property

$$d_{\alpha\beta}(\mathbf{Q}_{||};l_z\kappa,l'_z\kappa') = d_{\beta\alpha}(\mathbf{Q}_{||};l'_z\kappa',l_z\kappa)^*. \tag{5.11}$$

If the two-dimensional unit cell for the structure has been set up as described in the previous subsection, then associated with this is a two-dimensional Brillouin zone which can be constructed by means of methods standard in solid state physics [3]. We also have associated with the structure a set of reciprocal lattice vectors $\mathbf{G}_{||}$. By analyzing the structure of Eqs. (5.7) and (5.8), it follows that all distinct eigenfrequencies and eigenvectors of the structure are obtained by allowing $\mathbf{Q}_{||}$ to range over only the first Brillouin zone. The proof of this statement follows that presented in standard discussions of the lattice dynamics of crystals, and we omit it from the present text. Suppose we consider a semi-infinite crystal, with periodic boundary conditions applied to the two directions parallel to the surface. If there are N_s (two-dimensional) unit cells in the macroscopic area submitted to the periodic boundary conditions, it is again a standard result of solid state theory that there are precisely N_s values of $\mathbf{Q}_{||}$ allowed in the first Brillouin zone. Thus, the number of choices for $\mathbf{Q}_{||}$ equals the number of two-dimensional unit cells in the macroscopic crystal.

The fact that the wave vector $\mathbf{Q}_{||}$ parallel to the surface is a "good quantum number" greatly reduces the labor that must be expended in the analysis. We began with a solid that is, in effect, infinite in extent in the two directions parallel to the surface and semi-infinite in the third. The form of the eigenvector introduced in Eq. (5.7) eliminates two of these three dimensions. What remains in Eq. (5.10) is a problem isomorphic to the lattice dynamics of a one-dimensional linear array of molecules terminated at one end. One molecule is associated with each value of l_z, since we have possibly several atoms associated with each value of the index l_z.

The completes our formal description of the lattice dynamics of crystals in the presence of a surface, with an ordered adsorbate layer present. We must next examine the nature of the solutions to Eq. (5.10), and then we must relate these to electron energy loss spectra taken on surfaces. We turn first to the clean crystal, then to those covered with adsorbates.

5.2.2 The Eigenmodes of a Semi-Infinite Crystal

Here we examine the nature of the vibrational modes of a semi-infinite crystal with no adsorbate layer present, and with atoms in the surface located in the same position as those in the bulk. That is, suppose the surface is not reconstructed. The principles noted here will be directly applicable to the discussion of the effect of an adsorbate layer, or of the reconstructed surface. We begin with some general remarks, then turn to a simple example.

Before the influence of the surface on the lattice dynamics of the crystal can be examined, we need to review the lattice dynamics of the bulk. Here the normal modes of the crystal are the phonons familiar from solid state physics. (An authoritative and complete treatment of the lattice dynamics of crystals has been given by Maradudin *et al.* [4].) For each mode, the pattern of displacements is wavelike in all three spatial dimensions, not just in two of them as in Eq. (5.10). If there are z atoms per unit cell, then for each three-dimensional wave vector $\mathbf{Q}$, we have $3z$ phonon normal modes. We shall denote their frequencies by $\omega^{(j)}(\mathbf{Q})$, where the index j ranges from 1 to $3z$. Three of the modes, with $j = 1$, 2, and 3, are referred to as the acoustical branch of the phonon spectrum. These branches have the property that as the wave vector $\mathbf{Q}$ vanishes, so does the frequency $\omega^{(j)}(\mathbf{Q})$. For the acoustical branches, in the limit of very small $\mathbf{Q}$, we have $\omega^{(j)}(\mathbf{Q}) = c_j(\hat{Q})Q$, where $c_j(\hat{Q})$ is the velocity of sound, which in general is a function of the direction but not the magnitude of the wave vector $\mathbf{Q}$, in the limit $\mathbf{Q} \to 0$.

A description of the lattice dynamics of the bulk crystal is contained implicitly in the previous section. Suppose the crystal does not have a surface, but has infinite extent in the z direction. Then Eq. (5.10) continues to apply and has solutions of a particularly simple form. We may seek solutions where the eigenvector $e_\beta^{(s)}(\mathbf{Q}_{||}; l_z\kappa)$ has a wavelike variation in the z direction. Upon rearranging subscripts for convenience, we make in Eq. (5.10) the replacement

$$e_\beta^{(s)}(\mathbf{Q}_{||}; l_z\kappa) \to e_\beta(\mathbf{Q}j; \kappa) \exp[iQ_z R_{0z}(l_{||}l_z\kappa)], \qquad (5.12)$$

where $\mathbf{Q} = \mathbf{Q}_{||} + \hat{z}Q_z$ is the three-dimensional wave vector discussed in the previous section. Note that $R_{0z}(l_{||}l_z\kappa)$ is independent of $l_{||}$. The subscript s on the left-hand side of Eq. (5.12) has, in effect, been replaced by the combination $(Q_z j)$, and the notation has been changed to conform more closely to that found in the literature on crystal lattice dynamics. With $\omega_s(\mathbf{Q}_{||})^2$ replaced by $\omega(\mathbf{Q}j)^2$, the eigenvalue equation becomes

$$\omega(\mathbf{Q}j)^2 e_\alpha(\mathbf{Q}j; \kappa) - \sum_{\beta\kappa'} d_{\alpha\beta}^{(\infty)}(\mathbf{Q}; \kappa\kappa')e_\beta(\mathbf{Q}j; \kappa') = 0, \qquad (5.13)$$

where $d_{\alpha\beta}^{(\infty)}(\mathbf{Q}; \kappa\kappa')$ is the dynamical matrix of bulk lattice dynamics, defined as

$$d_{\alpha\beta}^{(\infty)}(\mathbf{Q}; \kappa\kappa') = \sum_{l_z'\kappa'} d_{\alpha\beta}(\mathbf{Q}_{||}; l_z\kappa, l_z'\kappa')$$

$$\times \exp[-iQ_z(R_{0z}(l_{||}l_z\kappa) - R_{0z}(l_{||}'l'\kappa'))]. \qquad (5.14)$$

Our three-dimensional crystal is constructed from a sequence of planes parallel to each other.[1] Let their separation be d_0. One sees easily that two

---

[1] One frequently encounters crystals where adjacent planes are not equivalent. One may have a staked arrangement such as $ABABAB\cdots$, where A and B are topologically inequivalent, for example. Such a case may be fit into the present discussion by using the index l_z to refer to an AB pair, where $\kappa = 1$ refers to the A plane of atoms and $\kappa = 2$ refers to a B plane. More complex arrangements can be handled in a similar fashion.

values of Q_z that differ by $2\pi/d_0$, or an integral multiple of this quantity, produce precisely the same pattern of atomic displacements. Thus, we may confine our attention to only values of Q_z that lie between two planes parallel to the surface direction ($\mathbf{Q}_{\|}$ lies within such a plane), and which are separated by the distance π/d_0. It is conventional to choose one of these at $Q_z = \pi/d_0$, and the second at $Q_z = -\pi/d_0$.

With Q_z chosen as just described, and $\mathbf{Q}_{\|}$ confined to the two-dimensional Brillouin zone described earlier, the three-dimensional Brillouin zone chosen here has the shape of a truncated cylinder with height $2\pi/d_0$, and with cross-section shape determined by the choice of the two-dimensional zone. While this is a perfectly fine choice of a Brillouin zone, and it is the one most convenient for the present discussion, in many cases it also differs from the choice conventional in the lattice dynamics of crystals in three dimensions. The latter is the Wigner–Seitz unit cell of the reciprocal lattice, which in general is not a truncated cylinder, but a polyhedron of complex form. We illustrate this in Fig. 5.2, where for the fcc lattice we give the Brillouin zone constructed here, and the choice conventional in solid state physics. With a_0 the distance between nearest-neighbor sites in the fcc lattice, we have $d_0 = a_0/\sqrt{2}$.

We now turn to the nature of the normal modes of the semi-infinite lattice. After some general remarks, we turn to a study of a simple case, the fcc lattice with (100) surface and nearest-neighbor interactions of central force character.

There are two distinctly different classes of normal modes of the semi-infinite crystal lattice. (A discussion of the normal modes of a semi-infinite lattice more complete than that provided here has been given by Wallis [5].) The first are clearly related to the bulk phonons just discussed. A bulk phonon will propagate up to the surface and be reflected from it. Mathematically, this means that there are eigenvectors of the semi-infinite crystal which describe bulk phonons that propagate up to the surface, and the outgoing waves that reflect off the surface. In general, there are several reflected waves. We may see this as follows. Let $\mathbf{Q}$ be a three-dimensional wave vector, which lies in the three-dimensional Brillouin zone, constructed as in Fig. 5.2a. Consider a mode associated with branch $j = j_0$ of the bulk phonon spectrum incident on the surface, and we write its frequency as $\omega(\mathbf{Q}_{\|}, Q_z, j_0)$ for the moment. Quite generally we will have $\omega(\mathbf{Q}_{\|}, +Q_z, j_0) = \omega(\mathbf{Q}_{\|}, -Q_z, j_0)$, where the wave with wave vector $\mathbf{Q}_{\|} - \hat{z}Q_z$ is the reflected wave. The surface may be viewed as a perturbation that mixes together these two bulk waves, each of which is degenerate. But there are in general other phonon branches which contain modes degenerate with the incident wave. For $j \neq j_0$, we may find $\omega(\mathbf{Q}_{\|}, Q_z, j_0) = \omega(\mathbf{Q}_{\|}, Q_z', j)$, $Q_z' \neq Q_z$. Reflected off the surface will be several waves; all degenerate modes of the bulk phonon

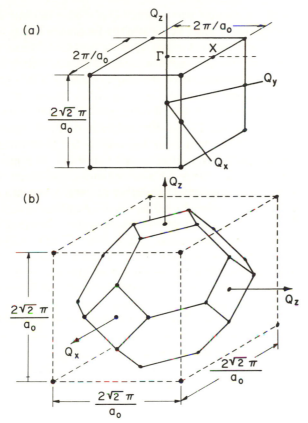

Fig. 5.2   Two choices of three-dimensional Brillouin zone for the fcc lattice. Here a_0 is the distance between nearest-neighbor atoms. (a) The choice convenient for use in the analysis of the semi-infinite geometry. (b) The choice conventional in solid state theory.

spectrum (with the same wave vector parallel to the surface) will be mixed by the perturbation provided by the surface. In the long-wavelength limit, where the crystal may be approximated by a continuum and the theory of elasticity serves to describe the vibrational modes of it, an excellent compact discussion of the reflectivity problem has been given by Landau and Lifshitz [6]. The discussion proceeds along very similar lines for the discreet lattice.

The second class of mode encountered in the discussion of surface lattice dynamics has no counterpart in the lattice dynamics of the bulk crystal. These are surface phonons, where in the two directions parallel to the surface, the atomic displacements have a wavelike nature as in Eq. (5.7), but as one moves into the crystal along a direction perpendicular to the surface, the eigenvectors fall to zero exponentially with l_z. If we consider a surface phonon with some particular wave vector $\mathbf{Q}_{||}$, then it is clear on physical grounds that its frequency must lie *outside* the frequency bands allocated to the bulk phonons with wave vectors $\mathbf{Q}$ with projection onto the surface plane equal to $\mathbf{Q}_{||}$. If this were not the case, then upon setting atoms in or near the surface into motion, with the displacement pattern phased from one two-dimensional unit cell to the next according to the rule in Eq. (5.7), one would find that the energy would propagate off into the crystal interior in the form of bulk phonons. The vibrational excitation will remain localized near the surface only if its frequency lies *outside* the bulk phonon bands, unless the symmetry of the excitation is such that it fails to excite vibrations of the substrate. There are diverse classes of surface phonons that are encountered, and we shall encounter a number of them in the discussion that follows. Once again, we refer the reader to the excellent review article by Wallis [5], which presents a discussion more complete than that found here.

We now turn to a specific example, the fcc crystal with nearest-neighbor forces. For this case, there is only one atom per unit cell, so in fact the indices κ and κ' are not required. If we have nearest-neighbor coupling only, then the only nonzero matrices $d_{\alpha\beta}(\mathbf{Q}_{||}; l_z, l'_z)$ defined in Eq. (5.9) are those with $l'_z = l_z \pm 1$. If we take the nearest-neighbor interactions to have central force character, then the form of $D_{\alpha\beta}(l_{||}l_z, l'_{||}l'_z)$ is given in Eq. (4.15); here the first derivative $\varphi'(a_0)$ vanishes, if the nearest-neighbor separation a_0 is such that the potential energy of the model crystal vanishes. Then if we introduce $k_0 = 2\varphi''(a_0)/M$, we have for the nonzero matrices that enter our discussion, with $q_x = Q_x a_0/\sqrt{2}$ and $q_y = Q_y a_0/\sqrt{2}$, and also with $l_z = 1$, the surface layer and the crystal placed in the upper half-space,

$\mathbf{d}(\mathbf{Q}_{||}; l_z l_z)$

$$= k_0 \begin{pmatrix} 2-\cos(q_x)\cos(q_y)-\tfrac{1}{2}\delta_{l_z,1} & +\sin(q_x)\sin(q_y) & 0 \\ +\sin(q_x)\sin(q_y) & 2-\cos(q_x)\cos(q_y)-\tfrac{1}{2}\delta_{l_z,1} & 0 \\ 0 & 0 & 2-\delta_{l_z,1} \end{pmatrix},$$

$$(5.15a)$$

$\mathbf{d}(\mathbf{Q}_{||}; l_z, l_z+1)$

$$= -\frac{k_0}{2} \begin{pmatrix} \cos(q_x) & 0 & +i\sin(q_x) \\ 0 & \cos(q_y) & +i\sin(q_y) \\ +i\sin(q_x) & +i\sin(q_y) & \cos(q_x)+\cos(q_y) \end{pmatrix},$$

$$(5.15b)$$

$$\mathbf{d}(\mathbf{Q}_{||}; l_z, l_z-1) \equiv \mathbf{d}^*(\mathbf{Q}_{||}; l_z, l_z+1).$$

$$(5.15c)$$

In Eq. (5.15a), the matrix elements on the diagonal of the matrix assume values different when $l_z = 1$ and we are in the surface layer. This is because the atoms in the surface layer have only eight nearest neighbors, rather than the 12 appropriate to a lattice site in the bulk of the crystal.

Even for the simple model described in the matrices just set down, a discussion of the vibrational spectrum of the crystal, most particularly that of the surface region, requires resort to numerical methods. However, we shall consider a special point in the Brillouin zone for which simple analytic solutions can be obtained. This is the point X of Fig. 5.2a, where we have $q_x = q_y = \frac{1}{2}\pi$. We write out Eq. (5.13) explicitly for this point, in slightly more schematic notation that supresses reference to $\mathbf{Q}_{||}$ and the mode index s. It is convenient to write the equations not in terms of the eigenvector amplitudes $e_x(l_z)$ and $e_y(l_z)$, but instead the combinations $e_{\pm}(l_z) = e_x(l_z) \pm e_y(l_z)$, which describe motions parallel to or perpendicular to the line from Γ to X in Fig. 5.1a.

For $l_z \geq 2$, i.e., for atoms within the second layer of the crystal or deeper, the equations for the X point of the two-dimensional zone read

$$3k_0e_+(l_z) + ik_0[e_z(l_z + 1) - e_z(l_z - 1)] = \omega^2e_+(l_z) \qquad (l_z \geq 2), \quad (5.16a)$$

$$2k_0e_z(l_z) + i(k_0/2)[e_+(l_z + 1) - e_+(l_z - 1)] = \omega^2e_z(l_z) \qquad (l_z \geq 2), \quad (5.16b)$$

and then simply

$$k_0e_-(l_z) = \omega^2e_-(l_z) \qquad (l_z \geq 2). \tag{5.17}$$

For the surface layer $l_z = 1$, we have the special equations

$$\tfrac{5}{2}k_0e_+(1) + ik_0e_z(2) = \omega^2e_+(1), \tag{5.18a}$$

$$k_0e_z(1) + (i/2)k_0e_+(2) = \omega^2e_z(1), \tag{5.18b}$$

and

$$\tfrac{1}{2}k_0e_-(1) = \omega^2e_-(1). \tag{5.19}$$

The first task is to outline, for our particular choice of $\mathbf{Q}_{||}$, the frequency bands alloted to the bulk crystal phonons. This is done by studying the solutions of Eq. (5.16) deep in the bulk of the crystal. We begin by solving these.

Note that Eqs. (5.16a) and (5.16b) form a coupled pair of equations in the variables $e_z(l_z)$ and $e_+(l_z)$. We seek solutions of these in the form $e_z(l_z) = e_z \exp(iq_zl_z)$, $e_+(l_z) = e_+ \exp(iq_zl_z)$, where $q_z = Q_za_0/\sqrt{2}$, and consultation with Fig. 5.1a shows q_z ranges from $-\pi$ to $+\pi$. If these forms are inserted into Eqs. (5.16a) and (5.16b), we find two bands of bulk phonons, with frequencies $\omega_+^2(q_z)$, $\omega_-^2(q_z)$ given by

$$\omega_{\pm}^2(q_z) = \tfrac{5}{2}k_0 \pm \tfrac{1}{2}k_0[1 + 8\sin^2(q_z)]^{1/2}. \tag{5.20}$$

As q_z is varied through its allowed range, $\omega_+(q_z)$ assumes its maximum value of $2\sqrt{k_0}$ when $q_z = \pm\frac{1}{2}\pi$. This is in fact the maximum vibrational frequency of the model crystal. The minimum value of $\omega_+(q_z)$ is $\sqrt{3k_0}$, and this occurs when $q_z = 0$, or $\pm\pi$. Similarly, $\omega_-(q_z)$ assumes a maximum value of $\sqrt{2k_0}$ and a minimum value of $\sqrt{k_0}$.

We are left with Eq. (5.17), which shows that each plane vibrates in the direction perpendicular to the line from Γ to X, independently of the remaining atomic planes. These eigenmodes are analogous to those of a set of independent oscillators, each uncoupled to its neighbors. The frequency of the oscillation is $\sqrt{k_0}$.

So when $\mathbf{Q}_{\parallel}$ is chosen to be at the X point of the Brillouin zone, we have two frequency bands covered by bulk phonons, separated by a gap. The first band extends from $\sqrt{k_0}$ to $\sqrt{2k_0}$, with the oscillation of e_- character equal in frequency to the lower bound of this band, and we then have a high-frequency band that extends from $\sqrt{3k_0}$ to the maximum frequency of the crystal, $2\sqrt{k_0}$. Our next step is to explore the equations for solutions that describe the surface phonons discussed earlier.

From our earlier general discussion, we argued that any surface phonons which are associated with this value of $\mathbf{Q}_{\parallel}$ must lie outside the bulk phonon bands of the host crystal. Thus, we have three possible frequency regimes within which surface phonons may be found. These can be frequencies higher than $2\sqrt{k_0}$, in the gap between $\sqrt{2k_0}$ and $\sqrt{3k_0}$, or below $\sqrt{k_0}$.

In fact, on very general grounds, we may forget the frequency regime above $2\sqrt{k_0}$. This is so for the following reason. We may imagine, formally, forming our model crystal in the following fashion: Suppose we begin with a large, three-dimensional crystal, with periodic boundary conditions applied in all three directions. We then set to zero all force constants which describe coupling between the atoms in the (100) plane at $l_z = 1$, and its neighboring (100) plane at $l_z = 0$. In this way, we create two semi-infinite crystals, each with a (100) surface plane. There is a theorem of harmonic vibrations of arrays of masses known as Rayleigh's theorem [4, p. 356 ff] which states the following. If we have an array of masses coupled by harmonic springs, and we decrease the spring constant of one or more springs, then the frequency of *each* vibrational normal mode must decrease. Hence, in our surface problem, we can find no vibrational modes with a frequency above the maximum of the infinitely extended crystal, since we can always create two semi-infinite crystals by setting to zero certain selected force constants, as already described. Thus, any surface phonons we find for our model with $\mathbf{Q}_{\parallel}$ chosen at the X point of the Brillouin zone must lie either in the gap between $\sqrt{3k_0}$ and $\sqrt{2k_0}$, or they must lie below $\sqrt{k_0}$.

In practice, the formation of a surface may lead to substantial changes in electronic structure there. It is possible that some force constants in the surface can increase in value, compared to the equivalent bulk values. Model analyses of the surfaces of transition metals which allow for surface relaxation show how such increased force constants may come about, even in a model that treats the surface region as if a pair of atoms interact via the same two-body potential as in the bulk.[2] Thus, while Rayleigh's theorem forbids the appearance of surface phonons with frequency above the maximum bulk phonon frequency for the particular simple model explored here, changes in surface-force constants such as those already described may lead to surface modes above the maximum bulk vibrational frequency, for even the clean, unreconstructed surface. We return to the model under study, where this possibility cannot occur.

We first consider modes localized near the surface which emerge from the solution of the combination of Eqs. (5.16) and (5.17). Surface phonons have displacement localized near the surface, so we begin by seeking solutions to Eqs. (5.16), in which the pattern of displacements decays to zero exponentially as one enters the bulk of the crystal. Thus, for these modes, we have

$$e_+(l_z) = e_+ \exp[-\alpha l_z], \tag{5.21a}$$

$$e_z(l_z) = e_z \exp[-\alpha l_z], \tag{5.21b}$$

where, upon substituting into Eqs. (5.16) we find that the decay constant α must be chosen so that

$$\sinh(\alpha) = (\sqrt{2}k_0)^{-1}(3k_0 - \omega^2)^{1/2}(\omega^2 - 2k_0)^{1/2}. \tag{5.22}$$

We shall shortly apply a constraint on the solution that will lead to a second relation between α and ω, and thus uniquely determine the frequency of the surface phonon. Note that Eq. (5.22) may be rearranged to read

$$\omega^2 = \tfrac{5}{2}k_0 \pm \tfrac{1}{2}k_0(1 - 8\sinh^2\alpha)^{1/2}. \tag{5.23}$$

For a given value of α, we have two possible frequencies. Finally, from Eq. (5.16a), there is a constraint on the ratio of the amplitudes e_+ and e_z:

$$e_+/e_z = +2ik_0 \sinh(\alpha)/(3k_0 - \omega^2). \tag{5.24}$$

It is now convenient to divide the set of four equations in Eqs. (5.16) and (5.18) into two hierarchies. We may seek a solution for which $e_+(l_z)$ is nonzero only for odd numbered layers (including the surface layer $l_z = 1$), while $e_z(l_z)$ is nonzero in only the even numbered layers. Then Eq. (5.18b) is satisfied trivially, since both $e_z(1)$ and $e_z(2)$ vanish identically. A second distinctly and nondegenerate solution (nondegenerate only in the presence of the surface, which renders even- and odd-numbered layers inequivalent)

---

[2] A series of calculations of force constant changes in surfaces of model bcc lattices has been given by Cheng *et al.* [7]. These authors find some of the force constants are stiffened for certain surfaces.

has $e_z(l_z)$ nonzero only in odd-numbered layers, and $e_+(l_z)$ nonzero in the even-numbered layers.

We shall consider first the frequency regime between $\sqrt{2k_0}$ and $\sqrt{3k_0}$. Furthermore, we seek the solution of the hierarchy which has $e_+(l_z)$ nonzero in the surface, and in all odd-numbered layers. Then from Eq. (5.18a), we obtain a second constraint on the solution which reads, after a bit of manipulation,

$$e_+/e_z = +2ie^\alpha. \tag{5.25}$$

This relation combined with Eq. (5.24) provides us with a second relation between ω^2 and the attenuation constant α:

$$\omega^2 = \tfrac{5}{2}k_0 + \tfrac{1}{2}k_0 e^{-2\alpha}. \tag{5.26}$$

For this to be consistent with Eq. (5.23), we must necessarily choose the plus sign there, and furthermore α must be chosen as the solution of

$$e^{-2\alpha} = (1 - 8\sinh^2\alpha)^{1/2}. \tag{5.27}$$

The only nontrivial solution (i.e., solution with α positive and nonzero) of Eq. (5.27) is $\alpha = 0.2885$, and this gives for the frequency of the mode in question

$$\omega_{s_6} = 1.668\sqrt{k_0}, \tag{5.28}$$

which lies in the gap between $\sqrt{2k_0}$ and $\sqrt{3k_0}$, as desired. Following the earlier literature on surface lattice dynamics [8], we refer to this particular mode as the S_6 mode.

We next consider the frequency region below $\sqrt{k_0}$, where our general considerations once again allow a surface phonon to exist. In this regime, the right-hand side of Eq. (5.22) is in fact purely imaginary, so we have no solutions with α real. However, if we set

$$\alpha = i(\pi/2) + \chi, \tag{5.29}$$

then the equation for χ becomes

$$\cosh\chi = 2^{-1/2}(3k_0 - \omega^2)^{1/2}(2k_0 - \omega^2)^{1/2}. \tag{5.30}$$

For $\omega < \sqrt{k_0}$, the right-hand side of Eq. (5.30) is real and greater than unity, so we have one unique choice for χ for each frequency below $\sqrt{k_0}$.

We may carry out the analysis of the surface-phonon frequency in parallel with our earlier discussion. It turns out that the only solution of Eqs. (5.16) and (5.18) that describes a disturbance localized near the surface is the hierarchy which allows $e_z(l_z)$ to be nonzero in the surface and in all odd-numbered layers, with $e_+(l_z)$ nonzero in those that are even numbered. The consistency condition for χ becomes

$$\exp[2\chi] = (1 + 8\cosh^2 \chi)^{1/2}, \tag{5.31}$$

and we then have

$$\chi = 0.6351 \tag{5.32}$$

while the frequency of this mode, called the S_4 mode, is then

$$\omega_{S_4} = 0.8481\sqrt{k_0}. \tag{5.33}$$

So at this point, we have two surface-phonon modes of the structure when $Q_\parallel$ is at the X point of the Brillouin zone in Fig. 5.2a, one in each frequency regime allowed by our general considerations. There is in fact a third mode that emerges easily from the foregoing analysis. We saw that for our choice of $Q_\parallel$, the normal coordinate $e_-(l_z)$ describes an independent oscillation of each of the atomic layers, with all others at rest when one particular layer is excited. From Eq. (5.19), we see that the surface layer oscillates with frequency $\sqrt{\tfrac{1}{2}k_0}$, below that $\sqrt{k_0}$ appropriate to the bulk layers. This oscillation frequency of the surface layer lies in the gap below $\sqrt{k_0}$, and this mode is in fact a true surface phonon also. Thus, we have a third mode with frequency

$$\omega_{s_1}^2 = \sqrt{\tfrac{1}{2}k_0} = 0.7071\sqrt{k_0}. \tag{5.34}$$

In Fig. 5.3, we show the pattern of atomic displacements in the surface for each of the three modes just examined. As we have seen, for each, the displacements are confined to the outermost few atomic layers, in the sense that the displacement amplitudes decay to zero exponentially as one moves into the bulk of the crystal. As our solution shows, the surface motion associated with the S_6 mode excites motion in the second layer *perpendicular*

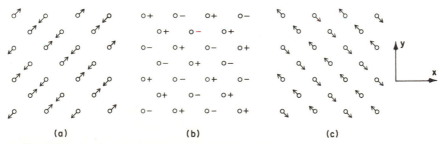

      (a)   (b)   (c)

Fig. 5.3   The displacements in the surface layer for the three surface modes at the X point of the two-dimensional Brillouin zone (Fig. 5.1a) for the fcc crystal with nearest-neighbor central force interactions and a (100) surface: (a) the S_6 mode; (b) the S_4 mode; (c) the S_1 mode. In (b) the displacements are perpendicular to the surface, so plus and minus signs are used to denote upward and downward displacements.

to the surface, and the second layer motion then excites parallel motion in the third layer that is in phase with that in the surface layer but which has amplitude smaller by the factor $\exp(-2\alpha) = 0.5616$. For the S_4 mode, there is a similar alternation of perpendicular and parallel motion, but with a $180°$ phase shift between the motion in the first and third layer [see the factor of $i\pi/2$ in Eq. (5.29)].

We have just examined the surface modes present for the model of the fcc crystal when $q_x = q_y = \frac{1}{2}\pi$. Quite clearly these modes are degenerate with those at the point $q_x = -\frac{1}{2}\pi$, $q_y = \frac{1}{2}\pi$. For the moment, we call these two points the X_1 and the X_2 points. Any linear combination of the eigenvector of a mode at X_1 with its equivalent at X_2 is also necessarily an eigenvector of the dynamical matrix, with eigenfrequency equal to that of the two equivalent modes. It will prove useful later to consider the eigenmotion produced by the sum and difference of the eigenvectors at X_1 and X_2, for the S_4 mode. These are illustrated in Fig. 5.4. We see breathing motions of alternate squares of Ni atoms, and these breathing motions drive the atoms in the layer just above into vertical motion.

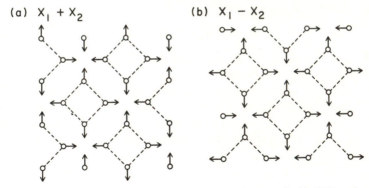

Fig. 5.4   The eigenmotion in the surface layer of the fcc crystal with (100) surface found by (a) adding and (b) substracting the eigenvectors of the S_4 mode at $q_x = q_y = \pi/2$ with that at $q_x = -\pi/2, q_y = \pi/2$.

Our discussion of surface phonons on the (100) surface of the fcc crystal has so far been confined to a single value of $\mathbf{Q}_\parallel$ in the two-dimensional Brillouin zone of the structure. We next discuss the dependence of the frequency of each mode with $\mathbf{Q}_\parallel$. It is possible to extend the analytic approach away from the special point we have considered, but in general the solution of the problem is cumbersome and not very enlightening, save in one limit to be described. Thus, we continue the discussion by presenting the results of numerical calculations, along the line from Γ to X, in the two-dimensional zone. Before the results are presented, it is useful to discuss the method that has been used, since it is one commonly employed in surface lattice dynamics.

The treatment presented here directly addresses the nature of the surface phonons in a semi-infinite crystal lattice. In the end, we found surface modes with displacement largely confined to other outermost few atomic layers. This suggests that from the point of view of a numerical analysis, it is not necessary to consider the entire semi-infinite crystal to generate such modes, as long as the outermost atomic layers are reliably modeled. Thus, one may consider a finite slab constructed from a finite number of layers N. The wave vector $\mathbf{Q}_{||}$ remains a "good quantum number" for such a slab, and in fact the eigenfrequencies and eigenvectors may be generated by solving the same equations, Eqs. (5.10), that formed the basis of the discussion of the present section. One simply allows l_z to range over a finite set of values, N in number.

If, as in the simple example explored here, there is one atom per unit cell in each plane of atoms, then to find all the eigenvectors and eigenvalues associated with a particular value of $\mathbf{Q}_{||}$, it is necessary to diagonalize a $3N \times 3N$ matrix. With present-day computers, this may be readily done for N in the range 20–25, or larger if desired. The surface modes may be identified simply from a visual inspection of the eigenvectors generated by the computer. Finally, the dispersion relation of the surface waves may be generated by scanning the two-dimensional Brillouin zone. The slab method has been widely applied in the study of surface lattice dynamics, since as one can see from our simple model of the fcc crystal, analytical methods are of limited utility in this area. We shall discuss other schemes for generating information on the lattice dynamics of crystal surfaces as we go along.

In Fig. 5.5, we show a breakdown of the modes of the fcc crystal with (100) surface, along the line from Γ to X. These results have been generated by the slab method, for a central force model dominated by nearest-neighbor couplings [8].[3] At the X point, we see the three surface phonons, S_1, S_4, and S_6, obtained earlier in the present section. The gap at X within which the S_6 mode lies does not extend throughout the zone, but closes as one moves toward Γ from X. The S_6 mode thus does not have a dispersion curve that extends throughout the zone, but is split off from the bulk phonon continuum only over a fraction of it. We see a similar behavior of the S_4 mode, and it is only the low frequency S_1 mode that extends throughout the two-dimensional Brillouin zone, including the region of very small $\mathbf{Q}_{||}$.

We turn next to a discussion of the properties of the S_1 mode in the limit $\mathbf{Q}_{||} \to 0$.

---

[3] Black, Campbell, and Wallis have examined the surface lattice dynamics of several transition metals, including Ni with a (100) surface, for a model with central force and also angle-bending force constants [9].

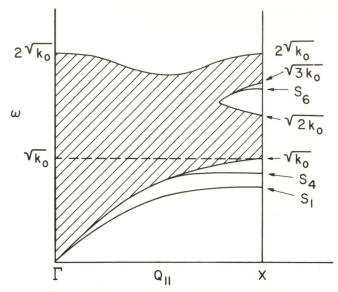

Fig. 5.5 The phonon dispersion curves for the fcc crystal with (100) surface, and nearest-neighbor central force interactions. We show the dispersion curves in the two-dimensional Brillouin zone (Fig. 5.1a) along the line from Γ to X. The cross-hatched region indicates the frequency bands alloted to the bulk phonons.

5.2.3 Surface Acoustical Phonons in the Long-Wavelength Limit; Rayleigh Waves

In the previous section, in our discussion of surface phonons, we saw that for wave vectors at the Brillouin zone boundary, the surface phonons we encountered described displacement patterns localized to only the outermost atomic layers. Quite clearly, a microscopic theory is required to describe such excitations.

In Fig. 5.5, we noted that one mode, the S_1 mode, survives as an identifiable surface mode even for small values of $Q_{||}$. While this mode describes a displacement pattern localized near the surface, in the long-wavelength limit $Q_{||}a_0 \ll 1$, the mode penetrates many layers into the crystal, quite in contrast to the behavior exhibited near the zone boundary. For such waves, where the displacement varies slowly on the scale of the lattice constant, a microscopic model is not required. Here the theory of elasticity if fully applicable, and the behavior of the mode is controlled by the elastic constants that may be obtained from bulk measurement. Such low-frequency, long-wavelength, and deeply pentrating surface acoustic waves have been well known in the literature on elasticity theory since the nineteenth century. These are frequently called Rayleigh waves, and recent years have seen the application of these waves in a number of practical devices. We begin by showing how our microscopic model yields a set of equations identical to those obtained from elasticity theory in the long-wavelength limit, then we discuss some general features of these modes. We again refer the reader to the review article by Wallis [5] for a more complete discussion. For the isotropic elastic continuum of elasticity theory, Landau and Lifshitz have provided a derivation of the dispersion relation of Rayleigh waves that is particularly compact. We now summarize this.

We proceed by examining the equation for the eigenvector amplitudes $e_x(l_z)$, $e_y(l_z)$, and $e_z(l_z)$ in the limit $Q_{||} \to 0$, for the model of the fcc crystal with central force interactions explored in the previous section. It will be convenient to express these in terms of Q_x and Q_y rather than the dimensionless wave-vector components q_x and q_y used previously. For $e_x(l_z)$, we have in the bulk of the crystal ($l_z \geq 2$), when $Q_x a_0 \ll 1$ and $Q_y a_0 \ll 1$,

$$k_0[1 + \tfrac{1}{4}a_0^2(Q_x^2 + Q_y^2)]e_x(l_z) + \tfrac{1}{2}k_0 a_0^2 Q_x Q_y e_y(l_z)$$
$$- \tfrac{1}{2}k_0[1 - \tfrac{1}{4}a_0^2 Q_x^2][e_x(l_z + 1) + e_x(l_z - 1)]$$
$$- i(a_0 k_0 Q_x/2\sqrt{2})[e_z(l_z + 1) - e_z(l_z - 1)] = \omega^2 e_z(l_z), \qquad (5.35)$$

while for the surface layer $l_z = 1$, we have

$$\tfrac{1}{2}k_0[1 + \tfrac{1}{2}a_0^2(Q_x^2 + Q_y^2)]e_x(1) + \tfrac{1}{2}k_0 a_0^2 Q_x Q_y e_y(1)$$
$$- \tfrac{1}{2}k_0[1 - \tfrac{1}{4}a_0^2 Q_x^2]e_x(2) - i(k_0/2\sqrt{2})a_0 Q_x e_z(2) = \omega^2 e_x(1). \quad (5.36)$$

Similar equations follow for $e_y(l_z)$ and $e_z(l_z)$; it will be sufficient for our purposes to write only the equations for $e_x(l_z)$ in explicit form.

Now if the disturbance varies slowly with l_z, we may regard $e_x(l_z)$ to be a smooth and slowly varying function of l_z, with a similar assumption applicable also to $e_y(l_z)$ and to $e_z(l_z)$. Then if we change variables from l_z, to the distance z measured from the surface, with

$$z = (a_0/\sqrt{2})(l_z - 1),$$

we have for the Cartesian component $e_\alpha(l_z \pm 1)$ the Taylor expansion

$$e_\alpha(l_z \pm 1) \to e_\alpha(z) \pm \frac{a_0}{\sqrt{2}}\frac{\partial e_\alpha}{\partial z} + \frac{a_0^2}{4}\frac{\partial^2 e_\alpha}{\partial z^2} + \cdots, \qquad (5.37)$$

so the difference equation in Eq. (5.35) may be replaced with the differential equation

$$\tfrac{1}{2}k_0 a_0^2 Q_x^2 e_x(z) + \tfrac{1}{2}k_0 a_0^2 Q_x[Q_y e_y(z) - i\,\partial e_z/\partial_z]$$
$$+ \tfrac{1}{4}k_0 a_0^2(Q_y^2 - \partial^2/\partial z^2)e_x(z) = \omega^2 e_x(z). \qquad (5.38)$$

While it may not be evident at first glance, in fact Eq. (5.38) is identical in form to the differential equation obeyed by the displacement vector **u(x)** in an elastic medium of cubic symmetry, with appropriate choice of the elastic constants c_{11}, c_{12}, and c_{44} that characterize the material. Consider such a crystal, and suppose its mass density is ρ. Then the x component of displacement, $u_x(\mathbf{x})$, obeys

$$\rho \ddot{u}_x = \frac{\partial}{\partial x}\left[c_{11}\frac{\partial u_x}{\partial x} + c_{12}\left(\frac{\partial u_y}{\partial y} + \frac{\partial u_z}{\partial z}\right)\right]$$

$$+ c_{44}\frac{\partial}{\partial y}\left(\frac{\partial u_x}{\partial y} + \frac{\partial u_y}{\partial x}\right) + c_{44}\frac{\partial}{\partial z}\left(\frac{\partial u_z}{\partial x} + \frac{\partial u_x}{\partial z}\right). \tag{5.39}$$

If we seek a solution of Eq. (5.39) for which

$$u_x(\mathbf{x}) = u_x(z)\exp[i\mathbf{Q}_\| \cdot \mathbf{x} - i\omega t], \tag{5.40}$$

with a similar form for the remaining Cartesian components, then after some rearrangement, Eq. (5.39) reads

$$\frac{c_{11}}{\rho}Q_x^2 u_x(z) + \frac{(c_{12} + c_{44})}{\rho}Q_x\left(Q_y u_y - i\frac{\partial u_z}{\partial z}\right)$$

$$+ \frac{c_{44}}{\rho}\left(Q_y^2 - \frac{\partial^2}{\partial z^2}\right)u_x(z) = \omega^2 u_x(z). \tag{5.41}$$

Equations (5.41) and (5.40) are identical, if we make the identification

$$c_{11} = (\rho/2)k_0 a_0^2 = (M/4\sqrt{2}a_0)\omega_M^2, \tag{5.42a}$$

$$c_{12} = c_{44} = (\rho/4)k_0 a_0^2 = (M/8\sqrt{2}a_0)\omega_M^2. \tag{5.42b}$$

The second statement on the right-hand side of Eqs. (5.42a) and (5.42b) follows upon noting the maximum phonon frequency $\omega_M = 2\sqrt{k_0}$, as we saw in Section 5.2.2.

A comparison of Eqs. (5.40) and (5.38) shows that in the limit of displacement patterns that vary slowly on the scale of a lattice constant, our microscopic lattice dynamical model passes smoothly over to the phenomenological theory of elasticity, with Eqs. (5.42a) and (5.42b) the expressions for the elastic constants provided by the microscopic model. It may be demonstrated on very general grounds that the equations of physically sensible lattice models pass smoothly over to the theory of elasticity in this limit [4]. Here we have examined only one very simple special case. It may be demonstrated that any model which utilizes a central force model of the interatomic couplings necessarily yields the relation $c_{12} = c_{44}$ for cubic crystals, provided each atom sits at a site of inversion symmetry and also that the crystal is subject to no external stress. In lattice dynamics, this is known as the Cauchy relation [4].

Since the nearest-neighbor central force model is frequently used to provide a crude, but quite adequate description of the phonon spectrum of a number of metals, most particularly the transition metals, it is useful to pause for a moment in our formal development to see how the predictions of Eqs. (5.42) compare with experiment. For Ni metal, $\omega_M = 295$ cm^{-1} and $a_0 = 1.76$ Å. Equation (5.42a) gives $c_{11} = 2.95 \times 10^{12}$ dyn/cm^2, in reasonable agreement with the measured value 2.50×10^{12} dyn/cm^2, considering the simplicity of the model. We have $c_{12} = c_{44} = 1.48 \times 10^{12}$ dyn/cm^2, while the measured values are 1.60×10^{12} dyn/cm^2 and 1.19×10^{12} dyn/cm^2, respectively.

We now have explicit contact between our microscopic model and the theory of elasticity, but only for $l_z \geq 2$. Any solution generated must also satisfy Eq. (5.36) which applies right at the surface. Now upon using Eq. (5.37) and rearranging, Eq. (5.36) becomes

$$\tfrac{1}{2}k_0 a_0^2 Q_x^2 e_x(1) + \tfrac{1}{2}k_0 a_0^2 Q_x\left(Q_y e_y(z) - i\frac{\partial e_z}{\partial z}\right) + \tfrac{1}{4}k_0 a_0^2\left(Q_y^2 - \frac{\partial^2}{\partial z^2}\right)$$

$$- k_0 a_0\left\{\frac{1}{2\sqrt{2}}\left(iQ_x e_z(1) + \frac{\partial e_x}{\partial z}\right)\right.$$

$$\left. + \tfrac{1}{8}a_0\left(Q_x^2 e_x(1) + \frac{\partial^2 e_x}{\partial z^2} - 2ia_0 Q_x\frac{\partial e_z}{\partial z}\right)\right\} = \omega^2 e_x(1). \qquad (5.43)$$

In Eq. (5.43), all derivatives are understood to be evaluated at the surface, $l_z = 1$ in the microscopic model.

Now, if the terms within braces on the left of Eq. (5.43) could be ignored, then Eqs. (5.43) and (5.39) are identical in mathematical form. We can proceed by requiring that Eq. (5.39) be satisfied *everywhere* in the crystal, including at the surface $l_z = 1$, and ensure that Eq. (5.43) is satisfied by appending to Eq. (5.39) the boundary condition

$$[iQ_x e_z(1) + \partial e_x/\partial z] + \tfrac{1}{8}(a_0 Q_x^2 e_x(1) + \cdots) = 0. \qquad (5.44)$$

In the limit $a_0|\mathbf{Q}_\||| \ll 1$, the terms in the second parentheses in Eq. (5.44) are very small compared to the first term, so to excellent approximation we may drop these and write the boundary condition as

$$iQ_x e_z + \left.\frac{\partial e_x}{\partial z}\right|_{z=0} = 0. \qquad (5.45)$$

The boundary condition stated in Eq. (5.45) again has a direct and straightforward interpretation in the theory of elasticity. To see this, recall the elements of the stress tensor are given by (in terms of the displacement **u** of an elastic continuum)

$$\sigma_{xz} = c_{44}\left(\frac{\partial u_x}{\partial z} + \frac{\partial u_z}{\partial x}\right), \tag{5.46a}$$

$$\sigma_{yz} = c_{44}\left(\frac{\partial u_y}{\partial z} + \frac{\partial u_z}{\partial y}\right), \tag{5.46b}$$

$$\sigma_{zz} = c_{11}\frac{\partial u_z}{\partial z} + c_{12}\left(\frac{\partial u_x}{\partial x} + \frac{\partial u_y}{\partial y}\right). \tag{5.46c}$$

Thus, Eq. (5.45) is equivalent to the boundary condition that σ_{xz} vanish at the surface. If we were to take the long-wavelength limit of the equation for $e_y(l_z)$ and $e_z(l_z)$, we would obtain from the equations for $l_z = 1$ the additional condition that $\sigma_{yz} = \sigma_{zz} = 0$ at the surface. Physically, these three conditions just state that the crystal surface is stress-free; these are in fact the standard boundary conditions that apply to the free surface of an unstressed crystal, in the theory of elasticity.

Now that we have made full contact with the theory of elasticity within the framework of our microscopic model treated in the long-wavelength limit, we conclude this subsection by sketching the derivation of the Rayleigh wave-dispersion relation for a special case. The (100) surface of an elastic solid with cubic symmetry is actually a somewhat complicated case to analyze. The direction from Γ to X involves some subtle questions discussed fully by Wallis [5]. It will be more instructive and useful to consider instead a simpler case, which is the propagation of Rayleigh wave on the surface of a fully isotropic elastic medium. Such a limiting case is realized if the elastic constants obey the isotropy relation

$$c_{11} - c_{12} = 2c_{44}, \tag{5.47}$$

and it is convenient to express all the relations in terms of the velocities c_l, c_t of longitudinal and transverse sound waves. These are independent of direction in the isotropic case, and given by $c_l = (c_{11}/\rho)^{1/2}$, $c_t = (c_{44}/\rho)^{1/2}$.

Since the dispersion relation of the Rayleigh wave is isotropic, one may choose its wave vector along the $\hat{x}$ direction with no loss of generality. It is also the two components u_x and u_z of the displacement that are nonvanishing, These obey, when a solution with the form in Eq. (5.40) is assumed,

$$\left[c_l^2 Q_{\parallel}^2 - c_t^2 \frac{\partial^2}{\partial z^2} - \omega^2\right]u_x(z) - i(c_l^2 - c_t^2)Q_{\parallel}\frac{\partial u_z}{\partial z} = 0, \tag{5.48a}$$

$$-iQ_{\parallel}(c_l^2 - c_t^2)\frac{\partial u_x}{\partial z} + \left[c_t^2 Q_{\parallel}^2 - \omega^2 - c_l^2 \frac{\partial^2}{\partial z^2}\right]u_z(z) = 0. \tag{5.48b}$$

We thus have two coupled equations for the two functions $u_x(z)$ and $u_z(z)$. If we assume, as in the discussion of the previous section, that we have solutions that describe displacements localized to the surface,

$$u_{x,z}(z) = u_{x,z} \exp(-\alpha z),$$

then α is found to obey

$$[c_l^2(Q_{||}^2 - \alpha^2) - \omega^2][c_t^2(Q_{||}^2 - \alpha^2) - \omega^2] = 0. \tag{5.49}$$

For a given choice of $\mathbf{Q}_{||}$ and ω, we thus have two distinctly different choices of α, denoted by α_t and α_l. These are

$$\alpha_l = [Q_{||}^2 - (\omega^2/c_l^2)]^{1/2} \tag{5.50a}$$

and

$$\alpha_t = [Q_{||}^2 - (\omega^2/c_t^2)]^{1/2}. \tag{5.50b}$$

For α_l and α_t to be real, quite clearly we must have $c_t Q_{||} > \omega$ and also $c_l Q_{||} > \omega$ will be obeyed because $c_l > c_t$ quite generally.

In contrast to the surface modes considered in the previous section, the Rayleigh wave does not consist of a disturbance that decays to zero as a simple exponential, but instead the solution is a superposition of the two decaying exponentials:

$$u_x(z) = u_x^{(t)} \exp(-\alpha_t z) + u_x^{(l)} \exp(-\alpha_l z) \tag{5.51a}$$

and

$$u_z(z) = u_z^{(t)} \exp(-\alpha_t z) + u_z^{(l)} \exp(-\alpha_l z). \tag{5.51b}$$

The ratio $u_x^{(t)}/u_z^{(t)}$, and also $u_x^{(l)}/u_z^{(l)}$ is fixed by Eqs. (5.48a) and (5.48b), so in fact the solution in Eq. (5.51a) contains two and not four free parameters. One has

$$u_x^{(t)}/u_z^{(t)} = -i\alpha_t/Q_{||} \tag{5.52a}$$

and also

$$u_x^{(l)}/u_z^{(l)} = +iQ_{||}/\alpha_l. \tag{5.52b}$$

The final step is to require the solution to satisfy the stress free boundary conditions given in Eqs. (5.46). When this condition is imposed on the solution, one finds that for a given choice of $\mathbf{Q}_{||}$, the boundary condition is satisfied for only one particular frequency. We have

$$\omega = c_R Q_{||}, \tag{5.53}$$

where for the isotropic elastic continuum, the Rayleigh wave velocity is

$$c_R = c_t \xi \tag{5.54}$$

with ξ a real number that satisfies (see [6] for details)

$$\xi^6 - 8\xi^4 + 8\xi^2(3 - 2c_t^2/c_l^2) - 16(1 - c_t^2/c_l^2) = 0. \qquad (5.55)$$

A common choice of parameters for the isotropic elastic continuum leads to the relation $c_t^2 = \frac{1}{3}c_l^2$. Then for this case the only root of Eq. (5.55) with $0 \le \xi \le 1$ is

$$\xi = 0.9491, \qquad (5.56)$$

which gives

$$\alpha_l = 0.8475 Q_{\|}, \qquad (5.57a)$$

$$\alpha_t = 0.3933 Q_{\|}. \qquad (5.57b)$$

The principal point to notice is that both attenuation constants in Eqs. (5.57) are proportional to $Q_{\|}$, so the displacement field of the Rayleigh surface wave is not confined to the outermost few atomic layers as was the case for the short-wavelength surface phonons considered in the previous section, but instead it penetrates a distance the order of $Q_{\|}^{-1}$, which is many atomic layers in the limit that $\mathbf{Q}_{\|}$ vanishes.

The behavior found for the specific examples explored in the previous and present section is typically encountered in the theory of surface phonons on a semi-infinite crystal. Near the boundaries of the two-dimensional Brillouin zone of the structure, a rich and complex spectrum of modes is encountered. These will have their displacements confined to the outer few atomic layers, and many of the modes have dispersion curves which extend over only a fraction of the two-dimensional Brillouin zone, to merge with the phonon continuum as one moves from the zone edge to the zone center. One mode will survive as $\mathbf{Q}_{\|} \to 0$, and this becomes the Rayleigh surface acoustic wave of elasticity theory. The displacement field associated with Rayleigh surface wave penetrates deeply into the crystal in this limit, as one sees from the preceding example.

5.2.4   Surface Optical Phonons in the Long-Wavelength Limit; the Case of Ionic Crystals

In the discussion of surface phonons on the clean crystal surface just presented, as already remarked, two classes of surface modes have been encountered. The low-frequency Rayleigh surface waves have displacement fields that penetrate deeply into the crystal as $\mathbf{Q}_{\|} \to 0$, while the remaining modes are localized on the outermost atomic layers.

In crystals with more than one atom per unit cell, we encounter surface optical phonons, i.e., surface modes which have finite, nonzero frequency in the limit $\mathbf{Q}_{\|} \to 0$. These modes also come in two distinctly different classes. If we consider a crystal lattice with two or more atoms per unit cell, and each may be regarded as electrically neutral so only forces of short range enter the lattice dynamics of the crystal, then in general as $\mathbf{Q}_{||} \to 0$ the atomic displacement associated with the surface optical phonon remains localized in the outermost few atomic layers [10]. Such modes may be encountered in the homopolar semiconductors silicon and germanium, and at the surface of metals of the appropriate crystal structure.

In ionic crystals, where the Coulomb interactions between the ions lead to couplings that extend over large distances, we encounter a new class of surface optical phonons which penetrate deeply into the crystal as $\mathbf{Q}_{||} \to 0$. These modes may be described by a macroscopic theory just as the Rayleigh waves can; in fact, in the present text, these surface optical phonons and their study by electron energy loss spectroscopy has been discussed already in Section 3.3.4. In Chapter 3, these waves were examined within the context of the theory of fluctuations in dielectric media. Here we reexamine their properties within a lattice dynamical framework. The present treatment ultimately yields results identical to those in Chapter 3, although the language is rather different.

For simplicity, consider an alkali halide crystal, and let $\mathbf{u}_0(l)$ denote the normal coordinate that describes the relative motion of the ions in the unit cell located at l, i.e., $\mathbf{u}_0(l)$ is the normal coordinate that describes motion of the two sublattices $180°$ out of phase, with relative amplitude such that the center of mass of the unit cell remains at rest. The coordinate $\mathbf{u}_0(l)$ then obeys the equation of motion

$$\ddot{\mathbf{u}}_0(l) + \omega_0^2 \mathbf{u}_0(l) = (e^*/M_r)\mathbf{E}(l), \tag{5.58}$$

where e^* is the transverse effective charge associated with the unit cell, M_r the reduced mass of the ions in it, and ω_0 the mechanical restoring force provided by the short-ranged interactions between the two sublattices. In effect, all consequences of the long-ranged interaction between the ions is included in the electric field $\mathbf{E}(l)$ generated by the ion motion. Now each unit cell has the dipole moment

$$\mathbf{p}(l) = e^*\mathbf{u}_0(l), \tag{5.59}$$

and as a consequence we have

$$E_\alpha(l) = \frac{e^*}{\varepsilon_\infty} \sum_\beta \sum_{l' \neq l} \left\{ \frac{3\hat{n}_\alpha(l,l')\hat{n}_\beta(l,l') - \delta_{\alpha\beta}}{R(l,l')^3} \right\} u_{0\beta}(l'), \tag{5.60}$$

where we assume ε_∞ is the background dielectric constant of the medium, $\hat{n}(l,l')$ a unit vector directed from l to l', and $R(l,l')$ the distance from l to l'. For the moment, our attention is confined to the infinitely extended crystal. In the limit that $\mathbf{u}_0(l)$ and $\mathbf{E}(l)$ vary slowly on the scale of the lattice constant, we may pass to the continuum limit, so Eqs. (5.58) and (5.60) may be combined into the single statement

$$u_{0\alpha}(\mathbf{x}) + \omega_0^2 u_{0\alpha}(\mathbf{x}) = \frac{ne^{*2}}{M_r \varepsilon_\infty} \int' d^3x' \sum_\beta \left\{ \frac{3\hat{n}_\alpha \hat{n}_\beta - \delta_{\alpha\beta}}{|\mathbf{x} - \mathbf{x}'|^3} \right\} u_{0\beta}(\mathbf{x}'), \qquad (5.61)$$

where $\hat{n}_\alpha$ is a unit vector directed from $\mathbf{x}$ to $\mathbf{x}'$, n the number of unit cells per unit volume, and the prime on the integral on the right-hand side of Eq. (5.61) indicates that a small spherical volume (infinitesimal in radius) centered at $\mathbf{x}$ is to be excluded from the integration.

It is useful to rearrange the integral on the right-hand side of Eq. (5.61), using the identity

$$\frac{3\hat{n}_\alpha \hat{n}_\beta - \delta_{\alpha\beta}}{|\mathbf{x} - \mathbf{x}'|^3} = \frac{\partial}{\partial x_\beta'} \left\{ \frac{(x_\alpha - x_\alpha')}{|\mathbf{x} - \mathbf{x}'|^3} \right\}. \qquad (5.62)$$

Then

$$\int' d^3x' \sum_\beta \left\{ \frac{3\hat{n}_\alpha \hat{n}_\beta - \delta_{\alpha\beta}}{|\mathbf{x} - \mathbf{x}'|^3} \right\} u_{0\beta}(\mathbf{x}')$$

$$= \sum_\beta \int' d^3x' \frac{\partial}{\partial x_\beta'} \left\{ \frac{(x_\alpha - x_\alpha')}{|\mathbf{x} - \mathbf{x}'|^3} u_{0\beta}(\mathbf{x}') \right\} - \int' d^3x' \frac{(x_\alpha - x_\alpha')}{|\mathbf{x} - \mathbf{x}'|^3} \nabla' \cdot \mathbf{u}_0(\mathbf{x}')$$

$$= \int d^2\mathbf{S}' \cdot \mathbf{u}_0(\mathbf{x}') \left\{ \frac{(x_\alpha - x_\alpha')}{|\mathbf{x} - \mathbf{x}'|^3} \right\} + \frac{\partial}{\partial x_\alpha} \int d^3x' \frac{\nabla' \cdot \mathbf{u}_0(\mathbf{x}')}{|\mathbf{x} - \mathbf{x}'|}. \qquad (5.63)$$

The surface integral in Eq. (5.63) covers an infinitesimal sphere which surrounds the point $\mathbf{x}$, and here the vector $d^2\mathbf{S}'$ is directed *toward* the point $\mathbf{x}$ from the surface of the sphere. Contributions from portions of the surface very distant from $\mathbf{x}$ will be negligible. Finally, on the surface of the infinitesimal sphere, the difference between $\mathbf{u}_0(\mathbf{x})$ and $\mathbf{u}_0(\mathbf{x}')$ may be overlooked, and

$$\int d^2\mathbf{S}' \cdot \mathbf{u}_0(\mathbf{x}') \left\{ \frac{(x_\alpha - x_\alpha')}{|\mathbf{x} - \mathbf{x}'|^3} \right\} = \mathbf{u}_0(\mathbf{x}) \cdot \int d^2\mathbf{S}' \frac{(x_\alpha - x_\alpha')}{|\mathbf{x} - \mathbf{x}'|^3}$$

$$\equiv \frac{4\pi}{3} u_{0\alpha}(\mathbf{x}), \qquad (5.64)$$

where the last step follows upon explicit evaluation of the integral left after $\mathbf{u}_0(\mathbf{x})$ is extracted.

Thus, for $\mathbf{u}_{0\alpha}(\mathbf{x})$ we have the equation of motion

$$u_{0\alpha}(\mathbf{x}) + \left[\omega_0^2 - \frac{4\pi ne^{*2}}{3M_r \varepsilon_\infty} \right] u_{0\alpha}(\mathbf{x}) = \frac{ne^{*2}}{M_r \varepsilon_\infty} \frac{\partial}{\partial x_\alpha} \int \frac{d^3x' \nabla' \cdot \mathbf{u}_0(\mathbf{x}')}{|\mathbf{x} - \mathbf{x}'|} \qquad (5.65)$$

On the right-hand side of Eq. (5.65), the prime on the integral is no longer required, since the near vicinity of the point $\mathbf{x} = \mathbf{x}'$ no longer leads to singular contributions.

We first apply Eq. (5.65) to the infinitely extended crystal. Here we have solutions of the form

$$\mathbf{u}_0(\mathbf{x}, t) = \mathbf{u}_0(\mathbf{Q}) \exp(i\mathbf{Q} \cdot \mathbf{x} - \omega t). \tag{5.66}$$

There are two distinct cases to examine:

(i)   Suppose that $\mathbf{u}_0(\mathbf{Q}) \perp \mathbf{Q}$: Then $\mathbf{V} \cdot \mathbf{u}_0(\mathbf{x})$ vanishes everywhere, the integral on the right-hand side of Eq. (5.65) is zero, and the frequency of the lattice oscillation equals $[\omega_0^2 - 4\pi n e^{*2}/3M_r\varepsilon_\infty]^{1/2}$. This solution is the long-wavelength transverse optical phonon of the crystal, and if we call its frequency ω_{TO}, then

$$\omega_{TO}^2 = \omega_0^2 - (4\pi n e^{*2}/3M_r\varepsilon_\infty). \tag{5.67}$$

The second term on the right-hand side is a local field correction to the frequency ω_0^2, which results from the restoring forces of short-range character.

(ii)   Suppose that $\mathbf{u}_0(\mathbf{Q}) \parallel \mathbf{Q}$: Then $\mathbf{V} \cdot \mathbf{u}_0(\mathbf{x}') = iQu_0(\mathbf{Q})$, so that

$$\frac{\partial}{\partial x_\alpha} \int \frac{d^3x' \, \mathbf{V}' \cdot \mathbf{u}_0(\mathbf{x}')}{|\mathbf{x} - \mathbf{x}'|} = iQu_0(\mathbf{Q}) \frac{\partial}{\partial x_\alpha} e^{i\mathbf{Q} \cdot \mathbf{x}} \int \frac{d^3x' \, e^{i\mathbf{Q} \cdot (\mathbf{x}' - \mathbf{x})}}{|\mathbf{x} - \mathbf{x}'|}$$

$$\equiv -4\pi \mathbf{u}_0(\mathbf{x}). \tag{5.68}$$

When this is inserted into Eq. (5.65), we find that when $\mathbf{u}_0(\mathbf{Q})$ and $\mathbf{Q}$ are parallel, the frequency of the lattice motion is given by

$$\omega_{LO}^2 = \omega_0^2 + \frac{8\pi n e^{*2}}{3M_r\varepsilon_\infty} = \omega_{TO}^2 + \frac{4\pi n e^{*2}}{M_r\varepsilon_\infty}. \tag{5.69}$$

This solution is the longitudinal optical phonon well known in the lattice dynamics of ionic crystals. If we identify the static zero frequency dielectric constant $\varepsilon(0)$ with the combination [see Eq. (3.35)]

$$\varepsilon(0) = \varepsilon_\infty + (4\pi n e^{*2}/M_r\omega_{TO}^2), \tag{5.70}$$

then Eq. (5.69) is equivalent to the Lyddane–Sachs–Teller relation

$$\omega_{LO} = [\varepsilon(0)/\varepsilon_\infty]^{1/2}\omega_{TO}. \tag{5.71}$$

Now let us consider a semi-infinite alkali halide crystal, with surface in the xy plane and the crystal in the half-space $z > 0$. We shall show that there is a solution of Eq. (5.65) that describes an optical phonon localized near the surface of the crystal. We begin by noting the identity

$$\nabla^2 |\mathbf{x} - \mathbf{x}'|^{-1} = -4\pi \, \delta(\mathbf{x} - \mathbf{x}') \tag{5.72}$$

which, after taking the divergence of both sides of Eq. (5.65), leads to the relation valid everywhere inside the crystal

$$\frac{\partial^2}{\partial t^2} [\mathbf{V} \cdot \mathbf{u}_0(\mathbf{x})] + \omega_{LO}^2 [\mathbf{V} \cdot \mathbf{u}_0(\mathbf{x})] = 0. \tag{5.73}$$

This means that *all* possible solutions of Eq. (5.65) except those with frequency equal to ω_{LO} must have

$$\nabla \cdot \mathbf{u}_0(\mathbf{x}) \equiv 0, \tag{5.74}$$

a statement true for a crystal of arbitrary shape.

Now we examine Eq. (5.65) for a solution that describes an excitation that propagates parallel to the x direction with wave vector $Q_{\parallel}$ and profile in the z direction unspecified at the moment:

$$\mathbf{u}_0(\mathbf{x}_{\parallel}, z) = \begin{cases} \mathbf{u}_0(z) \exp(iQ_{\parallel}x - i\omega t), & z \geq 0, \\ 0, & z \leq 0, \end{cases} \tag{5.75}$$

where the form of $\mathbf{u}_0(z)$ must be chosen so Eq. (5.73) is obeyed in the bulk of the crystal.

Then we have for this case

$$\nabla' \cdot \mathbf{u}_0(\mathbf{x}') = \delta(z)u_{0z}(0)e^{i(Q_{\parallel}x - i\omega t)}, \tag{5.76}$$

so there is a wavelike array of sources located *on* the crystal surface, and not within it. In this circumstance, we must pause to modify Eq. (5.65), which was derived for only a distribution of elementary dipoles [Eq. (5.59)] imbedded *within* a crystal of dielectric constant. We see from Eq. (5.65) that the combination $ne^*\nabla' \cdot \mathbf{u}(\mathbf{x}')$ plays the role of a charge density that generates a long-ranged electric field, according to the laws of electrostatics. When the charge density is imbedded in the dielectric, the electric field is screened, and thus we have the factor of ε_{∞} in the denominator of Eq. (5.65). If charge is placed right on the surface of a dielectric, then the field is only partially screened, and in place of the factor of ε_{∞} we have, instead, $2/(1 + \varepsilon_{\infty})$. To see that this is so, one may consider the electric field generated inside the dielectric by a point charge placed just outside its surface [11]. The local field correction $-4\pi ne^{*2}/3M_r\varepsilon_{\infty}$ comes from the surface of the infinitesimal sphere that surrounds the field point $\mathbf{x}$, and this remains unaltered. Thus, when $\nabla' \cdot \mathbf{u}_0(\mathbf{x}')$ is nonzero only on the surface and zero everywhere in the bulk as in Eq. (5.73), we have

$$\ddot{u}_{0\alpha}(\mathbf{x}) + \omega_{TO}^2 u_{0\alpha}(\mathbf{x}) = \frac{2ne^{*2}}{M_r(1 + \varepsilon_{\infty})} \frac{\partial}{\partial x_\alpha} \int \frac{d^3x' \nabla' \cdot \mathbf{u}_0(\mathbf{x}')}{|\mathbf{x} - \mathbf{x}'|}. \tag{5.77}$$

If Eq. (5.75) is inserted into Eq. (5.77), the integral may be evaluated in closed form to give, with $\hat{x}$ and $\hat{z}$ directions,

$$[\omega_{TO}^2 - \omega^2]\mathbf{u}_0(\mathbf{x}) = \frac{4\pi i ne^{*2}}{M_r(1 + \varepsilon_{\infty})Q_{\parallel}} u_{0z}(0)[Q_{\parallel}\hat{x} + iQ_{\parallel}\hat{z}]$$

$$\times \exp[iQ_{\parallel}x - Q_{\parallel}z - i\omega t], \tag{5.78}$$

or after a bit of rearrangement,

$$\mathbf{u}_0(\mathbf{x}) = \frac{4\pi i n e^{*2}}{M_r(1 + \varepsilon_\infty)} \frac{u_{0z}(0)[\hat{x} + i\hat{z}]}{(\omega_{TO}^2 - \omega^2)} \exp[iQ_{||}x - Q_{||}z - i\omega t]. \quad (5.79)$$

The expression in Eq. (5.79) describes a disturbance in the crystal of optical phonon character, driven by the polarization charge density localized on the surface [Eq. (5.76)]. If we equate $\hat{z}$ components on both sides of Eq. (5.79), then let $z \to 0$ after using Eq. (5.75), we obtain a self-consistency relation that allows us to find the frequency of this self-sustained oscillation of the dielectric medium:

$$u_{0z}(0) = + \frac{4\pi n e^{*2}}{M_r(1 + \varepsilon_\infty)} \frac{u_{0z}(0)}{(\omega^2 - \omega_{TO}^2)}, \quad (5.80)$$

or

$$\omega^2 \equiv \omega_s^2 = \omega_{TO}^2 + \frac{4\pi n e^{*2}}{M_r(1 + \varepsilon_\infty)}. \quad (5.81)$$

With $\varepsilon(0)$ the static dielectric constant of the crystal, we have

$$\omega_s = \omega_{TO} \left(\frac{\varepsilon(0) + 1}{\varepsilon_\infty + 1} \right)^{1/2}, \quad (5.82)$$

a result identical to Eq. (3.36).

In Chapter 3, the surface modes of a semi-infinite dielectric, and also of a dielectric slab were derived simply by examining the position of the poles of the appropriate electron energy loss expressions [see Eq. (3.36), the discussion just before Eq. (3.36), and Section 3.4.1]. For an ionic crystal, where the frequency-dependent dielectric function is given by Eq. (3.35), all these modes may be derived equally well within the lattice dynamical theory outlined here, which is easily extended to the slab geometry.[4] The two approaches seem very different in spirit, but in fact are identical in physical content. Because there are two distinctly different theoretical frameworks that may be employed to describe these surface modes, in the literature, two different sets of names are used to describe them. An author who utilizes the lattice dynamical approach will refer to them as surface optical phonons as we do here, or possibly Fuchs–Kliewer modes, after the authors who described them first with the formalism of lattice dynamics instead of the older description provided by electromagnetic theory. Where the modes are described within the framework of electromagnetic theory, most particularly when the

---

[4] In a similar fashion, the surface plasma oscillations discussed in Chapter 3 may be derived by direct appeal to an explicit description of the dynamics of the electron gas.

effect of retardation ignored throughout this volume is included,[5] the modes are called surface polaritons, or possibly surface electromagnetic waves. The two terminologies bring very different physical pictures of the waves to mind, but in fact each set of terminology describes an entity with both lattice motion of optical phonon character and macroscopic electric fields present simultaneously; an author may choose to emphasize one of the two features as of primary importance by the choice of terminology, and the result is a literature riddled with a wide variety of terms all of which describe this particular entity.

From Eq. (5.61), we see that just as in the Rayleigh acoustic wave, for small values of $\mathbf{Q}_{||}$, the displacement field of the surface optical phonon penetrates deeply into the crystal a distance of the order of $Q_{||}^{-1}$. Its frequency remains finite in contrast to the Rayleigh wave, and the surface charge density displayed in Eq. (5.76) sets up a macroscopic electric field in the vacuum above the crystal that also extends out a distance of $Q_{||}^{-1}$. As a consequence, these modes couple to electrons strongly, and lead to prominent features in electron energy loss spectra, as discussed in Chapter 3. The long-wavelength Rayleigh waves have frequency too low to be observed in electron spectroscopy, though they are studied by other forms of surface spectroscopy (light scattering, atom scattering).

5.2.5   Influence of the Surface on Bulk Phonon Eigenvectors; the Green's Function Method

The last three sections have been devoted to the study of a new class of vibrational normal modes that occur when a surface is present. These are the surface phonons, in which the atomic displacements are localized to the near vicinity of the surface. As discussed in general terms in Section 5.2.2, the surface also alters the eigenvectors of the bulk phonons, by mixing together those associated with bulk waves with a given wave vector $\mathbf{Q}_{||}$ that are degenerate in frequency. The purpose of this section is to examine this issue in more detail, because when any quantity is calculated that depends on an *average* over the whole spectrum of thermally excited phonons, it is a rule of thumb that at the crystal surface the contributions from bulk waves and from the surface phonons are roughly equal. An example is the mean square displacement $\langle u_\alpha(l_{||}l_z\kappa)^2 \rangle_T$ of an atom in the crystal surface, where

---

[5] Retardation effects are important for wave vectors $Q_{||} \approx \omega_s/c$, with c the velocity of light. For the surface modes on ionic crystals, and also for the surface plasma oscillations on metals, the values of $Q_{||}$ explored in electron energy loss spectroscopy satisfy $Q_{||} \gg \omega_s/c$, and retardation effects are quite unimportant. For a discussion of the inclusion of retardation on the properties of the surface waves, see Mills and Burstein [12].

the angular brackets with subscript T denote an average over the ensemble of thermally excited phonon modes of the finite crystal.

The simplest example to consider to begin a discussion of the effect of a surface on bulk phonon eigenvectors is that of a stretched string of length L, with mass density ρ subjected to tension T. The elementary excitations here are phonons with frequency $\omega(k) = ck$, with c the velocity of sound, given by $c = (T/\rho)^{1/2}$. If we apply periodic boundary conditions, and let $e^{(\infty)}(k_n; x)$ denote an eigenvector associated with a particular wave vector k_n normalized in the standard manner

$$\int_0^L dx |e^{(\infty)}(k_n; x)|^2 = 1, \tag{5.83}$$

then we have

$$e^{(\infty)}(k_n; x) = \exp(ik_n x)/\sqrt{L}, \tag{5.84}$$

where the periodic boundary conditions require $k_n = 2\pi n/L$, with n a positive or negative integer.

Now consider the same stretched string of length L, but let both surfaces be "free," i.e., we do not apply periodic boundary conditions but instead allow each end to vibrate unimpeded. If $u(x)$ describes any vibrational motion of the string, from elementary mechanics, it follows that the boundary conditions to be applied to $u(x)$ at each end of the string are

$$(\partial u/\partial x)_0 = (\partial u/\partial x)_L = 0, \tag{5.85}$$

where the string is assumed stretched from $x = 0$ to $x = L$.

The plane wave eigenvectors, Eq. (5.84), fail to satisfy the boundary conditions in Eq. (5.85) which, for the problem at hand, are the equivalent of the stress-free boundary conditions of elasticity theory. We now have, instead,

$$e(k_n; x) = (\sqrt{2}/\sqrt{L}) \cos(k_n x), \tag{5.86}$$

where $k_n = n\pi/L$ rather than $2\pi n/L$, and n is a positive integer.

In the stretched string, the two waves with wave vector $k = k_n = +n\pi/L$ and $k = k_{-n} = -n\pi/L$ are degenerate solutions of the underlying wave equation, and the eigenvector in Eq. (5.85) may be regarded as a linear combination of these two formed so that the boundary condition in Eq. (5.85) is satisfied. The role of the surfaces is to lead to new eigenvectors, each of which is a coherent superposition of each of the two degenerate modes that emerge from the basic wave equation, so simple plane waves of the form given in Eq. (5.84) no longer serve as proper bulk phonon eigenvectors. The calculation of any property of the string near its "surface" requires careful attention the actual form of the eigenvector; clearly the same is true near the surface of a crystal.

There is one simple feature of the preceding example not shared in even slightly more complicated situations. This is that for the finite string, the allowed wave vectors are all evenly spaced, i.e., $k_n \equiv n\pi/L$. In the limit that $L \to \infty$, in k space, the modes are thus uniformly distributed, with the number of modes between k and $k + dk$ given by $L\,dk/\pi$. More generally, in one-dimensional problems, the allowed wave vectors are shifted slightly by application of the surface boundary conditions, so one finds that k_n is given by a relation of the form [4, p. 356]

$$k_n = L^{-1}[n\pi - \varphi(k_n)], \tag{5.87}$$

where $\varphi(k_n)$ lies between 0 and π. One must solve Eq. (5.87) to find the allowed wave vectors, and we see that inside the phonon bands, the shift away from $n\pi/L$ is always less than π/L for any mode, a result well known from related problems.

The effect of the wave vector shifts that follow from Eq. (5.87) may be seen by supposing the relation between frequency and wave vector as given by a function $\omega(k_n)$, whose form need not be specified in detail here. Then as $L \to \infty$, with $\varphi(k_n) \equiv 0$, let $dn^{(\infty)}$ be the number of vibration modes with frequency between ω and $\omega + d\omega$. Then $k_n = n\pi/L$ and we have

$$\frac{dn^{(\infty)}}{d\omega} = \frac{dn^{(\infty)}}{dk^{(n)}}\frac{dk^{(n)}}{d\omega} = \frac{L}{\pi}\frac{1}{(d\omega/dk)}, \tag{5.88}$$

which is a standard expression for the density of states of a one-dimensional vibrating system.[6]

From Eq. (5.87), with $\varphi(k_n)$ nonzero because of the surface boundary condition, one finds in place of Eq. (5.88) the new expression (5.17)

$$\frac{dn}{d\omega} = \frac{dn^{(\infty)}}{d\omega} + \frac{1}{\pi}\frac{d\varphi}{d\omega}, \tag{5.89}$$

where it is assumed that since $\varphi(k_n)$ is a function of wave vector, and since $\omega = \omega(k_n)$, the phase angle can be expressed as a function of frequency.

The result in Eq. (5.89) shows that the presence of the surface not only alters the eigenvectors of the bulk waves, but their frequency distribution as well. The quantity $(dn^{(\infty)}/d\omega)$ is the phonon density of states of the one-dimensional system with periodic boundary conditions applied, and this is proportional to L, as we see in Eq. (5.88). The change in density of states, $(+d\varphi/\pi\,d\omega)$, is smaller than $(dn^{(\infty)}/d\omega)$ by a factor of $1/L$, but nonetheless the presence of this shift in the density of states must be acknowledged in

---

[6] Recall that here, the integer n in Eq. (5.86) is confined to positive values only, so with the wave vector k so confined, we have Eq. (5.87) rather than the form $L/2\pi(d\omega/dk)$ often quoted.

calculating the influence of the surface on the thermodynamic properties of the system. There are subtle issues that arise when this is done, incidentally.[7]

[7] Consider, for example, the calculation of the specific heat of the finite system. The internal energy is $U(T) = \hbar \sum_n \omega(k_n)[\bar{n}(\omega(k_n)) + \frac{1}{2}]$, where $\bar{n}(\omega)$ is the Bose–Einstein function. Differentiation with respect to temperature gives the specific heat.

Both surface and bulk waves contribute to $U(T)$. Consider the bulk wave part, call it $U_B(T)$, and let there be N values of k_n that contribute, $k_1, k_2, \ldots, k_N$. The behavior of $U_B(T)$ as $L \to \infty$ with first correction from finite L is obtained as follows. For any function $g(n)$, with n from 1 to N, one has the Euler–MacLaurin summation formula

$$g(1) + \cdots + g(N) = \int_1^N dn\, g(n) + \tfrac{1}{2}[g(1) + g(N)] + (B_2/2!)[g'(1) - g'(N)] + \cdots .$$

Applied to our case, one finds for any function of k_n,

$$\sum_{n=1}^N g(k_n) = \frac{L}{\pi} \int_{k_1}^{k_N} dk_n\, g(k_n) - \int_{k_1}^{k_N} dk\, \frac{\partial \varphi}{\partial k}\, g(k) + \tfrac{1}{2}[g(k_1) + g(k_N)] + \cdots ,$$

where as $L \to \infty$, the terms omitted give corrections of order L^{-1}. The terms retained generate, when we return to the discussion of the semi-infinite crystal, the surface specific heat discussed in the literature on surface lattice dynamics (e.g., [5]).

The important point to note is that the calculation of the surface specific heat, and other surface corrections to thermodynamic properties of the crystal, require one to take account not only the rearrangement in frequency of the bulk waves ($\partial \varphi/\partial k$ terms), but also the first correction to the procedure of replacing the sum over a discreet set of modes by an integral (the term $\tfrac{1}{2}[g(k_m) + g(k_M)]$).

An interesting case to consider is a monoatomic, one-dimensional line of masses each with identical mass. Take the lattice constant to be equal to unity, and let the masses be coupled by harmonic springs. Then as in the example of the stretched string in the text, we have again $k_n \equiv n\pi/L$, where n ranges from 0 to $(N - 1)$, for a finite string of N masses separated by the distance a_0. Here $L = Na_0$. Hence $\varphi(k) \equiv 0$ and we have

$$\sum_{n=0}^{N-1} g(k_n) = \frac{L}{\pi} \int_0^{\pi/a_0} dk\, g(k) + \tfrac{1}{2}[g(0) - g(\pi/a_0)] + \vartheta(L^{-1}),$$

where $g(\pi/a_0)$ is preceded by a minus sign because the upper limit of the integral has been taken to be π, and not $\pi(1 - 1/N)$, as required for N finite. For an integrand $g(k)$ that describes the contribution of modes of wave vector k to thermodynamic properties of the medium, the integral gives the thermodynamic property of interest in the limit $L \to \infty$, while the first surface correction is displayed explicitly. By a very different method, the formula just quoted has been generated by Dobrzynski and Leman [13].

The point is that a full description of surface contributions to the thermodynamic properties of crystals requires inclusion not only of the change in frequency distribution within the phonon bands, as described by Eq. (5.90), but also the "bandedge" corrections generated from the Euler–MacLaurin formula. As the example shows, these are tricky to generate. In Ref. [14], a sum rule approach is developed that allows these band edge corrections to be generated, once $(\partial \varphi/\partial \omega)$ is known.

In discussions of thermodynamic properties of the *whole crystal*, the "band edge" corrections must be included as just described. In analysis of *local* properties of the surface, such as the mean square displacement of an atom there, they are quite unimportant.

The foregoing examples explore one-dimensional systems, while the concern of the chapter is with semi-infinite crystals with infinite extent in two directions parallel to the surface. The examples are directly relevant to the surface problem because, as we have seen, when we make use of the wave vector $\mathbf{Q}_{||}$ parallel to the surface as a good quantum number, our equations of motion [Eq. (5.10)] become isomorphic to those of a one-dimensional line of entities, though in general the set of equations encountered will be substantially more complex than those in our example.

For quite a number of calculations, we need the explicit form of the eigenvectors associated with the bulk phonons, as modified by the presence of the surface. Consider once again the calculation of the mean square displacement of some particular atom. This is a quantity of direct experimental interest because it enters the Debye–Waller factor that controls the temperature variation of the intensity of features in the low-energy electron diffraction pattern of the crystal. This quantity was considered briefly in Section 4.2.2, where the form of the spectral density function $\rho_{i\alpha}(\omega)$ that enters Eq. (4.29) was described for an isolated oxygen adatom on the Ni(111) surface. An expression for the spectral density function may be obtained in terms of the eigenvectors $e_\alpha^{(s)}(\mathbf{Q}_{||}; l_z\kappa)$ that enter the eigenvalue problem defined in Eq. (5.10). If there are N_s unit cells in the basic crystal to which periodic boundary conditions are applied in the two directions parallel to the surface, and $a_{\mathbf{Q}_{||}s}, a_{\mathbf{Q}_{||}s}^+$ are the boson annihilation and creation operators that destroy and create quanta of the mode $(\mathbf{Q}_{||}s)$ of our finite crystal, the displacement operator $u_\alpha(l_{||}l_z\kappa)$ may be written in the form [4],

$$u_\alpha(l_{||}l_z\kappa) = \sum_{\mathbf{Q}_{||},s} \left(\frac{\hbar}{2M(l_z\kappa)\omega_s(\mathbf{Q}_{||})N_s}\right)^{1/2} e_\alpha^{(s)}(\mathbf{Q}_{||}; l_z\kappa)$$
$$\times \exp[i\mathbf{Q}_{||} \cdot \mathbf{R}_0(l_{||}l_z z)](a_{\mathbf{Q}_{||}s} + a^+_{\mathbf{Q}_{||}s}), \qquad (5.90)$$

where the eigenvector $e_\alpha^{(s)}(\mathbf{Q}_{||}; l_z\kappa)$ is chosen to satisfy the condition

$$\sum_{l_z\kappa\alpha} |e_\alpha^{(s)}(\mathbf{Q}_{||}; l_z\kappa)|^2 = 1.$$

From Eq. (5.87), it is easy to show that the component of the mean square displacement $\langle u_\alpha(l_{||}l_z\kappa)^2\rangle_T$ in the Cartesian direction α of the atom $(l_{||}, l_z\kappa)$ is given by

$$\langle u_\alpha(l_{||}l_z\kappa)^2\rangle_T = \sum_{\mathbf{Q}_{||},s} \frac{\hbar}{2M(l_z\kappa)\omega_s(\mathbf{Q}_{||})N_s} |e_\alpha^{(s)}(\mathbf{Q}_{||}; l_z\kappa)|^2[1 + 2\bar{n}(\mathbf{Q}_{||}s)], \quad (5.91)$$

with $n(\mathbf{Q}_{||}s)$ the Bose–Einstein function that gives the number of thermally excited modes present at temperature T.

From Eq. (5.91), we see that one requires the eigenvectors of the various normal modes, and not just the eigenfrequencies to calculate the mean square displacement of one particular atom. In fact, Eq. (5.91) may be cast in a form identical to Eq. (4.29), if we define the function

$$\rho_{l_z\kappa;\alpha}(\omega) = \frac{1}{N_s} \sum_{\mathbf{Q}_{||}s} |e_\alpha^{(s)}(\mathbf{Q}_{||}; l_z\kappa)|^2 \delta(\omega - \omega_s(\mathbf{Q}_{||})) \equiv \frac{1}{N_s} \sum_{\mathbf{Q}_{||}} \rho_{l_z\kappa;\alpha}(\mathbf{Q}_{||}, \omega), \quad (5.92)$$

where in the last line the partial spectral density $\rho_{l_z\kappa;\alpha}(\mathbf{Q}_{||}, \omega)$ defined by

$$\rho_{l_z\kappa;\alpha}(\mathbf{Q}_{||}, \omega) = \sum_s |e_\alpha^{(s)}(\mathbf{Q}_{||}; l_z\kappa)|^2 \delta(\omega - \omega_s(\mathbf{Q}_{||})) \qquad (5.93)$$

gives the contribution of the modes of wave vector $\mathbf{Q}_{||}$ to the frequency spectrum of thermal vibrations in Cartesian direction α of the atom in plane l_z, at site κ in the two-dimensional unit cell. Later in this chapter, such partial spectral density functions will be directly related to the electron energy loss spectrum of the surface in question. With the definition in Eq. (5.92), the quantity $\langle u_\alpha^2(l_{||}l_z\kappa)\rangle_T$ becomes

$$\langle u_\alpha^2(l_{||}l_z\kappa)\rangle_T = \frac{\hbar}{2M(l_z\kappa)} \int_0^\infty \frac{d\omega}{\omega}[1 + 2\bar{n}(\omega)]\rho_{l_z\kappa;\alpha}(\omega), \qquad (5.94)$$

a form identical to Eq. (4.29) except for a change in notation.

In Section 5.2.2, where our discussion of surface phonons was initiated, for crystal geometries where there is periodicity in the two directions parallel to the surface, both eigenfrequencies and eigenvectors may be generated by considering a slab of material with a finite number N of atomic layers. Through the processing of this information for selected values of $\mathbf{Q}_{||}$, one may construct the partial spectral densities defined in Eq. (5.93). Then by considering a net of values of $\mathbf{Q}_{||}$ distributed throughout the two-dimensional Brillouin zone, the spectral density $\rho_{l_z\kappa;\alpha}(\omega)$ itself may be built up along with the mean square displacement $\langle u_\alpha^2(l_{||}l_z\kappa)\rangle$ of selected atoms in the crystal. The method has also been used to calculate surface corrections to the specific heat of the crystal, by examining the variation of the total specific heat with the thickness of the model film. (The specific heat calculation does not require the eigenvectors, but only the eigenfrequencies.) A number of these analyses are discussed in the review article by Wallis [5]. Quite generally these "slab calculations" involve a substantial expenditure of computer time if accurate results are to be obtained, simply because one must sample a substantial number of normal modes for the sums on $\mathbf{Q}_{||}$ and s to be evaluated with precision.

There are a variety of methods by which one may evaluate quantities such as the surface specific heat or mean square displacement in selected temperature regimes without the need to diagonalize the full dynamical matrix of a finite sample, obtain the eigenvectors, and construct the spectral densities defined above. However, electron energy loss spectroscopy does directly sample the frequency spectrum of the atomic vibrations, and here we do need the partial spectral densities defined in Eq. (5.93), as we shall appreciate soon. A method that serves as an alternate to the slab method is the Green's function approach. For simple models, analytic formulas can often be obtained for the spectral densities from this approach, and for situations where analytic methods prove cumbersome, numerical schemes may be used that avoid the diagonalization of large matrices. We conclude this section with a summary of the Green's function approach, as it applies to the calculation of the partial spectral densities defined in Eq. (5.93).

In essence, in Chapter 4, we have already encountered a Green's function in Eq. (4.32), though we did not refer to this object by such a name, nor did we discuss how it was to be calculated. In the present context, we shall examine the following function of the complex variable z:

$$U_{\alpha\beta}(l_z\kappa, l_z'\kappa'; \mathbf{Q}_{||}z) = \sum_s \frac{e_\alpha^{(s)}(\mathbf{Q}_{||}; l_z\kappa)e_\beta^{(s)}(\mathbf{Q}_{||}; l_z'\kappa')^*}{z^2 - \omega_s^2(\mathbf{Q}_{||})}. \tag{5.95}$$

If methods can be developed to calculate this function, then partial spectral densities such as that defined in Eq. (5.93) may readily be constructed by noting the rule, with the limit $\varepsilon \to 0$ implied,

$$\rho_{l_z\kappa;\alpha}(\mathbf{Q}_{||}, \omega) = \frac{i\omega}{\pi} \left[U_{\alpha\alpha}(l_z\kappa, l_z\kappa; \mathbf{Q}_{||}, \omega + i\varepsilon) \right.$$

$$\left. - U_{\alpha\alpha}(l_z\kappa, l_z\kappa; \mathbf{Q}_{||}, \omega - i\varepsilon) \right]$$

$$\equiv \frac{2\omega}{\pi} \mathrm{Im}\{U_{\alpha\alpha}(l_z\kappa, l_z\kappa; \mathbf{Q}_{||}, \omega - i\varepsilon)\}. \tag{5.96}$$

Upon using the equation satisfied by the eigenvector $e_\alpha^{(s)}(\mathbf{Q}_{||}; l_z\kappa)$, one may readily establish that the Green's function obeys the equation

$$z^2 U_{\alpha\beta}(l_z\kappa, l_z'\kappa'; \mathbf{Q}_{||}z) - \sum_{\gamma l_z''\kappa''} d_{\alpha\gamma}(\mathbf{Q}_{||}; l_z\kappa, l_z''\kappa'')U_{\gamma\beta}(l_z''\kappa'', l_z'\kappa'; \mathbf{Q}_{||}z)$$

$$= \sum_s e_\alpha^{(s)}(\mathbf{Q}_{||}; l_z\kappa)e_\beta^{(s)}(\mathbf{Q}_{||}; l_z'\kappa')$$

$$= \delta_{\alpha\beta}\delta_{l_zl_z'}\delta_{\kappa\kappa'}. \tag{5.97}$$

The last equality in Eq. (5.97) follows because $e_\alpha^{(s)}(\mathbf{Q}_{||;||z}\kappa)$ obeys a completeness relation quite analogous to that stated in Eqs. (4.24b)–(4.24d).

The primary difference between Eqs. (5.4) and (5.97) is that the latter is an inhomogeneous rather than a homogeneous equation. Thus, even if the system cannot be solved in closed form, a variety of approximation schemes may be envoked. Consider, for example, the calculation of the diagonal element $U_{\alpha\alpha}(l_z\kappa, l_z\kappa; \mathbf{Q}_{||}z)$. Equation (5.97) may be rearranged to assume the form

$$U_{\alpha\alpha}(l_z\kappa, l_z\kappa; \mathbf{Q}_{||}z) = z^{-2} + z^{-2} \sum_{\gamma l_z''\kappa''} d_{\alpha\gamma}(\mathbf{Q}_{||}; l_z\kappa, l_z''\kappa'')$$

$$\times U_{\gamma\alpha}(l_z''\kappa'', l_z\kappa; \mathbf{Q}_{||}z). \tag{5.98}$$

Now upon eliminating $U_{\gamma\alpha}(l_z''\kappa'', l_z\kappa; \mathbf{Q}_{||}z)$ from the right-hand side of Eq. (5.98) by use of Eq. (5.97), one finds

$$U_{\alpha\alpha}(l_z\kappa, l_z\kappa; \mathbf{Q}_{||}z) = z^{-2} + z^{-4} d_{\alpha\alpha}(\mathbf{Q}_{||}; l_z\kappa, l_z\kappa)$$

$$+ z^{-4} \sum_{\gamma l_z''\kappa''} \sum_{\eta l_z'''\kappa'''} d_{\alpha\gamma}(\mathbf{Q}_{||}; l_z\kappa, l_z''\kappa'')$$

$$\times d_{\gamma\eta}(\mathbf{Q}_{||}; l_z''\kappa'' l_z'''\kappa''') U_{\eta\alpha}(l_z'''\kappa''', l_z\kappa; \mathbf{Q}_{||}z), \tag{5.99}$$

which after one more cycle of iteration becomes

$$U_{\alpha\alpha}(l_z\kappa, l_z\kappa; \mathbf{Q}_{||}z) = z^{-2} + z^{-4} d_{\alpha\alpha}(\mathbf{Q}_{||}; l_z\kappa, l_z\kappa)$$

$$+ z^{-6} \sum_{\gamma l_z''\kappa''} d_{\alpha\gamma}(\mathbf{Q}_{||}; l_z\kappa, l_z''\kappa'')$$

$$\times d_{\gamma\alpha}(\mathbf{Q}_{||}; l_z''\kappa'', l_z\kappa) + \cdots . \tag{5.100}$$

Quite clearly, if this procedure is continued, we generate the power series expansion of the Green's function in inverse powers of the variable z^2. Each coefficient may be evaluated by performing certain specified matrix multiplications with the dynamical matrices which enter Eq. (5.10). While the calculation of high-order terms requires many arithmetic operations and hence becomes very costly in computer time, several of the leading terms may be evaluated quite readily. We then have an asymptotic expansion of the Green's function for large values of z^2, without the need of dealing with the eigenvalue problem stated in Eq. (5.10). Given such an asymptotic expansion, there are mathematical methods that use this information to generate an approximation to the Green's function for all values of z, and which also matches the asymptotic expansion in the region of large z^2. This is done by approximating the Green's function as the ratio of two finite polynomials. In the mathematical literature, such a form is known as a Padé approximate.

A more satisfactory procedure, closely related to that just outlined, is the continued fraction scheme for approximating the Green's function. A clear and readable account of this method, applied to a problem closely related to that described here, has been given by Haydock et al. [15]. Its application to investigations in the lattice dynamics of crystal surfaces has been explored recently, for the direct construction of total spectral densities such as that defined in Eq. (5.92) [16, 17], and for the partial spectral densities. defined in Eq. (5.93) that are probed by electron energy loss spectroscopy [18]. Upon noting from Eq. (5.98) that the leading term in $U_{\alpha\alpha}(l_z \kappa l_z \kappa; \mathbf{Q}_{||} z)$ is simply the Green's function may always be represented as a continued fraction with the form

$$U_{\alpha\alpha}(l_z \kappa l_z \kappa; \mathbf{Q}_{||} z) = \left(z^2 - A_1 - \cfrac{|B_2|^2}{\left(z^2 - A_2 - \cfrac{|B_3|^2}{z^2 - A_3 - \cdots} \right)} \right)^{-1}. \quad (5.101)$$

The basic scheme allows one to calculate the coefficients $\{A_n, B_n\}$ that enter Eq. (5.101) by a matrix multiplication scheme not greatly different in computational labor than that required to generate the asymptotic expansion used in the Padé scheme. After a finite number of coefficients $A_1, A_2, \ldots, A_n$, and $B_1, B_2, \ldots, B_n$ are generated, the hope is that each sequence approaches a well-defined asymptotic limit A_∞, B_∞. Then one approximates the Green's function by using the actual values generated for the A_m, B_m coefficients for $m \le n$, but one replaces these by A_∞, B_∞ for $m > n$. The beauty of the scheme is that once this is done, the continued fraction may be summed out to infinity, to give the form

$$U_{\alpha\alpha}(l_z \kappa, l_z \kappa; \mathbf{Q}_{||} z) = \left(\left(z^2 - A_1 - \cdots - \frac{|B_n|^2}{z^2 - A_n - T(z^2)} \right) \right)^{-1} \quad (5.102)$$

with

$$T(z^2) = \tfrac{1}{2}(z - A_\infty) - \tfrac{1}{2}([z - A_\infty]^2 - 4B_\infty)^{1/2}. \quad (5.103)$$

The function $T(z^2)$ has a branch cut on the real z axis of the complex z plane which extends from $A_\infty + 2B_\infty^{1/2}$ to $A_\infty - 2B_\infty^{1/2}$. We shall appreciate shortly that such a branch cut exists in frequency regimes where, for a particular $\mathbf{Q}_{||}$, one encounters the bulk phonon continuum. In the example explored by analytic methods in Section 5.2.2, there must be one branch cut between $\sqrt{k_0}$ and $\sqrt{2k_0}$, and a second between $\sqrt{3k_0}$ and $\sqrt{2k_0}$. The surface modes associated with a particular value of $\mathbf{Q}_{||}$ appear as poles of the Green's function separated from the phonon continuum, as one can see from direct inspection of Eq. (5.95).

There are cases in which the sequence $A_1, A_2, \ldots$ and/or the corresponding sequence of the B_i do not converge to a simple limit, but to an alternate sequence of values $A_a^{(\infty)}, A_b^{(\infty)}$. In such a case, it is straightforward to derive a generalization of Eq. (5.99). This will happen in cases such as that just cited, where the nature of the bulk phonon spectrum is such that the Green's functions has not a single branch cut, as in Eq. (5.99), but a branch cut structure with two or more disconnected segments. If the convergence properties are unclear for the largest value of the index n handled conveniently by the computer at hand, the method breaks down as a practical tool. Finally, the technique is also limited to the study of models where the interatomic forces are of tolerably short range, so the required multiplications do not become too costly in computer time before a value of n has been reached sufficiently large for the convergence properties of the sequence to become clear. The slab method does not suffer from this difficulty, and has been applied to the study of surface lattice dynamics in alkali halide crystals by de Wette and his colleagues, with impressive results [19].

We conclude this section by returning to the example considered in Section 5.2.2, for which the relevant Green's functions may be obtained by analytic methods. The examination of these results will help the reader to understand the general mathematical structure of these objects.

Consider the fcc crystal with (100) surface explored earlier in this chapter, described within the simple lattice dynamical model in which all atoms are coupled by identical nearest-neighbor interactions of central force character. This is precisely the model used in Sections 5.2.2 and 5.2.3, where the basic properties of surface phonons are explored. If $l_z = 1$ again denotes the surface layer, and we recall that we have only one atom in each two-dimensional unit cell so reference to the index κ may be suppressed, then we shall study the partial spectral density $\rho_{1z}(\mathbf{Q}_{\|}, \omega)$, i.e., the contribution to the frequency spectrum of the modes of wave vector $\mathbf{Q}_{\|}$, to the vibrational motion of atoms in the surface layer perpendicular to the surface. For the construction of this spectral density, we require the function $U_{zz}(11; \mathbf{Q}_{\|}z)$. The equation satisfied by this function may be generated easily by using the dynamical matrices in Eq. (5.15), after noting Eq. (5.97). In the interest of simplicity, we shall confine our attention to the case where $\mathbf{Q}_{\|}$ lies at the point labeled X in the two-dimensional Brillouin zone that forms the upper face of the rectangular prism in Eq. (5.2a). This is in fact the same point in the Brillouin zone explored in detail in the analysis of surface phonons presented in Section 5.2.2 so the equations we utilize here have a structure nearly identical to and readily deduced from Eqs. (5.16a), (5.16b), and (5.18b). If we begin with $U_{zz}(1, 1; \mathbf{Q}_{\|}^{(x)}z)$, it satisfies

$$(z^2 - k_0)U_{zz}(1, 1; \mathbf{Q}_{\|}^{(x)}z) - (i/2)k_0 U_{+z}(2, 1; \mathbf{Q}_{\|}^{(x)}) = 1, \qquad (5.104)$$

where we define

$$U_{+z}(l_z, 1; \mathbf{Q}_{\|}^{(x)}z) = U_{xz}(l_z 1; \mathbf{Q}_{\|}^{(x)}z) + U_{yz}(l_z, 1; \mathbf{Q}_{\|}^{(x)}z). \qquad (5.105)$$

We are thus led to require the equation obeyed by $U_{+z}(2, 1; \mathbf{Q}_{||}^{(x)}z)$. In fact, for $l_z \geq 2$, one finds that

$$(z^2 - 3k_0)U_{+z}(l_z, 1; \mathbf{Q}_{||}^{(x)}z) - ik_0[U_{zz}(l_z + 1, 1; \mathbf{Q}_{||}^{(x)}z)$$
$$- U_{zz}(l_z - 1, 1; \mathbf{Q}_{||}^{(x)}z)] = 0, \tag{5.106}$$

while again for $l_z \geq 2$,

$$(z^2 - 2k_0)U_{zz}(l_z 1; \mathbf{Q}_{||}^{(x)}z) - (i/2)k_0[U_{+z}(l_z + 1, 1; \mathbf{Q}_{||}^{(x)}z)$$
$$- U_{+z}(l_z - 1, 1; \mathbf{Q}_{||}^{(x)}z)] = 0. \tag{5.107}$$

Note that we need $U_{zz}(l_z 1; \mathbf{Q}_{||}^{(x)}z)$ only for odd integers l_z, and $U_{+z}(l_z 1; \mathbf{Q}_{||}^{(x)}z)$ for even integers.

For convenience, in what follows, we shall suppress explicit reference to the combination $\mathbf{Q}_{||}^{(x)}z$ that appears in each function. The hierarchy of equations outlined in Eqs. (5.104)–(5.107) may be solved as follows.

Both Eqs. (5.106) and (5.107) are homogeneous difference equations and are solved by the forms

$$U_{zz}(l_z, 1) = \Lambda_z^{(\infty)} \exp(-\alpha l_z) \tag{5.108}$$

and

$$U_{+z}(l_z, 1) = \Lambda_+^{(\infty)} \exp(-\alpha l_z). \tag{5.109}$$

Substitution into Eqs. (5.106) and (5.107) then shows that α is given by

$$\sinh(\alpha) = (1/\sqrt{2k_0})(3k_0 - z^2)^{1/2}(z^2 - 2k_0)^{1/2}, \tag{5.110}$$

a relation virtually identical to Eq. (5.22), except that now in place of ω^2 on the right-hand side we have the complex variable z^2. From Eq. (5.96), we see that ultimately we wish to consider the case where $z \to \omega \pm i\varepsilon$. We note that with $z = \omega + i\varepsilon$, or $z = \omega - i\varepsilon$, we must always choose that solution of Eq. (5.110) for which $\text{Re}(\alpha) > 0$, since otherwise the Green's functions will diverge as we move into the crystal interior.[8]

In selected regimes of frequency, we have already explored solutions to Eq. (5.110) in Section 5.2.2. When $z = \omega$, a positive real number, we have solutions to Eq. (5.110) with α real and positive in the frequency regime between $\sqrt{2k_0}$ and $\sqrt{3k_0}$, while between zero frequency and $\sqrt{k_0}$, we have solutions with $\alpha = i(\pi/2) + \chi$, and χ real. Both of these frequency regimes

---

[8] Note that if some complex number α_0 is a solution of Eq. (5.106), so is $\alpha_0 + 2\pi in$ for any integer, since $\sinh(\alpha + 2\pi in) = \sinh \alpha$. But since for any integer $\exp[-(\alpha + 2\pi in)l_z] = \exp[-\alpha l_z]$, $\alpha_0 + 2\pi in$ and α_0 are not physically distinct solutions of the equation. Hence, we may confine our search for solutions to that strip of the complex α plane which includes the real axis, and extends up to but does not include the line $\text{Im}(\alpha) = 2\pi$. Within this strip there is one and only one solution to Eq. (5.106) with $\text{Re}(\alpha) > 0$.

lie *outside* the frequency bands allocated to bulk phonons with wave vector $Q_{||}^{(x)}$ parallel to the surface, as one sees from Fig. 5.5.

Within the bulk phonon bands at the X point, in the frequency regime between $\sqrt{k_0}$ and $\sqrt{2k_0}$ and that between $\sqrt{3k_0}$ and $2\sqrt{k_0}$, with z on the real axis, α is purely imaginary. In essence, Eqs. (5.105) and (5.106) describe a wave which is launched from the surface, and which propagates deeply into the bulk of the crystal. Here, we imagine ε to be finite in the relation $z = \omega \pm i\varepsilon$, and always select that root α with positive real part.

Finally, for the forms in Eqs. (5.108) and (5.109) to satisfy Eqs. (5.106) and (5.107), we are led to also require

$$\Lambda_+^{(\infty)}/\Lambda_z^{(\infty)} = +2ik_0 \sinh(\alpha)/(3k_0 - z^2), \tag{5.111}$$

a relation quite similar to Eq. (5.24). Thus, we have one free parameter in our solution so far.

This one free parameter is determined uniquely by requiring that the forms in Eqs. (5.108) and (5.109) also satisfy the inhomogeneous equation, Eq. (5.104). After a bit of algebra, from this requirement one finds

$$\Lambda_z^{(\infty)} = \frac{(3k_0 - z^2)e^{+\alpha}}{(\frac{5}{2}k_0 + \frac{1}{2}k_0 e^{+2\alpha} - z^2)}, \tag{5.112}$$

which yields the desired Green's function:

$$U_{zz}(1,1;Q_{||}^{(x)}z) = \frac{(3k_0 - z^2)}{(\frac{5}{2}k_0 + \frac{1}{2}k_0 e^{+2\alpha} - z^2)}, \tag{5.113}$$

with α and z^2 linked by Eq. (5.110) and the requirement that $\text{Re}(\alpha) > 0$.

We consider the various characteristic frequency regimes next. When we take $z = \omega - i\varepsilon$, then let $\varepsilon \to 0$ for ω in the frequency regime between $\sqrt{2k_0}$ and $\sqrt{3k_0}$, we have remarked earlier that Eq. (5.106) yields a solution with α positive and real. The right-hand side of Eq. (5.108) is real here, and furthermore we have $\frac{5}{2}k_0 + \frac{1}{2}k_0 e^{+2\alpha} > 3k_0$ always. Hence, there are no singularities on the right-hand side of Eq. (5.108) and the partial spectral density on the right-hand side of Eq. (5.93) vanishes identically.

Now below the frequency $\sqrt{k_0}$ we have, according to Eq. (5.29) solutions with $\alpha = i\pi/2 + \chi$, where χ is real and positive. Thus, with $z = \omega - i\varepsilon$ and the limit $\varepsilon \to 0$ again, we have

$$U_{zz}(11;Q_{||}^{(x)}\omega) = \frac{(3k_0 - \omega^2)}{(\frac{5}{2}k_0 - \frac{1}{2}e^{+2\chi} - \omega^2)}, \tag{5.114}$$

where χ is found from Eq. (5.30). A study of the denominator shows it vanishes when $\omega = 0.8481\sqrt{k_0}$, which is precisely the frequency of the S_4 surface phonon described in Section 5.2.2.

Thus, we see that the surface phonons manifest themselves as *poles* of the Green's function which lie outside the frequency regimes which are allocated to the bulk phonons of the wave vector $Q_{||}^{(x)}$. In Section 5.2.2, for the X point in the two-dimensional Brillouin zone, we found *three* surface modes, while only one of them produces a pole in $U_{zz}(11; Q_{||}^{(x)}\omega)$. The reason for this is that the S_4 mode is the only mode for which there is a displacement component perpendicular to the surface, in the surface layer itself. Hence this is the only one which contributes to the frequency spectrum of the atomic vibrations normal to the surface, and is thus the unique contributor to $U_{zz}(11; Q_{||}^{(x)}\omega)$. Now for frequencies in the near vicinity of the pole at $\omega = \omega_{s_4}$, we have

$$U_{zz}(11; Q_{||}^{(x)}, \omega - i\varepsilon) \cong \frac{r(\omega_{s_4})}{\omega - \omega_{s_4} - i\varepsilon}, \tag{5.115}$$

where $r(\omega_{s_4})$ is the residue at the pole which occurs when $\omega = \omega_{s_4}$. Thus, in the region of frequency between 0 and $\sqrt{k_0}$, we have

$$\rho_{1z}(Q_{||}^{(x)}, \omega) = 2\omega_{s_4} r(\omega_{s_4}) \delta(\omega - \omega_{s_4}). \tag{5.116}$$

The surface mode thus contributes a pole to the partial spectral density, with integrated strength controlled by the residue at the pole.

For the model explored here, it is possible to obtain closed form, analytic solutions for the behavior of the Green's function in the frequency regime between $\sqrt{k_0}$ and $\sqrt{2k_0}$, and also between $\sqrt{3k_0}$ and $2\sqrt{k_0}$. Here one finds the partial spectral density is a continuous function of frequency, with form determined in part by the density of states associated with the bulk phonons of wave vector projection onto the surface plane equal to $Q_{||}^{(x)}$, and partly the eigenvectors as influenced by the boundary conditions applicable at the crystal surface. For the case where the frequency lies between $\sqrt{k_0}$ and $\sqrt{2k_0}$ we find, after letting $\omega = \sqrt{k_0(1 + x)}$ with x ranging between 0 and 1,

$$U_{zz}(11; Q_{||}^{(x)}, \omega - i\varepsilon) = \frac{2(2 - x)^{1/2}}{(2 - x)^{1/2}(1 + x) - ix^{1/2}(1 - x)^{1/2}(3 - x)^{1/2}}, \tag{5.117}$$

a form that offers little insight into the physics of the surface vibrations, but, nonetheless, is a simple and compact form.

In conclusion, the purpose of this subsection has been to establish that not only are the surface phonons studied earlier of interest, but also we need to be concerned in the study of surface lattice dynamics with the influence of the surface on the bulk phonons. After our discussion of the theory of lattice dynamics at the surface is completed, and we turn our attention to the connection between the theory and electron energy loss spectroscopy, we shall appreciate that the bulk modes may be studied quite directly, in suitable experiments.

5.2.6 Surface Phonons in the Presence of Ordered
Adsorbate Layers: Lateral Interactions
between Adsorbates

In Chapter 4, we examined in detail the nature of the vibrational motion of an isolated atom or molecule adsorbed on the crystal surface. Here we turn to a discussion of the motion of an ordered overlayer of adatoms or molecules bound to the surface. As stated earlier in the present chapter, our attention will be confined exclusively to the case of an ordered overlayer commensurate in structure with the underlying substrate. Also, we shall discuss primarily an overlayer of atoms, since this is the simplest case to examine, yet all the essential principles are illustrated by it.

Consider first a single, light adsorbate atom bound tightly to the surface, i.e., chemisorbed rather than physisorbed. Then as we saw in Chapter 4, we encounter vibrational modes of the adsorbate–substrate complex which have frequencies that lie above the phonon bands of the substrate; the displacement field associated with these high-frequency modes is localized to the near vicinity of the adsorbate. If the adsorbate is sufficiently light, then its vibrational frequencies may be calculated accurately by supposing all the substrate atoms remain at rest when the vibrational mode is excited. The first corrections to this picture may also be treated simply, and in Table 4.1 we have summarized a series of formulas which give these first corrections, for the case where the adatom "sees" only the nearest-neighbor substrate atom through couplings of central force character. Very similar reasoning also applies to the high-frequency modes of molecules adsorbed on the surface, though the expressions one obtains are more complex.

Now consider an ordered layer of adsorbed atoms. It will prove illuminating to consider a specific system that is simple enough to be analyzed in quantitative terms, without great complexity. We shall once again examine the fcc crystal with a (100) surface. Suppose an ordered overlayer of atoms is present with the $c(2 \times 2)$ structure. The geometry is illustrated in Fig. 5.6, where in Fig. 5.6a we give the $c(2 \times 2)$ structure and in Fig. 5.6b we illustrate the $p(2 \times 2)$ structure. Both of these are encountered commonly for layers of hydrogen, oxygen, and sulfur atoms on the (100) surfaces of the fcc transition metals. In both ordered structures, each adsorbate atom sits on a site of fourfold symmetry. In the $c(2 \times 2)$ structure, one has two adsorbate atoms per unit cell, and the coverage $\Theta = 0.5$, while $\Theta = 0.25$ for the $p(2 \times 2)$ structure. A possible choice of unit cell is shown also in each figure; quite clearly there are two substrate atoms per unit cell in the $c(2 \times 2)$ structure, and four in the $p(2 \times 2)$.

The vibrational normal modes of a structure such as that described in Fig. 5.6a must be discussed within the framework of the phonon theory developed in this chapter. It is interesting to explore the same model used in

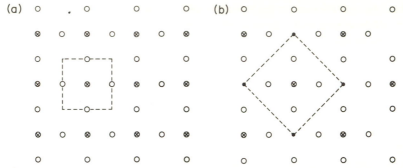

Fig. 5.6 (a) The $c(2 \times 2)$ overlayer on the (100) surface of an fcc crystal. The open circles are substrate atoms, and crossed circles are adsorbate atoms. The dashed square is a possible choice of the unit cell of the structure. (b) The $p(2 \times 2)$ layer on the (100) surface, with convention identical to that used in (a).

Section 4.2.2, where each adatom (mass M_A) interacts with each of its four neighbors in the substrate (mass M_s) by means of nearest-neighbor central force interactions.[9] In practice, in dense overlayers, there will also be direct interactions between the adsorbate atoms themselves. We turn to a discussion of the influence of these couplings later in the present section.

Suppose, as in Section 4.2.2, we begin with a description of the adatom motion with all substrate atoms "pinned" in place. Then each atom vibrates independently of all others. If we consider some particular normal mode of one atom, say that in which the displacement is perpendicular to the surface, then if there are N_s adatoms on the surface, we have N_s degenerate vibrational normal modes of the system. In the language of lattice dynamics, the adatom system forms an array of N_s Einstein oscillators, with frequency for vibration normal to the surface given by Eq. (4.38a), and that for the twofold degenerate modes parallel to the surface by Eq. (4.38b).

We should describe the motion of the adatom system in the phonon language developed in the present chapter, when the adsorbed layer is present. Consider the mode polarized perpendicular to the surface, and let $\xi_z(l_{||}; 1)$ be the displacement formed as in Eq. (5.3) of the adatom in the unit cell labeled by $l_{||}$. (The label 1 is used to indicate that the adsorbate layer is that with $l_z = 1$, and since there is only one adatom per unit cell, the label κ is dropped here and elsewhere in the present section.) Now if $\mathbf{Q}_{||}$ is any wave vector parallel to the surface, and we form the combination

$$e_z(\mathbf{Q}_{||}; 1) = N_s^{-1/2} \sum_{l_{||}} \exp[-i\mathbf{Q}_{||} \cdot \mathbf{R}_0(l_{||}; 1)]\xi_z(l_{||}; 1), \qquad (5.118)$$

[9] A full discussion of this model is given in Ref. [18]. There it is also argued that in the $c(2 \times 2)$ structure, electron energy loss data suggests that the oxygen overlayer lies much closer to the surface than the 0.88 Å assumed later in this section. A brief summary of these arguments will be presented later in the present chapter.

then $e_z(\mathbf{Q}_{||}; 1)$ is also an eigenvector of a mode of our vibrational problem with the frequency $\omega_\perp^{(0)}$ given in Eq. (4.38a), since it is a linear combination of eigenvectors associated with the N_s-fold degenerate vibrations of frequency $\omega_\perp^{(0)}$. The factor of $N_s^{-1/2}$ is inserted so that the eigenvector just defined is conveniently normalized.

Now if we construct the two-dimensional Brillouin zone for our structure, as described earlier in this chapter, we may form an eigenvector $e_z(\mathbf{Q}_{||}; 1)$ such as that in Eq. (5.113) for each wave vector $\mathbf{Q}_{||}$ in the zone. Each eigenvector describes a *surface phonon* with displacement localized to the adatom layer, i.e., a coherent, phased motion of the adatom layer. If we denote the dispersion relation of this mode by $\omega_\perp(\mathbf{Q}_{||})$, then we have $\omega_\perp(\mathbf{Q}_{||}) = \omega_\perp^{(0)}$ independent of $\mathbf{Q}_{||}$. The dispersion curve is then a flat curve in the two-dimensional Brillouin zone.

At the level of approximation that applies in the previous discussion, where each adatom vibrates independently of all others, it does not matter at all whether one regards the adatoms as an array of uncoupled oscillators with normal modes described by the set of displacement $\xi_z(l_{||}; 1)$, or whether one chooses to use the language of phonon theory with its Brillouin zones, wave vectors, and so on. From the mathematical viewpoint, the relation in Eq. (5.114) and its inverse

$$\xi_z(l_{||}; 1) = N_s^{-1/2} \sum_{\mathbf{Q}_{||}} \exp[+i\mathbf{Q}_{||} \cdot \mathbf{R}_0(l_{||}; 1)] e_z(\mathbf{Q}_{||}; 1) \qquad (5.119)$$

describe a unitary transformation between one basis set $\{\xi_z(l_{||}; 1)\}$ and a second equally acceptable set $\{e_z(\mathbf{Q}_{||}; 1)\}$, each of which describes an N_s-fold degenerate motion. The results of a calculation of any physically measurable quantity must necessarily be independent of which basis set as chosen.

As soon as coupling between the adsorbate atoms is introduced, then as we shall see shortly the dispersion relations of the new surface phonons of the structure introduced by the adsorbate layer will acquire a dependence on the wave vector $\mathbf{Q}_{||}$ parallel to the surface. Then it becomes imperative to use the language of phonon theory, since the displacement vectors $\xi_z(l_{||}; 1)$ are no longer by themselves eigenvectors of the vibrational Hamiltonian.

The experimental study of lateral interactions between absorbates remains in the very early stages, as far as the existence of quantitative experimental data is concerned. In Section 3.4.4, we have presented strong evidence of dipole–dipole interactions for layers of CO adsorbed on Ru. Also, in a beautiful study by electron energy loss spectroscopy of a $c(2 \times 2)$ adsorbed layer of CO on Cu(100), Andersson and Persson also find direct evidence for the importance of lateral interactions of dipolar character [20]. We shall discuss their work in the next section.

Even if direct interactions between adsorbate atoms or molecules are too weak to influence the surface lattice dynamics directly, there are indirect couplings between adsorbates mediated by the substrate. Suppose now the substrate atoms are allowed to vibrate, in response to the motion of the adsorbate. Then, as we have seen in Chapter 4, when an adsorbate vibrates with some frequency, assumed to lie above the maximum phonon frequency of the substrate at present, then motion of the substrate atoms in its near vicinity is excited. A nearby adsorbate senses this motion, and responds to it; we thus have indirect interactions between adsorbates through the inter-mediary of the substrate.

For the simple model of the adsorbate layer on the (100) surface of an fcc crystal, the influence of this indirect interaction on the surface-phonon dispersion curve may be explored in the limit $M_A \ll M_s$ by the method used to generate the formulas in Eqs. (4.39) and (4.41), and Table 4.1. We consider the $c(2 \times 2)$ adsorbate layer, once again with each adsorbate atom coupled to the nearest-neighbor atoms by means of interactions of central force character. Let $d_{\alpha\beta}(\mathbf{Q}_{||};11)$ be that portion of the dynamical matrix that describes the adatom motion, constructed as in Eq. (5.9), while $d_{\alpha\beta}(\mathbf{Q}_{||};1,l_z\kappa)$ and $d_{\alpha\beta}(\mathbf{Q}_{||};l_z\kappa,1)$ describes coupling between the adsorbate layer and the substrate. It is then straightforward to derive an effective dynamical matrix that describes motion of the adsorbate layer with effects of order M_A/M_s included in it. This may be done by following the derivation that led to Eq. (4.37), in our approximate description of the motion of an isolated adatom. The results are as follows. With the substrate atoms clamped in position, the motion of the substrate layer is governed by $d_{\alpha\beta}(\mathbf{Q}_{||}11)$:

$$\omega^2 e_\alpha(\mathbf{Q}_{||};1) = \sum_\beta d_{\alpha\beta}(\mathbf{Q}_{||};11)e_\beta(\mathbf{Q}_{||};1), \tag{5.120}$$

To first order in M_A/M_s, the same equation applies with $d_{\alpha\beta}(\mathbf{Q}_{||};11)$ replaced by the effective dynamical matrix $\tilde{d}_{\alpha\beta}(\mathbf{Q}_{||};11)$,

$$\tilde{d}_{\alpha\beta}(\mathbf{Q}_{||};11) \cong d_{\alpha\beta}(\mathbf{Q}_{||};11) + \omega^{-2} \sum_{l_z'\kappa'} \sum_\eta d_{\alpha\eta}(\mathbf{Q}_{||};1l_z'\kappa') d_{\eta\beta}(\mathbf{Q}_{||};l_z'\kappa',1). \tag{5.121}$$

For selected directions in the two-dimensional Brillouin zone, one may readily work out the surface-phonon dispersion curves to first order in M_A/M_s, for our model of the $c(2 \times 2)$ adlayer on the (100) surface of an fcc crystal. Suppose, once again, we consider the direction with $Q_x = Q_y$, and let $Q_x = Q_y = Q$ with a_0 the distance between nearest-neighbor substrate atoms. Then to first order in M_A/M_s, one finds, with $\omega_\perp^{(0)}$ given by Eq. (4.38a),

$$\omega_\perp^2(\mathbf{Q}_{||}) = \omega_\perp^{(0)2}\left[1 + \frac{1}{2}\frac{M_A}{M_s} - \frac{1}{2}\frac{M_A}{M_s}\frac{\cos(2\alpha)}{\cos^2(\alpha)}\sin^2(\tfrac{1}{2}Qa_0)\right] \tag{5.122}$$

for the dispersion relation of the mode polarized normal to the surface. We have two modes polarized parallel to the surface, one with displacement parallel to the wave vector $\mathbf{Q}_{||} = Q(\hat{x} + \hat{y})$, and one normal to it. If we refer to this pair of modes by $\omega^2_{\|_a}$ and $\omega^2_{\|_b}$, respectively, then with $\omega^{(0)2}_{\|}$ given by Eq. (4.38b),

$$\omega^2_{\|_a} = \omega^{(0)2}_{\|}\left[1 + \frac{M_A}{M_s}\frac{1}{\sin^2(\alpha)} + \frac{M_A}{M_s}\cot^2(\alpha)\sin^2(\tfrac{1}{2}Qa_0)\right] \qquad (5.123)$$

and

$$\omega^2_{\|_b} = \omega^{(0)2}_{\|}\left[1 + \frac{M_A}{M_s}\frac{1}{\sin^2(\alpha)} - \frac{M_A}{M_s}\cot^2(\alpha)\sin^2(\tfrac{1}{2}Qa_0)\right]. \qquad (5.124)$$

Strictly speaking, it is misleading to refer to the modes with the dispersion relations in Eqs. (5.122), (5.123), and (5.124) as perpendicular and parallel modes, because with coupling to the substrate included and $Q_{\|} \neq 0$, each eigenvector is a mixture of motion parallel and perpendicular to the surface. But when $M_A \ll M_s$, where Eqs. (5.122)–(5.124) are valid, one may show the component of motion parallel to the surface in the mode described by Eq. (5.122) is of order (M_A/M_s), and is thus small in the relevant limit. A similar statement holds for the modes described by Eqs. (5.123) and (5.124).

It is interesting to compare the frequencies associated with the ordered overlayer considered here, with those for the isolated adatom considered in Chapter 4. Consider an *isolated* oxygen atom absorbed on the Ni(100) surface, in the fourfold site appropriate to the structures illustrated in Fig. 5.6. If we assume, as in Chapter 4, the oxygen sits above the surface a distance of 0.88 Å, and the force constants are as described in Section 4.2.2, the frequency of the isolated adatom is 430 cm^{-1}, for the vibrational mode with motion normal to the surface (Table 4.3). A result very close to this follows from Eq. (4.39a). If we consider a $c(2 \times 2)$ layer of oxygen atoms placed above the (100) surface the same distance, and use the value of the oxygen–nickel coupling employed to generate Table 4.3, then Eq. (5.122) gives 396 cm^{-1} in good accord with the 401 cm^{-1} obtained from an exact solution of this model. At the Brillouin zone boundary of the $c(2 \times 2)$ structure, where $Q = Q_x = Q_y = \pi/a_0$, we have 464 cm^{-1} for the frequency of the surface phonon polarized normal to the surface. This example shows that even in the absence of *direct* coupling between adatoms, the indirect interaction through the substrate can influence the lattice dynamics of the adsorbed layer very strongly.

If we consider the $p(2 \times 2)$ adatom layer illustrated in Fig. 5.6b, then to first order in the ratio M_A/M_s, the frequencies of the surface phonons associated with the adatom layer are independent of $Q_{\|}$, and identical to the expressions given in Eqs. (4.39a) and (4.39b). Thus, for this less dense layer, indirect coupling between the adatoms is higher order in the ratio (M_A/M_s), and plays a minor role in this particular example.

The expressions just derived describe the dispersion relations of new surface-phonon branches induced by the adsorbate layer, with frequency higher than the maximum substrate phonon frequency ω_M. For the adsorbate layer, these new surface phonon branches are the analog of the high-frequency modes with frequency $\omega_v > \omega_M$, as discussed in the introductory paragraphs of Section 4.2.2. We also argued that it is possible to have frequencies ω_v of an adsorbate complex below ω_M. Here we cannot have well-defined vibrational modes of the adsorbate layer with infinite lifetime even in the harmonic approximation, since when vibrating, an adsorbate layer may "radiate" its energy down into the substrate layer. Any motion, then, has a finite lifetime and, strictly speaking, we may not have eigenmodes of the system localized at the surface, in the frequency regime between 0 and ω_M. This is so, unless we consider a mode decoupled from the substrate by symmetry.

Nonetheless, if a motion of an adsorbed molecule or atom is excited in the frequency regime between 0 to ω_M and if the coupling is weak, the vibrational excitation may be long lived, to appear as a "near-normal" mode with lifetime long compared to the basic vibrational period. In an experiment, it may prove difficult to distinguish between such long-lived resonances of the surfaces, and true normal modes. We conclude this section with an example, within a theoretical model, of such a resonance mode. We shall see another example when we discuss experimental studies of oxygen overlayers on the Ni(100) surface by the method of electron energy loss spectroscopy.

In their theoretical study of the interaction of an oxygen atom with a cluster of 20 Ni atoms arranged to mimic the Ni(100) surface, Upton and Goddard found two distinctly different equilibrium positions [21]. One has a distance $R_\perp = 0.88$ Å above the surface, and the second has $R_\perp = 0.26$ Å above it. While their calculations have been carried out for an isolated oxygen atom above the surface, they argue that the lower position may possibly be realized for the high density $c(2 \times 2)$ structure. We shall see in the next section that electron energy loss studies of this structure provide evidence that this picture is correct. For the moment, we use the parameters provided by their work to explore features that occur in that portion of the frequency spectrum of the surface vibrations sampled in electron energy loss spectroscopy.

In Fig. 5.7, we show the frequency spectrum $\rho_\perp(\omega)$ for fluctuations in the surface dipole moment of a model Ni(100) crystal with a $c(2 \times 2)$ layer of oxygen placed on it.[10] The spectral density is formed to describe the frequency spectrum of the *relative* motion of the oxygen and outermost Ni layer in the direction normal to the surface. For this, we require a spectral density defined as in Eq. (5.90) with $\mathbf{Q}_{||} = 0$, and $|e_\alpha^{(s)}(\mathbf{Q}_{||}, l_z\kappa)|^2$ replaced by the quantity

$$|e_s^{(s)}(\mathbf{Q}_{||}; 1) - \tfrac{1}{2}(e_z^{(s)}(\mathbf{Q}_{||}; 21) + e_z^{(s)}(\mathbf{Q}_{||}; 22))|^2,$$

[10] We are grateful to Dr. Talat S. Rahman for providing us with these results.

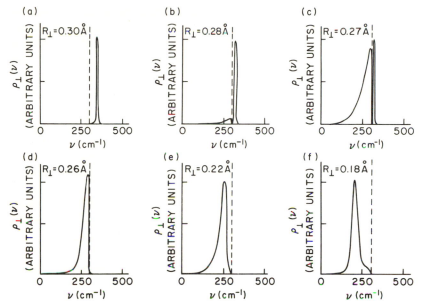

Fig. 5.7 (a)–(e). A series of calculations which show the frequency spectrum of surface dipole oscillations at $\mathbf{Q}_{||} = 0$, for the $c(2 \times 2)$ layer of oxygen on a Ni(100) surface. The figures are labeled with the distance $R_\perp$ of the layer of oxygen nuclei above the surface, and in each calculation $\varphi''(d_0)$ is the same and has been adjusted to fit the curvature of the potential energy curve calculated by Upton and Goddard, which places the $c(2 \times 2)$ structure 0.26 Å above the Ni surface. In each figure, the dashed line illustrates the maximum Ni phonon frequency of 295 cm^{-1}.

where $e_z^{(s)}(\mathbf{Q}_{||}; 1)$ describes the oxygen motion normal to the surface when the mode $(\mathbf{Q}_{||}, s)$ is excited, and $e_z^{(s)}(\mathbf{Q}_{||}; 2\kappa)$ that of the Ni atoms in site κ within the unit cell of the outermost Ni layer, which has $l_z = 2$. The spectral density has been constructed through use of the Green's function method described in the previous subsection [18].

The spectral densities in Fig. 5.7 have been constructed in an artificial way, but they serve to illustrate the point we wish to make. Upton and Goddard find the oxygen equilibrium position to have $R_\perp = 0.26$ Å, as mentioned earlier. The value of $\varphi''(d_0)$ required in the calculations has been fitted to the shape of the potential energy curve they calculate, with Ni atoms held rigidly fixed, through use of Eq. (4.38a). Then to generate Fig. 5.7, $\varphi''(d_0)$ has been held fixed while $R_\perp$ has been varied. As we see in Eq. (4.38a), as the oxygen atom is lowered toward the surface, the angle α increases toward $\frac{1}{2}\pi$, and the oxygen vibration frequencies drop, for motion normal to the surface. Quite clearly, the influence of the Ni vibrations must be incorporated into the description of the problem, and this has been done in Fig. 5.7. What is artificial is the fact that $R_\perp$ has been varied while $\varphi''(d_0)$ is held fixed, in the calculations.

In Fig. 5.7a, calculated for $R_\perp = 0.30$ Å, we see a sharp feature just above the maximum Ni phonon frequency of 295 cm^{-1}. This comes from a surface phonon very similar in nature to that described by Eq. (5.117). In the harmonic approximation of lattice dynamics, such surface phonons introduce features in the spectral density which are Dirac delta functions, as we see from Eqs. (5.115) and (5.116). In the numerical calculations, this feature has been artificially broadened. As the oxygen is lowered toward the surface, the surface phonon lowers in frequency to ultimately merge with and become undistinguishable from the Ni phonons. The spectra when this is about to occur, or just after it occurs (Figs. 5.7c and 5.7d) are complex in form and little can be said about them. By the time the oxygen restoring force has become so small that the characteristic frequencies of the adlayer are low compared to ω_M, we again see a clean, well-defined roughly symmetric feature in the spectral density function. This is a resonance mode as described in Section 4.2.2 and earlier in this section. The width in this figure is not artificially introduced, but is present even in the harmonic approximation of lattice dynamics. The adatom layer has a low-frequency resonance, but when this resonance is excited, its energy leaks off into the bulk of the crystal in the form of bulk phonon "radiation." To the experimentalist, such a relatively long-lived resonance of the surface region will appear as a line in an energy loss spectrum, indistinguishable in any qualitative sense from features introduced by true surface phonons, which are infinitely long-lived in the harmonic approximation. We shall encounter examples of such resonance modes in actual spectra in the next section, when we discuss studies of oxygen overlayers on the Ni(100) and Ni(111) surfaces by electron energy loss spectroscopy. We remind the reader that such low-frequency resonance modes are well known in the theory of bulk lattice dynamics of impure crystals [4, 22], though there has been rather little discussion of them in the context of surface lattice dynamics.

In Chapter 4, we discussed the vibrational spectra of molecules adsorbed in the surface at some length. Here we encounter low-frequency modes referred to often as the hindered rotations and translations of the molecular entity. When due account is taken of the substrate motion, these low-lying features necessarily become resonance modes of the sort illustrated in Fig. 5.7e. It is possible for such a resonance mode to be very long lived, if it has symmetry such that it couples weakly to bulk phonons in its frequency range or if the density of bulk phonon states is very small there; it may also be broadened to the point where it no longer exists as a well-defined linelike feature in the spectral density. Without resort to model calculations, it is difficult to decide in advance what the width of the feature will be.

We now turn to a summary of experimental studies of electron energy loss spectra of surfaces in which surface phonons or surface resonances have been studied.

In this section, we have not included direct interactions between adatoms in our formulation of the surface lattices dynamical problems. From a theoretical point of view, this can be done without difficulty. A number of authors have explored this question, with emphasis on the role of dipole–dipole interactions between adsorbates [23]. While inclusion of such effects into the theory is quite straightforward, when the dipole–dipole interactions are modeled from the microscopic point of view, delicate questions about the influences of image charges arise. We prefer not to address these issues here, but we shall assume when necessary that all self-image effects are included in the definition of the dipole moment effective charge introduced in Chapter 3. In the next section, we shall discuss recent electron energy loss studies of CO on the Cu(100) surface which provide direct evidence of the importance of dipole–dipole interactions between the adsorbed CO molecules [20].

5.3 THE STUDY OF SURFACE PHONONS BY ELECTRON ENERGY LOSS SPECTROSCOPY

In this section, we discuss certain electron energy loss experiments which explore the properties of surface phonons on crystal surfaces. The emphasis is not on the high-frequency vibrations of adsorbates, with frequency so high the substrate is weakly excited, but rather the energy loss spectra of either clean surfaces or those covered with ordered overlayers, where the modes studied involve substrate atom motion as an essential feature.

5.3.1 General Remarks

We wish to examine processes in which an electron with impact energy E_I strikes the surface, to emit a phonon with frequency $\omega_\alpha(\mathbf{Q}_{||})$. For the purposes of this section, it does not matter whether the mode is a surface phonon, or a bulk mode with eigenvectors modified by the surface, as discussed earlier. Thus, the subscript α refers to one of the modes of the structure with wave vector $\mathbf{Q}_{||}$, either a surface or bulk phonon. Quite clearly, energy is conserved in the scattering process, so if the electron emits the phonon in the scattering process, it emerges with energy E_S given by

$$E_S = E_I - \hbar\omega_\alpha(\mathbf{Q}_{||}). \tag{5.125}$$

Next consider conservation of wave vector components. Let us recall the discussion of the inelastic scattering of electron by phonons in crystal lattices of infinite spatial extent, since the same principles apply to the problem of present interest. An electron of wave vector $\mathbf{k}^{(I)}$ may emit a phonon of wave vector $\mathbf{Q}$ in the bulk of the crystal, to be scattered to a final state of wave vector $\mathbf{k}^{(S)}$. If $\mathbf{G}$ is one of the reciprocal lattice vectors of the (three-dimensional) crystal, then in addition to energy conservation, we have $\mathbf{k}^{(S)} = \mathbf{k}^{(I)} - \mathbf{Q} + \mathbf{G}$. Wave vector is not strictly conserved, but only to within a reciprocal lattice vector.

We may easily adapt this picture to the problem of the scattering of electrons from phonons at the crystal surface. As the electron approaches the crystal, it sees a structure with periodicity in the two directions parallel to the surface plane. We are assuming here, as elsewhere in the chapter, that any adsorbates are present as an ordered overlayer commensurate with the crystal lattice. Thus, quite clearly, if the electron emits a phonon with wave vector $\mathbf{Q}_{||}$ parallel to the surface, we have

$$\mathbf{k}_{||}^{(S)} = \mathbf{k}_{||}^{(I)} - \mathbf{Q}_{||} + \mathbf{G}_{||}, \tag{5.126}$$

where $\mathbf{G}_{||}$, possibly zero, is one of the reciprocal lattice vectors of the two-dimensionally periodic structure encountered by the electron. Here $\mathbf{k}_{||}^{(I)}$ and $\mathbf{k}_{||}^{(S)}$ are projections of the wave vector of the incident and scattered electron onto the plane parallel to the crystal surface. The proof of this relation proceeds along the lines of the derivation encountered in the theory of electron–phonon interactions in the bulk. One notes that the incoming and scattered electron wave functions have the Bloch property in the two directions parallel to the surface, so they may be written $\exp(i\mathbf{k}_{||} \cdot \mathbf{x}_{||})u_{\mathbf{k}_{||}}(\mathbf{x}_{||}, z)$, with $\mathbf{k}_{||}$ the appropriate wave vector and $u_{\mathbf{k}_{||}}(\mathbf{x}_{||}, z)$ is a periodic function of the projection $\mathbf{x}_{||}$ of the position vector of the electron onto a plane parallel to the surface. This combined with the general structure of the electron–phonon interaction leads to Eq. (5.126). The presence of the surface breaks down the translational symmetry of the crystal in the direction normal to the surface, so no condition analogous to Eq. (5.126) applies to wave vector components *normal* to the surface.

The relation in Eq. (5.126) is to be employed as follows. The wave vector $\mathbf{Q}_{||}$ of the phonon involved in the inelastic scattering necessarily lies within the first Brillouin zone of the crystal. Neither $\mathbf{k}_{||}^{(I)}$ or $\mathbf{k}_{||}^{(S)}$ are so constrained, but instead are determined by the scattering geometry. If, for example, we ignore the small change in energy $\hbar\omega_\alpha(\mathbf{Q}_{||})$ given on the right-hand side of Eq. (5.125), then the magnitude of the scattered electron wave vector equals that of the incident one, and $\mathbf{k}_{||}^{(S)}$ is then determined by the location of the detectors. Now the shortest reciprocal lattice vectors in the set $\{\mathbf{G}_{||}\}$ just span across the first Brillouin zone by construction, and the remainder have magnitude larger than its width. Then if the scattering geometry is such that $\mathbf{k}_{||}^{(S)} - \mathbf{k}_{||}^{(I)}$ lies *within* the first Brillouin zone, Eq. (5.126) can be satisfied only with the choice $\mathbf{G}_{||} \equiv 0$, since for any nonzero $\mathbf{Q}_{||}$ within the first zone and $\mathbf{G}_{||} \neq 0$, $\mathbf{Q}_{||} + \mathbf{G}_{||}$ necessarily lies *outside*. Suppose, instead, that $\mathbf{k}_{||}^{(S)} - \mathbf{k}_{||}^{(I)}$ lies *outside* the first Brillouin zone. Then clearly Eq. (5.126) cannot be satisfied with $\mathbf{G}_{||} \equiv 0$ and $\mathbf{Q}_{||}$ constrained to lie within the first zone. There is, however, one and only one reciprocal lattice vector $\mathbf{G}_{||}$ which has the property that $\mathbf{k}_{||}^{(S)} - \mathbf{k}_{||}^{(I)} - \mathbf{G}_{||}$ lies within the first zone. One sees this by noting that the first Brillouin zone is in fact one particular unit cell (the Wigner–Seitz unit cell) of the reciprocal lattice, and each $\mathbf{G}_{||}$ is a vector from the origin of $\mathbf{k}$ space to a point on the reciprocal lattice. Thus, if $\mathbf{k}_{||}^{(S)} - \mathbf{k}_{||}^{(I)}$ lies outside the first Brillouin zone, it lies within some other unit cell of the reciprocal lattice, and there is one and only one reciprocal lattice vector that can translate it back into the first Brillouin zone. It then follows that if $\mathbf{k}_{||}^{(S)} - \mathbf{k}_{||}^{(I)}$ lies outside the first Brillouin zone, Eq. (5.121) is satisfied for one unique choice of $\mathbf{Q}_{||}$ within it, and one unique value of $\mathbf{G}_{||}$.

Suppose we consider an incident electron with wave vector $\mathbf{k}^{(I)}$, and it emits a phonon with wave vector $\mathbf{Q}_{||}$ and frequency $\omega_\alpha(\mathbf{Q}_{||})$, as just described. To specify the direction $\mathbf{k}^{(S)}$ of the outgoing electron once $\mathbf{Q}_{||}$ and $\omega_\alpha(\mathbf{Q}_{||})$ are known, we require three conditions, since the wave vector $\mathbf{k}^{(S)}$ of the outgoing electron has three Cartesian components. Equations (5.125) and (5.126) are precisely three constraints, so the direction of the outgoing electron is uniquely determined by these three statements.

There is one useful approximation that greatly simplifies the analysis of experimental data, and we shall invoke this here. If, for example, we consider the scattering of electrons by substrate phonons, we are concerned with modes for which $\hbar\omega_\alpha(\mathbf{Q}_{||})$ is the order of 50 meV. In typical experiments, the electron impact energy is of the order of 5 eV, as we have seen. Thus, we make little error if we determine the direction of the outgoing electron by using Eq. (5.126), and dropping the term $\hbar\omega_\alpha(\mathbf{Q}_{||})$ from the right-hand side of Eq. (5.125). We then have a very simple result: all phonons associated with a particular wave vector $\mathbf{Q}_{||}$, whether they are surface phonons or bulk phonons, scatter the electron in the same outgoing direction. Examination of the energy spectrum of electrons scattered inelastically into particular directions provides information on the phonon modes associated with a particular wave vector $\mathbf{Q}_{||}$ determined, as we have seen, by the scattering geometry. From our discussion of the lattice dynamics of surfaces, we then expect the energy loss spectrum to consist of *lines* produced by surface phonons, continuous *bands* produced by scattering from the bulk phonons associated with the value of $\mathbf{Q}_{||}$ associated with the scattering geometry. From plots such as that given in Fig. 5.5, one can see the features to be expected in an energy loss spectrum. For example, if the geometry selects $\mathbf{Q}_{||}$ to be located at the X point of Fig. 5.5, in general we would expect to see lines in the spectrum at the frequencies of the three surface phonons, and energy loss bands between $\sqrt{k_0}$ and $\sqrt{2k_0}$, and also between $\sqrt{3k_0}$ and $2\sqrt{k_0}$. Which modes are seen depends, of course, on the magnitude of the electron phonon matrix element and on scattering geometries of high symmetry; also, selection rules may suppress some features. For the fcc crystal with (100) surface explored in this chapter, and a schematic form of the electron–phonon interaction, Roundy and Mills have calculated energy loss spectra which illustrate these points [25].

With these general principles in hand, we next turn our attention to specific examples of electron energy loss spectra, taken under circumstances where the motions of atoms in the substrate play a key role.

5.3.2 Electron Energy Loss Spectroscopy of Phonons on Crystal Surfaces: Experimental Examples

The first successful study of inelastic electron scattering by the surface vibrations of crystals was carried out on the crystal ZnO [26]. This is an ionic material which supports surface optical phonons that penetrate deeply into the crystal as the wave vector $\mathbf{Q}_{||} \to 0$. These modes were discussed from the lattice dynamical point of view in Section 5.2.4, and the manner in which they enter the dielectric response treatment of the electron energy loss problem was examined in Section 3.3.4.

Since these modes have a displacement field that penetrates deeply into the crystal as $\mathbf{Q}_{||} \to 0$, and since each unit cell has an oscillatory electric dipole moment, the electric field in the vacuum above the crystal is very strong. A consequence is that electrons suffer strong electric dipole scattering from these excitations, as they approach and exit from the crystal. In Fig. 4.14, we show an energy loss spectrum, for electrons scattered very near the specular direction by the dipole mechanism. Note that roughly 60% of the electrons that emerge from the crystal are contained in the one-phonon loss peak at 69 meV.

As we have seen in Section 3.3.4, for scattering from this particular mode, simple expressions for the energy-loss cross section are readily obtained. This involves only macroscopic parameters of the crystal, and details of the scattering geometry. This is true not only for the one-phonon losses, but also for the multiphonon losses described in Section 3.3.5. In the original experiments on ZnO, several basic features of the theory were illustrated by the data. A dependence of the one-phonon loss cross section on $E_0^{-1/2}$, after it is normalized to the intensity of the specular beam, is predicted by the theory, as we see from Eqs. (3.38) and (3.39) with the limit $\vartheta_c \to \infty$ taken. The energy variation of the loss cross section on impact energy, and the variation with angle of incidence θ_I were explored experimentally, and found in excellent accord with the theory. Also, in Fig. 4.14, we see an "anti-Stokes" loss feature in the spectrum, in which the electron emerges from the surface after absorption of a surface phonon, to have energy $E_S = E_I + \hbar\omega_s$ greater than the impact energy E_I. The one-phonon loss and gain peaks should have an intensity ratio of $\exp(-\hbar\omega_s/k_B T)$ from Eqs. (3.38) and (3.39); this is in accord with studies of the temperature variation of the features in the loss spectrum. Finally, multiquantum losses were studied up to $5\hbar\omega_s$, with peak intensities described by the Poisson distribution in Eq. (3.43) valid when $\hbar\omega_s \ll k_B T$. The case of electron energy loss from the surface of ZnO thus offers an example where several key features of the dipole scattering theory may be placed alongside data, and verified explicitly.

Quite recently, Matz and Lüth [27] have studied the electron energy loss spectrum of the (110) surface of GaAs where, once again, one expects strong dipole losses from surface optical phonons very similar to those explored on the ZnO surface [26]. The samples studied in this work were also doped, with electron concentrations in the range 10^{17}–10^{18} cm^{-3}. For such n-type materials, in addition to the surface optical phonon described in Section 5.2.4, one has a surface plasmon which, near $\mathbf{Q}_{\parallel} = 0$ has fields that extend deeply into the substrate, and also into the vacuum above the crystal. This mode, and inelastic scattering of electrons from it, was discussed in Section 3.3.2, where the emphasis was placed on the apparent broadening in energy of electrons in the specular beam that occurs at low carrier concentrations. In the experiments of Matz and Lüth, the carrier concentration was sufficiently high that the surface plasmon appears as a well defined loss peak, as in Eq. (3.26).

In fact, what is intriguing is that in the experiments, the surface-plasmon frequency was very close to that of the surface optical phonon. Under these conditions, one must not regard the two excitations as independent modes, but in fact they couple to form two new modes, each of which may be regarded as an admixture of a surface plasmon and a surface optical phonon. These coupled modes are easily described by noting from Eq. (3.24) the bulk losses occur when the real part of the frequency-dependent dielectric constant $\varepsilon(\omega)$ of the medium equals -1. Thus, the frequencies of the coupled modes are found by solving

$$\varepsilon(\omega) = -1, \tag{5.127}$$

where the dielectric constant in Eq. (3.36) is supplemented by the free-carrier contribution:

$$\varepsilon(\omega) = \varepsilon_{\infty} + \frac{\omega_p^2}{\omega_{TO}^2 - \omega^2} - \frac{\Omega_p^2}{\omega^2}, \tag{5.128}$$

where $\omega_p^2 = 4\pi n e^{*2}/M_r$, and $\Omega_p^2 = 4\pi n_e e^2/m^*$ is the electron plasma frequency, n_e the electron density, and m^* the effective mass. Equation (5.127) gives for the coupled pair of surface excitations the frequencies $\omega_{s\pm}$ which may be written

$$\omega_{s\pm}^2 = \tfrac{1}{2}[\omega_s^2 + \omega_{sp}^2] \pm \tfrac{1}{2}[(\omega_s^2 + \omega_{sp}^2)^2 - 4\omega_{TO}^2\omega_{sp}^2]^{1/2}, \tag{5.129}$$

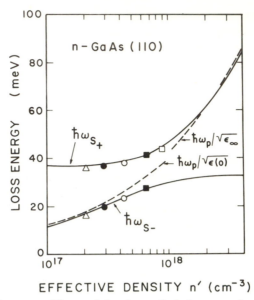

Fig. 5.8 The frequency of the coupled surface optical phonon–surface plasmon modes of the (110) surface of doped GaAs, as a function of carrier concentration. The data is taken from Matz and Lüth [27], used with permission.

where ω_s is the surface optical phonon frequency given in Eq. (3.36) [with $\varepsilon(0)$ the static dielectric constant of the undoped sample], and ω_{sp} is the surface plasmon frequency that appears in Eq. (3.26).

In Fig. 5.8, we reproduce a figure from the paper of Matz and Lüth, which gives the peak positions of the $\omega_{s\pm}$ modes measured by them as a function of the effective electron concentration. We see clear evidence for deviations from the behavior expected if the two surface modes were uncoupled. The solid lines are calculated from Eq. (5.128) and the various data points are indicated.

In Section 3.3.2, we saw that the electron "looks" into the substrate a depth roughly equal to the width of the depletion layer, in doped semiconductors such as that explored by Matz and Lüth. Thus, such experiments should prove a powerful probe of the dynamical properties of the nonuniform electron gas there. In fact, effective electron concentration used by Matz and Lüth to construct Fig. 5.8 in some cases differed from that in the bulk. It would be most useful to have in hand theoretical studies of the normal modes of the surface with the effect of the depletion layer included, with results directed toward explicit data such as this. This is an attractive and exciting area for further experimental and theoretical study.

The modes described above, for both the ZnO and the GaAs surface are surface excitations, but in the limit $\mathbf{Q}_{||} \to 0$, their amplitude extends deeply into the material, a distance the order of $Q_{||}^{-1}$. We now turn our attention to experiments that probe surface phonons that are truly microscopic in nature, i.e., modes in which the displacements are nonzero only within the outermost atomic layers of the crystal.

Shortly after the work on ZnO, the electron energy loss spectrum of electrons reflected from the (111) surface of silicon was examined for electrons scattered very near the specular direction [28, 29]. On the (2×1) reconstructed surface obtained after cleavage in ultrahigh vacuum, a well-defined loss peak at 56 meV was observed. Since silicon is a homopolar semiconductor, symmetry arguments require the dipole moment effective charge e^* to vanish identically in the bulk of the crystal. Thus, as $\mathbf{Q}_{||} \to 0$, this crystal surface cannot support surface optical phonons with the macroscopic character discussed for the examples of ZnO and GaAs described above. In fact, the 56-meV loss peak is clearly an excitation of microscopic character. That this is so is clear from the experimental observation that adsorption of monolayer quantities of oxygen on the surface causes the mode to vanish from the loss spectrum, and to be replaced by features from the oxygen vibrational motions. Also, if the surface is converted to the complex (7×7) reconstructed surface by heating, once again the signal from the 56 meV disappears. Since both treatments modify the environment of the outermost few atomic layers, the 56-meV mode must have displacements localized there.

As already pointed out, in the bulk of the silicon crystal, the dipole moment effective charge e^* must vanish by symmetry. Yet the 56-meV feature in the energy loss spectrum is very clearly an electric dipole active mode. The atoms in the surface of the crystal sit in environments of low symmetry, so they may possess a nonzero dipole moment effective charge. Evans and Mills developed the theory of electric dipole scattering by surface vibrations largely in response to this fascinating data [30]. This paper developed the theory of electric dipole scattering by an array of oscillating dipoles on the crystal surface, and also from the deeply penetrating modes of the sort observed on ZnO and GaAs. By a very different approach, the results of Evans and Mills have been recovered in Chapter 3 of the present work. When the theory is applied to the 56-meV loss on the silicon surface, it estimated that the surface atoms have a dipole moment effective charge $e^*/\varepsilon_\infty^{1/2} \cong 0.5e - 1.0e$. A dynamic effective charge of this magnitude can have a profound effect on the lattice dynamics of the surface region. In fact, Trullinger and Cunningham have explored the influence of dynamic effective charges of this magnitude on the lattice dynamics of the surface, to conclude that their presence may drive the unreconstructed surface unstable, and hence lead to reconstruction of the surface [31].

One may inquire into the origin of the 56-meV loss peak. The feature does not lie in any gaps in the bulk phonon spectrum, as do the surface phonons illustrated in Fig. 5.5 (see Fig. 9 of Ref. [29]). Thus, the feature is associated not with a true surface phonon, but with a resonance mode such as that displayed in Fig. 5.7f. Ludwig and Zimmerman have carried out studies of the lattice dynamics of the unreconstructed silicon surface [32]. They find a surface phonon on the boundary of the Brillouin zone of the *unreconstructed* (111) surface that agrees well in frequency with the 56-meV loss peak. They argue that when the surface reconstructs, and one forms the Brillouin zone appropriate to the reconstructed surface, the zone boundary surface phonon becomes folded back to the zone center, to become a (dipole active) $\mathbf{Q}_{\|} \cong 0$ mode of the reconstructed surface. We shall encounter modes of this sort shortly, when we examine the electron energy loss spectra of oxygen over-layers on the Ni(100) and the Ni(111) surface. This suggestion of Ludwig and Zimmerman is appealing because it is consistent with the observation that the loss appears on the surface only when the (2×1) reconstructed form is realized.

All of the foregoing examples explore surface phonons on the clean surface (sometimes reconstructed) of crystals. We now turn our attention to recent experiments which explore surface modes on adsorbate covered surfaces. Most particularly, recent experiments explore oxygen overlayers on both the Ni(100) surface [33], and the Ni(111) surface [34].

Consider, for example, the $p(2 \times 2)$ overlayer of oxygen on the Ni(100) surface. The geometry of this surface is illustrated in Fig. 5.6b, where the open circles give the positions of the Ni atoms in the outermost layer of the substrate, and the cross-hatched circles give the oxygen positions.

In the upper portion of Fig. 5.9, we show the near-specular energy loss spectrum taken by Lehwald and Ibach for this structure. The feature at 430 cm^{-1} lies very close to the frequency associated with vibration of the oxygen adatom against the Ni substrate, as estimated from Eq. (4.39a) with the parameters of Upton and Goddard as input [18, 21]. In Section 5.2.6, we saw that for the low-density $p(2 \times 2)$ structure, to first order in M_A/M_s, the high-frequency surface phonon branch associated with the motion of oxygen against the Ni substrate is dispersionless, with frequency equal to that of the isolated adatom, within the model that allows the oxygen atom to "see" only its four nearest neighbors on the surface. Thus, clearly, the 430 cm^{-1} feature has its origin in scattering from this surface phonon branch near $\mathbf{Q}_{\|} = 0$; the electric dipole activity has its origin in the stretching of the oxygen–Ni bond by the coherent motion of the oxygen overlayer against the substrate.

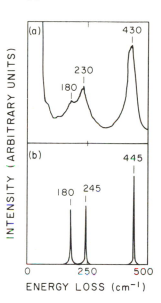

Fig. 5.9   (a) The electron energy loss spectrum from the $p(2 \times 2)$ layer of oxygen on the Ni(100) surface, from Lehwald and Ibach [33]. This is taken near the specular, where dipole scattering dominates. (b) A theoretical spectrum generated by Rahman *et al.* [18], with input data provided by the calculations of Upton and Goddard [21]. Both spectra and data are used with permission.

In addition, we see two clear features in the electron energy loss spectrum at 180 and 230 cm^{-1}, well below the maximum Ni phonon frequency of 295 cm^{-1}. The lower part of the figure gives a theoretical electron energy loss spectrum generated for the $p(2 \times 2)$ structure by Rahman *et al.* [18]. We see the doublet also in the theoretical spectrum. The origin of these loss features may be appreciated by comparing the oxygen positions in the $p(2 \times 2)$ layer, as given in Fig. 5.6b, with the pattern of displacements associated with the surface phonon illustrated in Fig. 5.4b, for the (100) surface of the fcc crystal. When the $p(2 \times 2)$ adlayer is placed over the substrate surface, with the surface phonon excited, the "breathing motion" of the nearest neighbors of each oxygen are such that the oxygen adlayer is set in coherent vertical motion. The adlayer thus induces dipole activity in the S_6 surface phonon in Fig. 5.5. In fact, from Fig. 5.4, we see that the S_6 mode on the clean surface is twofold degenerate. One of these modes, the one with breathing motion in the square just below the oxygen, couples to the adlayers, which the second fails to do, since motion of the four nearest-neighbor Ni atoms leaves the O–Ni bond length unchanged. In the language of solid state physics, when the phonon dispersion curves of the free surface are folded back into the Brillouin zone associated with the surface with $p(2 \times 2)$ adlayer present, the surface phonons at the X point of the original Brillouin zone appear at the Γ point of the new zone. The presence of the adlayer renders one of them dipole active, since its displacement has symmetry such that the oxygen adlayer is driven by it. This accounts for the 230 cm^{-1} mode of the structure, while the 180 cm^{-1} feature may be regarded as a

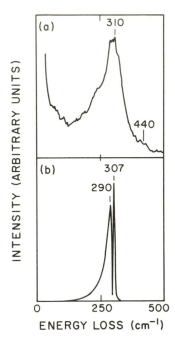

Fig. 5.10 (a) The electron energy loss spectrum from the $c(2 \times 2)$ layer of oxygen on the Ni(100) surface, from Lehwald and Ibach [33]. (b) A theoretical spectrum generated by Rahman *et al.* [18], for the case where the oxygen layer sets only 0.27 Å from the Ni surface. The force constants are taken from a fit to an *ab initio* calculation by Upton and Goddard. Spectra and data are used with permission.

translation of the O–Ni$_4$ normal to the surface, hindered by coupling to the substrate. The result is a second resonance mode.

The theoretical calculations in Fig. 5.9 assume, following Upton and Goddard, that the oxygen adlayer sits 0.88 Å above the surface, a position very close to that determined by low-energy electron diffraction studies of the $p(2 \times 2)$ oxygen overlayer on the Ni(100) surface. It is intriguing to examine Fig. 5.10, where the energy loss spectrum is presented for (dipole scattering) from the $c(2 \times 2)$ oxygen overlayer on the Ni(100) surface. We illustrate this in Fig. 5.10, where we see a spectrum that differs qualitatively from that seen for the $p(2 \times 2)$ structure. If, as in the theory illustrated in Fig. 5.9, one assumes the $c(2 \times 2)$ layer is 0.88 Å above the substrate surface, it is found that the calculated loss spectrum contains a single feature at 401 cm^{-1} (essentially from the surface mode described by Eq. (5.117) with $\mathbf{Q}_{\parallel} = 0$), with no features near or below 295 cm^{-1}, as seen in the data. It has been argued that for the dense $c(2 \times 2)$ structure, the oxygen overlayer should lie very close to the Ni(100) surface [21]. If these parameters are employed in the theory, the spectra given in the lower portions of Fig. 5.10 is generated. These agree very well with the electron energy loss data.

These studies provide an example of a situation where the study of the electron energy loss spectrum leads to information on basic geometrical features of the adsorbate geometry. The *qualitative* differences in the experimental spectra in Figs. 5.9 and 5.10 show that features in the energy loss spectrum can be very sensitive to surface geometry. We have the good fortune here to have the *ab initio* calculations of Upton and Goddard in hand, while a simple lattice dynamical model provides a quite satisfactory description of the substrate.

This discussion has examined in some detail the electron energy loss spectra of ordered oxygen adlayers on the Ni(100) surface. Ibach and Bruchman have also studied ordered layers on the Ni(111) surface, again to find distinct features well below the maximum Ni phonon frequency of 295 cm^{-1} [33]. They have identified particular substrate phonons, forbidden to be dipole active on the clean surface, which are "activated" by the presence of the oxygen overlayer. Black has calculated the spectral density function associated with the motion of an isolated oxygen adatom on the Ni(111) surface, and compared features in the spectral density with the data [35].

The preceding data provide no direct information on the nature of lateral interactions between adsorbates. In very elegant experimental and theoretical studies of the $c(2 \times 2)$ CO layer on the Cu(100) surface, Andersson and Persson [20], and Persson and Ryberg [36] have extracted quantitative information on the nature of lateral interactions between the CO molecules, to conclude they are dipolar in character. Certain features of their analysis are very similar to the study of CO adsorbed on the Ru surface that we have presented in Section 3.4.3. The variation of infrared absorption intensity with coverage, and the coverage dependence of the CO stretching frequency were the focus of our attention. The work of Andersson and Persson explores the absolute value and dependence of the electron energy loss cross section in the dipole dominated regime with wave vector transfer $\mathbf{Q}_{\|}$, to deduce values for the parameter α_v and α_e discussed in Section 3.4.3. They also find that dipole–dipole interactions with origin in the electronic polarizability play an important role in controlling the magnitude of the electron energy loss cross section. In our formulation, this enters through the factor ε_∞^2 in Eq. (3.64). The treatment of the dipole scattering problem presented by Andersson and Persson provides a generalization of this form to finite wave vector transfers $\mathbf{Q}_{\|}$. From their analysis of the variation of dipole scattering intensity with wave vector transfer, these authors conclude that for the adsorbed CO, α_v is larger than the value appropriate to the gas phase by roughly a factor of 5. Recall that in Section 3.4.3, we also found α_v substantially larger than the gas phase value, so for these two different substrates a similar result follows. Finally, Andersson and Persson measure the dependence of the frequency of the CO stretching mode with wave vector transfer $\mathbf{Q}_{\|}$, to find appreciable dispersion well accounted for by the assumption that the lateral interactions between the CO molecules is purely dipolar in nature, with strength as deduced from the energy loss data.

Earlier in this chapter, it will be recalled that for ordered adsorbate layers of atoms, we found dispersion in the surface phonon branch with origin in the indirect coupling through the substrate, even in the absence of direct lateral interactions between adsorbates [Eqs. (5.117)–(5.110)]. For the case of oxygen on Ni, the dispersion introduced into the surface-phonon branch was appreciable. If, however, the same picture is applied to the CO stretching mode of the adsorbed layer, the dispersion introduced through this mechanism is very small, since the CO stretching frequency is so very much higher than the maximum vibration frequency of the Cu substrate. It would be most intriguing to see studies very similar to those carried out by Andersson and Persson completed for adsorbates that lead to surface-phonon branches split off from the substrate phonon bands by a modest amount, since here the indirect coupling through the substrate may play an important role.

From this section, the reader may appreciate that by means of electron energy loss spectroscopy, surface phonons may be studied on a diverse array of clean crystal surfaces, and adsorbate covered surfaces. All of the experiments discussed here examine near-specular scattering, where the dipole mechanism provides the dominant coupling to the vibrational motions of the surface. Quite clearly, if large-angle scatterings could be routinely studied to enable exploration of vibrational excitations throughout the two-dimensional Brillouin zone, there would be a significant expansion in the quantitative study of surface phonons. Hopefully, such studies will be undertaken in the near future.

5.4 ANHARMONICITY AND DOUBLE LOSSES

The entire discussion of the present chapter and most of that of the previous one is based on the harmonic approximation of lattice dynamics. When equations of the potential energy of the array of vibrating masses have been expanded in powers of the atomic displacements, as in Eq. (4.11), we have terminated the expansion after the quadratic terms. Then, at least in the formal sense, the vibrational Hamiltonian may be diagonalized to yield the phonon normal modes of the semi-infinite crystal, possibly with adsorbates present. Chapters 4 and 5 have been devoted to the analysis of the nature of these phonon modes, and how they may be studied by electron energy loss spectroscopy.

If the next terms in the expansion in Eq. (4.11) are retained, such as those cubic and quartic in the displacements, then one finds interactions between the phonons that lead to temperature-dependent frequency shifts, lifetime effects, and so on. Very little theoretical work has been done on this topic in the lattice dynamics of surfaces, and direct experimental studies are also sparse. In this section, we shall consider scattering processes in which, by virtue of anharmonicity, overtone excitations of an adsorbed molecule may be excited. We also comment briefly on multiple losses. These topics have been touched upon earlier. In Section 3.3.5, we examined multiple losses seen in spectra taken on ionic crystal surfaces, where surface optical phonons are present. Also, in Section 4.4.5, group theoretical considerations were applied to the excitations of overtones, with comments once again on multiple losses. Here, our emphasis will be on a simple model description of dipole active overtone excitations, with brief comments on multiple losses from dense layers of adsorbed molecules.

The simplest case to consider is a molecule adsorbed on a surface, with an internal mode that has frequency very high compared to the maximum phonon frequency ω_M of the substrate. Then, as we have seen in Chapter 4, the atomic displacements associated with the mode will remain highly localized to the molecule. We may introduce a normal coordinate u, and as this mode is excited we have a potential energy function $V(u)$ that for small excursions from the equilibrium configuration $u = 0$ of the molecule has the form $\frac{1}{2}M_r\omega_e^2 u^2$, with ω_e the frequency of small-amplitude vibrations about the equilibrium position and M_r the effective mass associated with the vibrational motion. Here we use ω_e for the vibrational frequency calculated in the harmonic approximation, a notation used in the literature on molecular vibrations. In the simple limit considered here, the only difference between the adsorbed molecule and the gas-phase species is the form of $V(u)$. Within this schematic picture, we can obtain a feeling for the role of anharmonicity in electron energy loss spectroscopy.

While $V(u)$ is necessarily quadratic for small excursions about the equilibrium position, for excitations of larger amplitude, deviations from the small-amplitude quadratic behavior influence the spectrum. In molecular physics, it is common to use an empirical form for the potential $V(u)$ called the Morse potential,

$$V(u) = D_e[1 - \exp(-au)]^2, \qquad (5.130)$$

whose form is sketched in Fig. 5.11. For small values of the displacement amplitude u, we have $V(u) \cong a^2 D_e^2$, so in the harmonic approximation, the vibrational frequency is given by

$$\omega_e = a(2D_e/M_r)^{1/2}. \qquad (5.131)$$

The Morse potential is widely used in the literature on molecular vibrations, because its form correctly mimics a real molecular potential, and also because the energy levels and wave functions may be found exactly. The

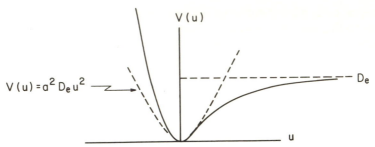

Fig. 5.11   A sketch of the Morse potential defined in Eq. (5.125).

bound energy levels have energies $\hbar\omega_n$ given by

$$\hbar\omega_n = \hbar\omega_e(n + \tfrac{1}{2})[1 - x_e(n + \tfrac{1}{2})], \qquad (5.132)$$

where

$$x_e = \hbar\omega_e/4D_e. \qquad (5.133)$$

The parameter x_e provides a dimensionless measure of the anharmonicity present in the potential, in that in the limit $x_e \to 0$, the vibrational spectrum becomes that of the simple harmonic oscillator. The quantum number n assumes the values $n = 0, 1, 2, \ldots$ up to the integer just below $\tfrac{1}{2}(1 - x_e)/x_e$; presumably for physically realistic potentials, the parameter x_e will be less than unity.

In an electron energy loss study of an adsorbed molecule described by the foregoing picture, the electron will typically encounter the molecule in the vibrational ground state $n = 0$ (when $k_B T \ll \hbar\omega_e$), and excite it to the first vibrational level $n = 1$. This is, in the localized picture described here, the analog of the one-phonon loss process. With $x_e \neq 0$, the excitation energy associated with this transition is not $\hbar\omega_e$, but rather $\hbar(\omega_1 - \omega_0) \equiv \hbar\omega^{(1)}$. From Eq. (5.132), one has

$$\omega^{(1)} = \omega_e(1 - 2x_e), \qquad (5.134)$$

so the measured one-phonon loss appears at a frequency shifted below that of the vibrational frequency calculated in the harmonic approximation. The oscillatory part of the electric dipole moment of the molecule may be written, as earlier,

$$p = e^*u \qquad (5.135)$$

and the intensity of the one-phonon loss is controlled by the matrix element $\langle 0|p|1\rangle$, with $|0\rangle$ and $|1\rangle$ the ground- and first-vibrational wave functions of the oscillator. A convenient and usable tabulation of this and related matrix elements for the Morse potential oscillator has been given by Rosenstock [37]. In the limit $x_e \ll 1$, the dipole-moment matrix element is well approximated by the form appropriate to the simple harmonic oscillator:

$$\langle 0|p|1\rangle \cong e^*(\hbar/2M_r\omega_e)^{1/2}. \tag{5.136}$$

Earlier in the present volume, as remarked at the beginning of this section, we have discussed multiphonon excitation processes, in which the incoming electron creates two or more quanta of excitation as it scatters from the crystal. If we consider the isolated molecule examined above, then such a scattering is one in which the electron excites the molecule from the ground $n = 0$ level to the second excited vibrational state with $n = 2$. This is referred to, as in the earlier discussions, as an overtone excitation. In the absence of anharmonicity, the excitation energy associated with such a transition is simply given by $2\hbar\omega_e$. With anharmonicity included, as one readily sees from Eq. (5.132), one has

$$\hbar(\omega_2 - \omega_0) = 2\hbar\omega_e - 6\hbar\omega_e x_e = 2\hbar\omega^{(1)} - 2\hbar\omega_e x_e. \tag{5.137}$$

Thus, there is a downward shift from $2\hbar\omega^{(1)}$ produced by the anharmonic nature of the vibrational motion equal in magnitude to $2\hbar\omega_e x_e$. Thus, one does not see the overtone at $2\hbar\omega^{(1)}$ for such an excitation, but at a downshifted frequency instead.

For a harmonic oscillator, the two-quantum transition just described is dipole forbidden, since the matrix element of the normal coordinate u between the $n = 0$ and $n = 2$ level vanishes. In the presence of anharmonicity, this is no longer the case. When x_e is small compared to unity, one has [37]

$$\langle 0|p|2\rangle \cong e^*(x_e/2)^{1/2}(\hbar/2M_r\omega_e)^{1/2}. \tag{5.138}$$

A comparison between Eqs. (5.134) and (5.137) shows that frequency shift of the overtone excitation away from $2\hbar\omega_e$ and its strength relative to the "one-phonon" peak are both controlled by the parameter x_e. These two features of the energy loss spectrum thus should be correlated, within the framework of the present simple model. If one wishes to pursue this question, there is one difficulty. This is that as the molecule is excited to large amplitude, the dynamic electric dipole moment p of the molecule is no longer simply proportional to u. We will have a relation of the form

$$p(u) = e^*u(1 - \gamma au + \cdots), \tag{5.139}$$

where the dimensionless parameter γ is independent of x_e, and is controlled by charge flow which occurs between the constituents as the molecule is stretched. This parameter may have either sign. The matrix element in

Eq. (5.138) must then be modified to read

$$\langle 0|p|2\rangle = e^*\left(\frac{x_e}{2}\right)^{1/2}\left(\frac{\hbar}{2M_r\omega_e}\right)^{1/2}\left[1 - 2\gamma a\left(\frac{\hbar}{aM_r\omega_e x_e}\right)^{1/2}\right], \quad (5.140)$$

where in general the first and second terms are quite comparable in magnitude. We may see this as follows. If $\langle u^2\rangle_0$ is the mean square displacement of the oscillator in its ground state, one has (when $x_e \ll 1$)

$$\langle u^2\rangle_0 = \hbar/2M_r\omega_e. \quad (5.141)$$

From Eq. (5.141), it follows that

$$a^2\langle u^2\rangle_0 \equiv x_e, \quad (5.142)$$

a relation that provides one with an intuitive understanding of the meaning of the anharmonicity parameter x_e. Equation (5.140) then reduces to

$$\langle 0|p|2\rangle = e^*\left(\frac{x_e}{2}\right)^{1/2}\left(\frac{\hbar}{2M_r\omega_e}\right)^{1/2}(1 - 2\gamma), \quad (5.143)$$

and with γ comparable to unity in typical situations, the nonlinear variation of dipole moment with displacement and anharmonicity play a comparable role in inducing dipole activity into the double quantum transition.

The foregoing discussion applies to "two-phonon," or overtone excitations of an isolated adsorbed molecule by the incoming electron. At finite coverage, of course, such overtone excitations still occur, as illustrated schematically in Fig. 5.12a. But now a second kind of "two-phonon" excitation, the multiple loss, may occur as illustrated in Fig. 5.12b. For the case where lateral interactions between the adsorbates are ignored, the two quantum multiple loss will occur at precisely $2\hbar\omega^{(1)}$, without the shift displayed in Eq. (5.137). This multiple loss is, for the adsorbed molecule, the analog of the multiple losses examined in Section 3.3.5, where the incoming electron may emit two or

(a)   (b)

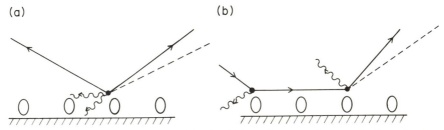

Fig. 5.12   (a) A two-phonon transition in which an electron strikes a molecule, and excites it from the $n = 0$ to the $n = 2$ level in a single scattering event. (b) The double scattering in which an electron excites one molecule from $n = 0$ to $n = 1$, then from $n = 0$ to $n = 1$.

more phonons by a sequence of scatterings with one phonon emitted in each basic interaction. Such multiple losses, for adsorbed molecules, were examined again in Section 4.4.5. For a layer of adsorbed molecules, the following criterion may be used to distinguish between overtone excitation, as illustrated in Fig. 5.12a and the multiple loss shown in Fig. 5.12b. The intensity of the overtone excitation will exhibit a linear variation with coverage, while the multiple loss shown in Fig. 5.12b will show a quadratic variation. Examples for the observation of overtones will be discussed in more detail in Section 6.4.

REFERENCES

1. G. A. Sonorjai, "Principles of Surface Chemistry," p. 38 ff. Prentice-Hall, Englewood Cliffs, New Jersey, 1972.
2. L. Dobrzynski and A. A. Maradudin, *Phys. Rev. B* **7**, 1207 (1973); erratum, *Ibid.*, **12**, 6006 (1975).
3. C. Kittel, "Introduction to Solid State Physics," 5th ed., p. 50 ff. Wiley, New York, 1976.
4. A. A. Maradudin, E. W. Montroll, G. H. Weiss, and I. P. Ipatova, "Lattice Dynamics in the Harmonic Approximation," 2nd ed. Academic Press, New York, 1971.
5. R. F. Wallis, *Progr. Surface Sci.* **4**, 233 (1973).
6. L. D. Landau and E. M. Lifshitz "Theory of Elasticity," p. 101 ff. Addison-Wesley, Reading, Massachusetts, 1959.
7. D. J. Cheng, R. F. Wallis, and L. Dobrzynski, *Surface Sci.* **43**, 400 (1974).
8. R. E. Allen, G. P. Alldredge and F. W. de Wette, *Phys. Rev. B* **4**, 1661 (1971).
9. J. E. Black, D. A. Campbell, and R. F. Wallis, *Surface Sci.* (to be published).
10. R. F. Wallis, D. L. Mills, and A. A. Maradudin, *in* "Localized Excitations in Solids" (R. F. Wallis, ed.), Figs. (1) and (2). Plenum, New York, 1968.
11. J. D. Jackson, "Classical Electrodynamics," p. 110 ff. Wiley, New York, 1962.
12. D. L. Mills and E. Burstein, *Rep. Progr. Phys.* **37**, 817, Section X (1974).
13. L. Dobrzynski and G. Leman, *J. Physique* **30**, 116 (1969).
14. D. L. Mills, *Phys. Rev. B* **1**, 264 (1970).
15. R. Haydock, V. Heine, and M. J. Kelly, *J. Phys. C* **5**, 2845 (1972); **8**, 2591 (1975).
16. J. E. Black, B. Laks, and D. L. Mills, *Phys. Rev. B* **22**, 1818 (1980).
17. M. Mostoller and U. Landman, *Phys. Rev. B* **20**, 1755 (1979).
18. T. S. Rahman, J. E. Black, and D. L. Mills, *Phys. Rev. Lett.* **46**, 1469 (1981); *Phys. Rev. B* (to be published).
19. T. S. Chen, G. P. Alldredge, F. W. de Wette, and R. E. Allen, *J. Chem. Phys.* **55**, 3121 (1971).
20. S. Andersson and B. N. J. Persson, *Phys. Rev. Lett.* **45**, 1421 (1980).
21. T. H. Upton and W. A. Goddard III, *Phys. Rev. Lett.* **46**, 1635 (1981).
22. A. S. Barker, Jr., and A. J. Sievers, *Rev. Mod. Phys.* **47**, Suppl. 2, p. 51 (1975).
23. G. D. Mahan and A. A. Lucas, J. Chem. Phys. **68**, 1344 (1978).
24. J. M. Ziman, "Electrons and Phonons," Chapter 5. Oxford Univ. Press, London and New York, 1963.
25. V. Roundy and D. L. Mills, *Phys. Rev. B* **5**, 1347 (1972).
26. H. Ibach, *Phys. Rev. Lett.* **24**, 1416 (1970).
27. R. Matz and H. Lüth, *Phys. Rev. Lett.* **46**, 500 (1981).

28. H. Ibach, *Phys. Rev. Lett.* **27**, 253 (1971).
29. H. Ibach, *J. Vac. Sci. Technol.* **9**, 713 (1971).
30. E. Evans and D. L. Mills, *Phys. Rev. B* **5**, 4126, (1972); **7**, 853 (1973).
31. S. Cunningham and S. E. Trullinger, *Phys. Rev. B* **18**, 1898 (1978).
32. W. Ludwig, *Proc. Int. Conf. Solid Surfaces, 2nd, Kyoto, 1974* [*Japan J. Appl. Phys., Suppl.* 2, 879 (1974)].
33. S. Andersson, *Surface Sci.* **79**, 385 (1979); S. Lehwald and H. Ibach, "Vibration at Surfaces". (R. Caudano, R. Gilles, A. A. Lucas, eds.), p. 137. Plenum, New York, 1982.
34. H. Ibach and D. Bruchmann, *Phys. Rev. Lett.* **44**, 36 (1980).
35. J. E. Black, *Surface Sci.* **100**, 555 (1980).
36. B. N. J. Persson and R. Ryberg, *Phys. Rev. B* (to be published).
37. H. B. Rosenstock, *Phys. Rev. B* **9**, 1963 (1974).

CHAPTER 6

APPLICATIONS OF VIBRATION SPECTROSCOPY IN SURFACE PHYSICS AND CHEMISTRY

6.1 GENERAL REMARKS

At the time of writing this book, surface vibration spectroscopy is enjoying increasing attention, and a growing number of papers dealing with the subject experimentally and theoretically are appearing every month. Several reviews have been written [1–4]. In this chapter, rather than updating existing reviews or covering the entire experimental literature, we wish to focus on a number of selected examples that may serve as an illustration of present and future applications of vibration spectroscopy. As with other surface analysis tools, a microscopic understanding of surfaces, adsorbates, and surface reactions is seldom achieved using vibrational spectroscopy alone. In combination with other techniques, however, frequently rather detailed pictures of surface structure and bonding have emerged. The combination of electron energy loss spectroscopy, electron diffraction, and flash desorption has proved to be particularly powerful.

In order to be illustrative, one necessarily tends to focus on examples where the interpretation of the experimental data is relatively straightforward, and one is led to disregard the more controversial issues. This might lead to the (false) impression that interpreting vibrational data is always easy and unambiguous. To balance this view, we also add a case study at the end of this chapter which has been (and still is) controversial, despite existing vibrational data.

All examples considered in the following deal with adsorption on single-crystal surfaces, although electron energy loss spectroscopy, as well as the other techniques for probing surface vibrations, are not confined to single-crystal substrates. The reason for choosing single crystals is that the chances of obtaining an understanding of surface structures and a successful analysis of surface reactions and of new surface species are significantly higher on these surfaces, which have a limited number of adsorption sites. The spectral range of vibration spectroscopy is roughly 3000 cm^{-1}. With a resolution of 50 to 100 cm^{-1} typically for current electron spectrometers, 30–60 vibrational losses can be discerned within the spectral range of the apparatus. This puts an upper bound to the complexity of systems that can be analyzed. For molecules in a single site, the limit is to molecules no larger than those with between 10 and 20 atoms. The number becomes smaller when different adsorption sites are present, so that even a simple diatomic molecule such as CO occupying three to five surface sites simultaneously exhausts the analyzing capacity of electron spectroscopy. Essentially the same holds for optical techniques in vibrational spectroscopy, since their advantage of higher resolution is balanced by a smaller accessible frequency range, or limited sensitivity. Therefore adsorption studies on single-crystal surfaces with their limited and well-known selection of available surface sites together with a comparison of studies on different crystal faces are naturally the first steps in the exploration of the capabilities of the technique. With increasing experience and an atlas of vibrational frequencies of various molecules adsorbed on different crystal faces polycrystalline, "real" surfaces might be investigated successfully as well.

6.2 VIBRATION SPECTROSCOPY AND THE DETERMINATION OF ADSORPTION SITES

6.2.1 Adsorbed Atoms

Localized vibrations of adsorbed atoms and their interaction with the phonon spectrum of the substrate have already been considered in detail theoretically and experimentally for a few selected systems in Chapters 4 and 5. There, force constants and equilibrium positions calculated from a parameter-free theory of the electronic eigenstates of a cluster had been fed into lattice dynamical models, and a nearly perfect agreement between theoretical and experimental vibration spectra was found. With this agreement, the assumed site and position of the adsorbed atom is also confirmed. Obviously such elaborate studies are not going to be carried out for more but a few examples. This raises the question of whether qualitative structural information may be obtained from the vibrational data without elaborate theories. To address this question we take a close look at the entire range of experimental material, as far as it is available now, in order to establish some semiempirical rules or simplified procedures by which structural information is extracted from the vibrational spectrum.

In Table 6.1, we have listed frequencies of A_1-type vibration modes for adatoms in various sites together with the vibration frequencies of the di-

TABLE 6.1

Vibrational Frequencies of Atoms Perpendicular to the Surface for
Different Ligancies and (Fundamental) Frequencies of Equivalent
Diatomic Molecules[a]

System	Ligancy			Diatomic molecule	Ref.	Comment
	4	3	2			
O–W	610		740	1047	5	
O–Ni	430/310	580/450		~615	6–8	Two binding states
O–Pt		490		841	9	
O–Fe		400	500	870	10	Fe atoms not in contact
O–Ru		515		865	11	
O–Cu	340				12	
H–Ni	600 (590)	(1210)	(1370)	~1926	13, 19	Values in parentheses are calculated [19]
H–Pt		550		~2200	14	Only ω_e known for PtH
C–Ni	390	520			15	
S–Ni	360				16	

[a] Cf. Appendix B. For oxygen on nickel, two different binding states are observed.

atomic molecules. Ligancy 4 here means adsorption in a fourfold hollow site on a (100) surface of an fcc lattice, ligancy 3 the threefold site on a (111) surface, and ligancy 2 a bridging site. It is well established now from electron diffraction studies [17] that absorbed atoms prefer the site of highest possible coordination. For the systems quoted in Table 6.1 the adsorption sites had been determined by electron diffraction for oxygen on W, Ni, Cu, and sulfur on Ni [17]. For hydrogen on Ni, the site is independently confirmed by theoretical studies on the binding energy of hydrogen in various sites [18, 19]. For hydrogen on platinum, this site is inferred from additional information on the vibrational frequency for the motion parallel to the surface which we shall discuss later. The site assignment for oxygen on Ru(001) and Fe(110) and for carbon on Ni(100) and Ni(111) is merely based on analogy, in order to comply with the general pattern of preferred adsorption in sites of high coordination.

Data are still rather scarce; nevertheless, they allow several conclusions: Vibrations of atoms perpendicular to the surface in sites of high ligancy have a significantly lower frequency than the corresponding diatomic molecule. In cases where adsorption in different sites is observed for the same metal–adatom combination, the frequency is lower for higher ligancy. The concept that the ligancy is the important factor that accounts for the difference in the frequencies between the diatomic molecule and the adsorbed atoms is supported by the fact that high-frequency modes (in the 2000 cm^{-1} range) have been observed for weakly bound hydrogen on dispersed Pt, Ir, Ni, Fe, Co, Rh, and Pd [20]. These modes have been assigned to hydrogen bound to a single metal surface atom. Typical symmetric stretching vibrations for hydrogen bridging two metal atoms in transition metal complexes range between 800 and 1200 cm^{-1} [20]. The effect of higher ligancy on the frequency reduction for oxygen is smaller than for hydrogen. This seems to be mainly related to the size of the adatom: When a small adatom vibrates perpendicular to the surface in a fourfold adsorption site of the (100) and (111) surface of an fcc lattice, respectively, the angle between the vibrational motion and the nearest-neighbor bonds [α in Table 4.1, see also Eqs. (4.38)–(4.41)] is relatively high and therefore the stretching force constants are less effective.

Within the nearest-neighbor force constant model discussed in Chapter 4, the latter statement can also be rendered more precisely: In Fig. 6.1 we have plotted the frequency of the parallel and perpendicular vibration in reduced units as a function of the ratio of the adatom–substrate atom radius for three different surface sites. Assuming for the moment the force constant to each nearest neighbor to remain the same in all three sites (by the term force constant, we are referring to the quantity f in Table 4.1), the perpendicular vibration frequency of an adatom is approximately the same for the twofold and threefold sites. It drops, however, for all but the largest adatoms when going to the fourfold site. For oxygen on nickel, the ratio of frequencies between the threefold and fourfold sites is 1.33 for the low coverage state, while 1.25 would be predicted from Fig. 6.1, with the assumption already made and assuming that oxygen has its covalent radius. The oxygen–nickel bond length for the state characterized by the set of higher frequencies given in Table 6.1 for the threefold and fourfold sites, respectively, is approximately equal to the sum of the covalent atomic radius as determined from LEED investigations and theoretical calculations. The nearest-neighbor force constant required to provide the best fit to the experimental frequency ratio is therefore a little higher for the threefold than for the fourfold site (1.9 vs 1.65 10^5 dyn cm^{-1}). This is in accordance with the intuitive notion that "stronger" bonds are established with each nearest neighbor the smaller the ligancy. This effect would further tend to emphasize the trend of a reduction of the perpendicular frequency with higher ligancy of the adsorption site. It is therefore reasonable to assume that as a general rule the perpendicular vibrations follow the order

$$\omega_1 > \omega_2 > \omega_3 > \omega_4 \tag{6.1}$$

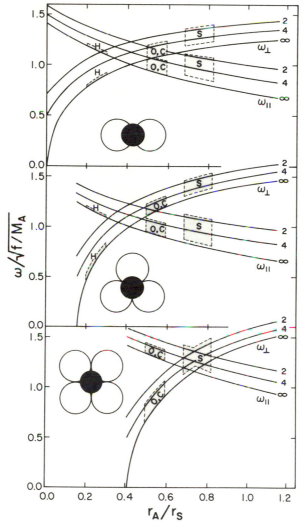

Fig. 6.1 Frequency (in reduced units) of the perpendicular and parallel vibration in the bridging site, the threefold and the fourfold hollow site predicted from the nearest-neighbor force constant model as a function of the ratio of the adatom to substrate atom radius.

with the index denoting the ligancy. We shall return to this subject with more quantitative estimates in Section 6.3.2.

Once the adatom radius or bond distance to the nearest neighbors is chosen, the nearest-neighbor force constant model also predicts the ratio of the parallel to the perpendicular vibration frequency. Since, at least in

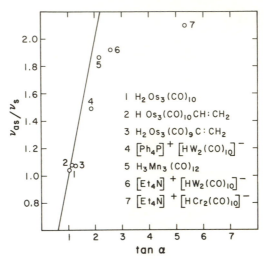

Fig. 6.2 Ratio of the asymmetric to symmetric stretching frequency versus tan α, where α is half the bond angle of the hydrogen atom bridging two metal atoms (after Howard *et al.* [21] with permission). The solid line represents the prediction of the nearest-neighbor force constant model.

principle, parallel vibrations of adsorbates can be measured by the off-specular technique (Section 3.5), one may turn the argument around and use the experimentally observed frequencies to determine the bond angle and bond length. The accuracy of the surface bond length obtained by this procedure can be estimated with the help of a study by Howard *et al.* [21]. Howard *et al.* tabulated the symmetric (corresponds to the "perpendicular" vibration) and the asymmetric hydrogen–metal stretching vibrations (the "parallel" vibration) for several cluster compounds for which the bond angle had been determined by neutron or x-ray diffraction. As we can see from Table 4.1, the ratio ν_{as}/ν_s should be equal to tan α for a light atom such as hydrogen. The actual data according to Howard *et al.* are shown in Fig. 6.2, together with the linear relationship predicted by the nearest-neighbor force constant model. At first glance deviations appear to be large, suggesting that angle-bending forces cannot be neglected. In terms of the structural analysis from vibrational data as proposed above, the deviation between the bond length calculated in the nearest-neighbor model and the real value is within the limits of error for other surface structure analysis technique for all but data point 7 in Fig. 6.2. Let us take data point 6, e.g., where the bond angle α is 68.6°. From the frequency ratio using the nearest-neighbor model one would obtain 62.5°. Converting these angles into a metal–hydrogen bond length using the metal–metal distance would yield a difference of about 0.06 Å between the two values.

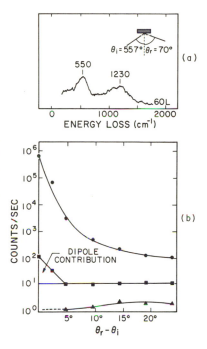

Fig. 6.3 (a) Electron energy loss spectrum and (b) angular profiles for hydrogen on the platinum (111) surface. (a) $T = 90°K$, $E_p = 6.5$ eV. (b) $E_p = 5$ eV; ●, elastic intensity; ■, inelastic intensity (550 cm^{-1}); ▲, inelastic intensity (1230 cm^{-1})

Unfortunately, up to now, parallel vibrations of adsorbed atoms have been reported in two cases only, H on Pt(111) and H on W(100). We discuss the result on platinum first. An energy loss spectrum together with an angular profile is shown in Fig. 6.3. Clearly only the 550 cm^{-1} loss exhibits a dipolar lobe near specular reflection while the 1230 cm^{-1} loss becomes invisible in specular reflection. On the other hand, it has been shown [14] that both energy losses grow simultaneously with coverage. From these two pieces of evidence it was inferred that the 550 cm^{-1} and 1230 cm^{-1} losses correspond to the perpendicular and parallel motion of hydrogen in the same site, respectively. By applying the relations of Table 4.1 and assuming the threefold site to be the binding site a bond angle $\alpha = 72.45°$ is deduced. Assuming, further, that the surface is not reconstructed, a Pt–H bond distance of 1.68 Å is calculated. This is a reasonable value since the bond length of the diatomic molecule PtH is 1.53 Å and the sum of the covalent radius is 1.75 Å.

The system W(100) plus H is complicated by the reconstruction that takes place on this surface. Nevertheless, the same kind of analysis as already outlined was carried out by Willis [22], assuming a certain model for the reconstruction which places the tungsten atoms together in close contact for the reconstructed $c(2 \times 2)$ phase. Then a W–H bond length of 2.15 Å was

TABLE 6.2

Vibrational Frequencies (Dipole Active)
of Oxygen or Ni, Pt, and Fe for Various
States of Oxidation[a]

	Adsorbed	Intermediate	Bulk oxide
Ni(111)	580	950[b]	545[c]
Ni(100)	435	910[c]	545[c]
Pt(100)	490	720[d]	
Fe(110)	500[e]	910[e]	660 Fe_2O_3[f]

[a] For Ni see also Table 6.1.
[b] 950 cm^{-1} is enhanced when oxygen is brought on to the surface by decomposing NO [24].
[c] Ref. [25].
[d] Produced by high-temperature treatment in O_2 [26].
[e] Produced by high-temperature treatment in O_2 [10].
[f] Highest LO-phonon frequency of Fe_2O_3 [27].
Data for [a-f] are used with permission.

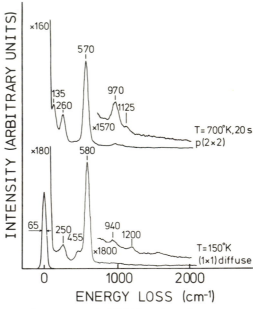

Fig. 6.4 Spectrum of oxygen on the Ni(111) surface after exposure at low temperatures at 2.0 langmuirs. $E_0 = 2$ eV, $\theta_i = 78° = \theta_r$. In addition to the characteristic 580 cm^{-1} loss due to oxygen in a threefold site, a loss at 455 cm^{-1} is observed. The intensity of this loss depends reversibly on the temperature and disappears above $\sim 250°$K. The loss may be associated with the occupation of the second threefold site of a (111) surface. The spectrum taken at 150°K also contains traces of high-frequency modes. After warming the crystal to 700°K, losses in the high-frequency range appear with higher intensity. These losses are probably related to subsurface oxygen.

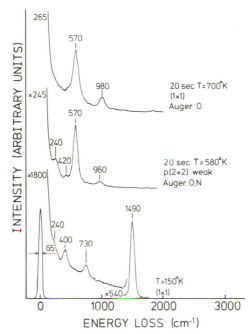

Fig. 6.5 Decomposition of NO by annealing the surface to 580 and 700°K at 0.5 langmuir. After annealing to 700°K the surface is nitrogen free since nitrogen desorbs. Again the remaining oxygen spectrum exhibits a peak at about 980 cm^{-1}, this time with an even greater intensity than in Fig. 6.4.

derived. For the unreconstructed $p(1 \times 1)$ structure at higher coverage a W–H bond distance of 2.05 was deduced. Both values seem rather high since the sum of covalent radius is only 1.76 Å.

Before turning to adsorbed molecules it may be worthwhile to pay attention to some additional data on vibrations of atoms regarding oxygen adsorption. For three different substrates (possibly four when an earlier investigation of oxygen on tungsten [5] is included which, however, rendered a rather complicated spectrum) a high vibrational frequency was observed for conditions of large exposures at elevated temperatures (Table 6.2). An example referring to the Ni(111) surface is shown in Figs. 6.4 and 6.5. The frequencies are roughly equal to the vibration frequencies of the diatomic M–O molecules which might be taken as an indication of the establishment of a bond where the oxygen sits on top of a metal atom. On the other hand, an on-top site must have a lower binding energy than sites of higher ligancy since the corresponding losses had not been observed at low coverages. The fact that the high-frequency losses seem to occur as an intermediate to the formation of bulk oxides suggests one might be dealing with oxygen below the first metal layer, probably in an octahedral interstitial site. As we have discussed in Section 4.4.6, such atoms should still give rise to inelastic electron scattering. The high frequency of these oxygen atoms compared to adsorbed atoms or bulk oxide modes may relate to the fact that the oxygen is not at an oxygen–metal equilibrium distance from the metal but rather squeezed into a site of a smaller volume dictated by the still remaining metal–metal bond. Such subsurface oxides might have a number of rather unique physical and chemical [23] properties which would warrant further investigations.

6.2.2   Adsorbed Diatomic Molecules (CO and NO)

Vibration spectra of diatomic species adsorbed on single-crystal surfaces have been reported for CO, NO, O_2, N_2, and C_2. As in the preceding section, we concentrate for now upon the question whether and how information about the binding sites of the adsorbed molecules may be obtained from vibration spectroscopy. Other questions of concern such as conditions for dissociative adsorption shall be discussed in Section 6.4.

Adsorbed CO.   By far the most vibrational data available for any adsorbate is on adsorbed CO. The stretching frequency of CO adsorbed on metal surface has been the workhorse of infrared reflection spectroscopy ever since this technique was introduced [28]. The reason that the CO stretching vibration was so popular in infrared reflection spectroscopy is partly because the frequency falls into a convenient range and partly because of the high dynamic dipole moment associated with the CO stretching vibration (Section 3.4.4). For the latter reason CO has also been investigated frequently with electron energy loss spectroscopy. But even before the advent of these single-crystal vibration spectroscopies, substantial experience with the vibration spectrum of adsorbed CO was available from infrared transmission spectroscopy. There the sample consists of a porous SiO_2 or Al_2O_3 support covered with finely dispersed metal particles which are usually cleaned by an oxidation-reduction cycle before they are brought into contact with adsorbates. This technique of IR transmission spectroscopy was introduced by Eischens et al. in 1954 [29]. Eischens and his colleagues also interpreted the different CO-stretching frequencies on these supported metal particles arising from CO species bonded to different sites. This assignment was based on the analogy to metal carbonyls where CO occurs bonded to a single metal atom or bridging between metal atoms. The entire spectroscopic material regarding CO adsorption on dispersed metals has been reviewed recently by Nguyen and Sheppard [30]. According to this review, the CO-stretching vibration is in the spectral region

$2130-2000$ cm^{-1}	for terminal CO	(M–CO),
$2000-1880$ cm^{-1}	for twofold bridged CO	(M$_2$–CO),
$1880-1800$ cm^{-1}	for threefold bridged CO	(M$_3$–CO),
< 1800 cm^{-1}	for fourfold bridged CO	(M$_4$–CO).

The intuitive picture one tends to associate with the frequency shift as a function of ligancy is the following: Bonding of the CO molecule to metal atoms involves mainly the two highest electron orbitals, the 5σ orbital and the unoccupied, antibonding $2\pi^*$ orbital. In order to establish a chemical bond, charge is donated from the doubly occupied 5σ orbital to the metal and accepted from the metal into the $2\pi^*$ orbital. The latter process would tend to weaken the CO bond and therefore also reduce the frequency of the CO stretching vibrations. For steric reasons, the overlap of the $2\pi^*$ orbital with the ligands is higher when the ligancy is higher, and therefore higher ligancy tends to result in a downward shift of the stretching frequency. This picture may be rather crude, but it serves the purpose of rationalizing a large body of experimental material.

Infrared transmission spectroscopy is limited by the fact that the support absorbs light below roughly 1300 cm^{-1}. The frequency regime below 1300 cm^{-1} was therefore inaccessible and nearly nothing was known about the metal carbon frequencies of adsorbed CO and the hindered rotations until recently. Although infrared reflection spectroscopy does not suffer from absorption of a support and in principle is capable of probing into the low-frequency regime of metal–adsorbate vibrations, technical difficulties hitherto have prevented its application in the far infrared. Thus, the use of electron energy loss spectroscopy was a major breakthrough.

The metal–carbon vibration is just as easily probed as the CO stretching vibration using electron energy loss spectroscopy. For the first time, metal–carbon frequencies and CO stretching frequencies can be correlated. The data available up to now are summarized in Table 6.3. Frequencies quoted in this table refer to low coverage with the exception of the last two entries. On the Pt(111) surface a bridging CO is observed for higher exposures only, and on the W(100) surface CO tends to adsorb dissociatively at room temperature. The molecular form of CO is thus not observed until the surface is covered with carbon and oxygen. Not included are CO data on a stepped surface, which we shall discuss in a different context. The agreement between the frequencies reported from electron energy loss spectroscopy and IR reflection spectroscopy applied to the same surface is within the error margin of the electron spectroscopy, which is about 10 cm^{-1}. This is not unexpected

TABLE 6.3

CO Stretching and Metal Carbon Frequencies
of Adsorbed CO on Various Crystal Surfaces[a]

Substrate	ν_{M-C}	ν_{CO}	Site	Ref.
Fe(110)	455	1890	Top	31
Ni(100)	480	2065	Top	32
	360	1930	Bridge	
Ni(111)	400	1810	Threefold bridge	33
Cu(100)	340	2090	Top	34
Ru(001)	445	1990	Top	35
Rh(111)	480	1990	Top	36
Pt(111)	480	2105	Top	37
	380	1870	Bridge	101
W(100)	365	2080	Top	38

[a] Data refer to low coverage except for the last two entries.

in view of the similar theoretical description of both techniques (Section 3.4.2). For higher CO converges, substantial frequency shifts, up to 100 cm^{-1}, have been reported as discussed and explained in Section 3.4.4.

Inspection of Table 6.3 shows that the assignment to certain adsorption sites is in accordance with the rules stated by Nguyen and Sheppard, with the exception of the Fe(110) surface. There the lower frequencies of terminal CO on Fe–carbonyls and a study of frequency shifts for CO isotopes on Fe–particles have led Erley [31] to propose the on-top (terminal) position despite the comparatively low frequency. This assignment is in agreement with the relatively high metal–carbon stretching vibration which falls into the range of frequencies characteristic for terminally bonded CO. If one excludes the two cases of weakly bonded CO on Cu(100) and W(100), the metal–carbon vibration for terminally bonded CO falls into a relatively narrow frequency range of only 35 cm^{-1} width, which is quite remarkable. As expected from the discussion in the previous section, the frequencies of the metal–carbon vibrations of bridging CO are lower.

So far the correlation between the frequencies of adsorbed CO and the adsorption sites had to be based entirely upon the result of vibrational spectroscopy itself. The feasibility of vibration spectroscopy on single-crystal surfaces, however, allows one to make contact with other surface techniques. The correlation between vibrational spectroscopy and electron diffraction from ordered overlayers of adsorbates has been particularly useful. These days, dynamical calculations of low-energy electron diffraction [39] have reached a stage of maturity where binding sites of small molecules can be determined from the diffraction intensities, just as we have already discussed for the case of atoms. Therefore the assignment of CO species to certain binding sites based on vibration spectroscopy can be tested with electron diffraction. An example is the $c(2 \times 2)$ overlayer of CO on Ni(100) where (after some controversy) it is now well established [40, 41] from electron diffraction that the CO is terminally bonded. Similarly, the terminal bonding for CO on Cu(100) was confirmed [41].

In some cases, the binding site can also be deduced from the number and sequence of CO species in correlation with the size and shape of the adsorbate unit mesh which is obtained from the diffraction pattern without an analysis of intensities. An example is CO on the Ni(111). In Fig. 6.6 the intensity of two CO stretching losses are plotted versus exposure. Except for small coverages and a singular point where a sharp $c(4 \times 2)$ diffraction pattern occurs, two losses are always observed. The spectrum characteristic of the $c(4 \times 2)$ layer has already been shown in Figs. 4.11 and 4.12 and is again

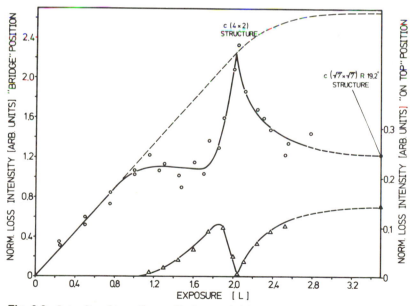

Fig. 6.6 Intensity of two discernable losses of CO on Ni(111) versus exposure to CO assigned to on-top ($\triangle$) (terminally) bonded CO and CO in threefold and twofold bridges ($\bigcirc$). $T = 140°K$.

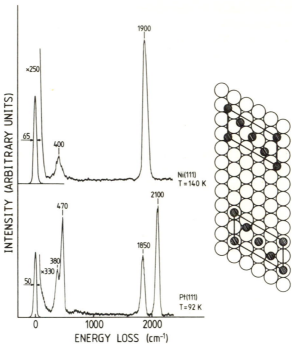

Fig. 6.7   Electron energy loss spectra of the Ni(111) and Pt(111) surfaces, each covered with half a monolayer of CO which orders into a $c(4 \times 2)$ overlayer. On the nickel surface the vibration spectrum indicates only a single CO species in a site of high symmetry. The only possibility for positioning the two-dimensional CO lattice on the surface consistent with the single type of adsorption site is to place all CO molecules into twofold bridges. By similar reasoning, half the CO molecules must occupy on-top sites on the Pt(111) surface. This example shows how powerful the *in situ* comparison of vibrational spectra and diffraction pattern can be, since a qualitative structure analysis is achieved without analyzing diffraction intensities.

reproduced in Fig. 6.7 together with the $c(4 \times 2)$ unit mesh. The absence of any second CO stretching loss and other losses than a single metal–carbon loss indicates that all CO molecules are adsorbed in a single kind of site of high symmetry. The only way to position the $c(4 \times 2)$ unit mesh in such a way that all atoms are in one site is to place the CO molecules into twofold bridges. The frequency of the CO stretching vibration of 1900 cm^{-1} (Fig. 6.7) also seems to be consistent with the rule Nguyen and Sheppard established for CO adsorbed on finely divided metals. Such a comparison must, however, be taken with some precaution. We have learned already in Section 3.4.4 that the frequency may shift upward with increasing coverage from dipole–dipole coupling by as much as 60–80 cm^{-1} on flat surfaces (Fig. 3.14). As we have seen, the magnitude of the frequency shift depends on the ionic and electronic polarizability of the CO–metal complex which can be determined from data on the adsorbate intensity and a measurement of the frequency shift as a function of coverage. From data of Campuzano and Greenler [42], who applied infrared reflection spectroscopy to the same system, a total dipole shift of roughly 35 cm^{-1} can be inferred. If we subtract this shift from the measured frequency for the $c(4 \times 2)$ structure, the resulting value would be rather on the lower boundary of the limits Nguyen and Sheppard placed on characteristic frequencies for the bridging site. This seems to be in accord with the general trend of the data on single-crystal surfaces as they are listed in Table 6.3. We may therefore conclude that the CO stretching frequencies on flat surfaces at low coverage, or after subtracting the dipole–dipole shift, tend to be somewhat lower than on dispersed metals.

While the analysis just given has allowed us to determine the binding site of CO for the $c(4 \times 2)$ adsorbate lattice, nothing is inferred about the binding site at low coverage. Both the authors of Refs. [33] and [42] observed an upward jump in the frequency by 30 to 50 cm^{-1} when the $c(4 \times 2)$ structure was reached. Since dipole–dipole interactions tend to be relatively structure insensitive, this jump can be associated with a change of the binding site from a threefold bridge to a twofold bridge. This interpretation is supported by the fine experiment involving infrared spectra of CO on Pd(111) published by Bradshaw and Hoffman [43] which shows the same frequency jump quite clearly (Fig. 6.8).

On the densely packed surfaces of nickel and palladium, CO seems to prefer sites of higher coordination. On the contrary, on Rh(111) and Pt(111), CO is terminally bonded at low coverage with additional bridging CO at higher coverage. In the case of the Pt(111) surface, two different metal–carbon frequencies at 480 cm^{-1} and approximately 380 cm^{-1} could be resolved for the terminally- and bridge-bonded CO (Table 6.3 and Fig. 6.7). As in the case of Ni(111), the position of the $c(4 \times 2)$ adsorbate lattice relative to the substrate can be deduced (Fig. 6.7).

A consistent model satisfying the picture provided by vibrational spectroscopy and of electron diffraction is not easily achieved for all CO–substrate systems. For the systems CO on Cu(100) and Cu(111) and CO on Ru(0001) the data appeared to be contradictory. On Cu(100), in both electron energy loss spectroscopy and infrared spectroscopy, a single stretching frequency (at 2080 cm^{-1}) was observed [28, 34, 44] at all coverages. As mentioned before, no problem arises in matching this observation to the existing structure analysis for the $c(2 \times 2)$ superlattice, at half a monolayer coverage. For slightly higher coverages the diffraction pattern converts into $c(7\sqrt{2} \times \sqrt{2})R\,45°$. Assuming that the repulsive forces between the CO molecules tend to keep them at approximately equal distance, unit cells as

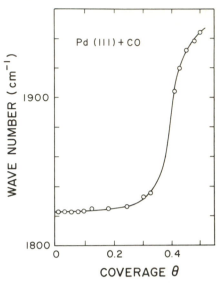

Fig. 6.8 CO stretching frequency versus coverage for Pd(111) at room temperature after Bradshaw and Hoffmann [43], used with permission. The jump in the frequency by ~ 60 cm^{-1} is caused by the change of the adsorption site from a threefold bridge to a twofold bridge. The coverage scale was calibrated at $\theta = 0.33$. The scale need not to be linear for high coverages since the coverages apparently were calculated from the infrared intensities (see Section 3.4.4).

depicted in Figs. 6.9a and 6.9b were proposed [45, 46]. The distances of the CO molecules in both structures were approximately equal to the van der Waals distance of CO, which seemed to provide a straightforward explanation of the coverage saturated with exposure at this point. The structures are, however, at variance with the site specificity of the CO stretching frequency encountered in so many other systems. Recently an alternative structure was proposed by Biberian and van Hove [47], which employs on-top sites only (Fig. 6.9c). The authors could show for this structure (and a few similar problematic cases) that their proposed structure produced the correct diffraction patterns in laser simulation experiments. Biberian and van Hove also argued that the van der Waals distance cannot play a crucial role in determining the distance between adsorbed CO molecules, since CO molecules are more densely packed also in carbonyls. Nevertheless, the CO molecules adsorbed on nearest-neighbor surface atoms should tend to repel each other and assume a tilted position. In this respect, experimental results of Andersson [34] on the same system and of Erley [31] for CO on Fe(110) become rather significant. Both authors observed an additional energy loss at 275 and 360 cm^{-1}, respectively, despite the fact that there was only a single CO stretching vibration (Fig. 6.11). This loss could well be a bending vibration (hindered rotation) which would become dipole active when the molecule is tilted.

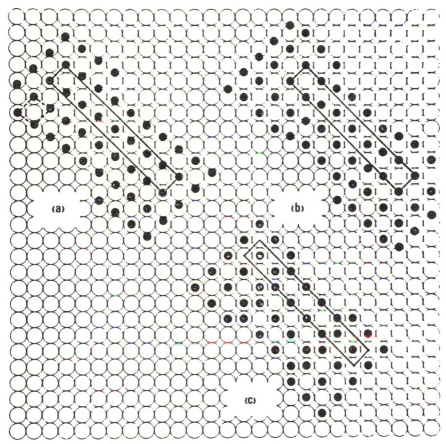

Fig. 6.9   Structure models to account for the $c(7\sqrt{2} \times \sqrt{2})R\,45°$ diffraction pattern for CO on Cu(100) after Biberian and van Hove [47] used with permission. O, Cu surface atoms; ●, adsorbed CO molecules. (a, b) [45, 46] used with permission. The CO molecules are placed into different sites, which is at variance with the single CO stretching frequency observed for this structure. (c) The new structure model proposed by Biberian and van Hove is consistent with the diffraction pattern, the single CO stretching frequency around 2080 cm^{-1}, and with the observation of a dipole active CO bending mode [34] (see also Fig. 6.11), used with permission.

Biberian and van Hove also reinterpreted the $c(4 \times 2)$ structure on Cu(111). By rearranging the atoms in the unit cell so that they are no longer at equal distances (Fig. 6.10), all CO atoms can be placed in on-top positions. Some of them are again on nearest-neighbor surface atoms, and one might expect them to tilt. This model allows one to retain the proposition that high vibration frequencies [~ 2080 cm^{-1} for CO on Cu(111)] indicate terminal bonding. At the same time it seems to invalidate our reasoning with

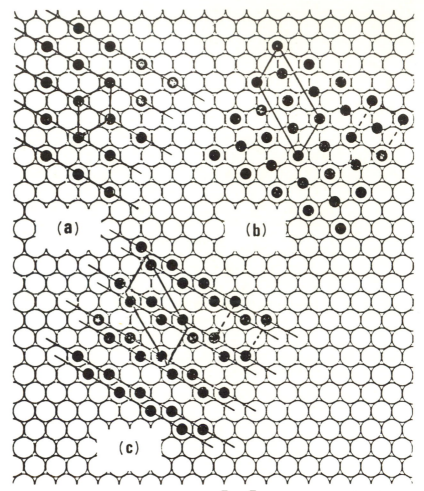

Fig. 6.10 Structure models for (a) the $(\sqrt{3} \times \sqrt{3})R\,30°$ pattern and (b, c) the $c(4 \times 2)$ pattern. O, Cu surface atoms; ●, adsorbed CO molecules. (b) Two types of sites are employed. An alternative would be to use bridging sites only as depicted in Fig. 6.7. (c) The new proposal by Biberian and van Hove allow use of top sites only which would be consistent with the high frequency of the CO stretching vibration at all coverages. At the same time the model places the CO atoms very close together and one would expect them to tilt away from each other again. Vibration spectra in the low-frequency regime are not yet reported for this system, yet Fig. 6.11 shows an example for a similar situation where the tilting is clearly evident from the spectrum ([47], used with permission).

the $c(4 \times 2)$ structure of CO on Ni(111), where we have argued that the observation of a single CO stretching vibration concomitant with the $c(4 \times 2)$ structure is evidence for a bridging site. Still, the reasoning there was backed by additional arguments: First, the spectrum of CO on Ni(111)

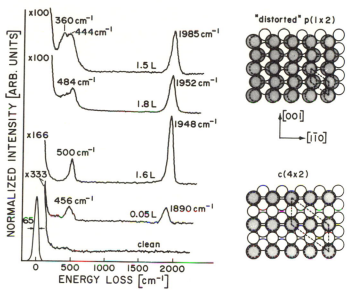

Fig. 6.11   A series of vibration spectra for CO on Fe(110) at $T = 120°$K after Erley [31] (with permission), and structural models. The proposed structure for high coverage is consistent with the following observations: The diffraction appears as $p(1 \times 2)$ in low-order beams, the single stretching vibration indicates terminally bonded CO for all CO molecules, the molecules must be tilted in order to let an additional bending mode become dipole active.

did not show any signs of a bending mode which one would expect to be dipole active in a structure like in Fig. 6.10c, just as was observed for Cu(100) and Fe(110) (Fig. 6.11). Second, on the Ni(111) surface, additional CO stretching vibrations of higher frequency (2050 cm^{-1}) were observed for smaller and higher coverages. Therefore, if one were to propose the structure of Fig. 6.10c for the Ni(111) surface, one would likewise have to assume that stretching vibrations around 2050 cm^{-1} could arise from nonterminally bonded CO. Such a reversed assignment seems not acceptable in view of the ample experience with carbonyls.

In conclusion, one might say that the conventional assignments of CO stretching vibrations to different adsorption sites is endorsed by the majority of single-crystal surface results. It is furthermore rather likely that the few cases in which the assignment has led to some difficulties are also about to be resolved in a way which is consistent with the conventional relation between stretching frequency and ligancy.

Adsorbed NO. As with CO, a large number of coordination complexes with NO ligands have been studied by vibration spectroscopy [48–50]. The stretching frequency of NO in those compounds covers a much wider range from 1300 to 2000 cm^{-1}. Nevertheless, an analysis of the vibration

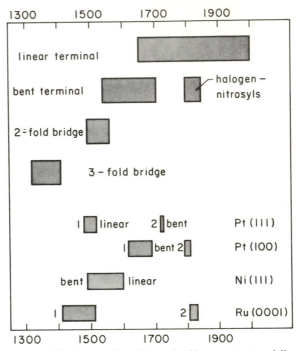

Fig. 6.12   NO stretching frequencies of nitrosyls of known structure following Pirug *et al.*
[51] and of adsorbed NO ([51–54] used with permission). The numbers 1 and 2 refer to the
sequence in which the species appear on the surface as a function of exposure.

frequencies of NO for coordination compounds where structure is known
again shows the general trend of lower stretching frequencies with higher
ligancy (Fig. 6.12). Unlike CO, NO tends to form a bent configuration, by
employing the extra single electron in the $2\pi^*$ orbital for establishing a
chemical bond (note CO_2 is linear, NO_2 bent). The stretching frequency for
terminal bent configuration is in general somewhat lower than for linear
species.

The experimental material on single-crystal surfaces is rather scarce. The
existing data on the NO stretching frequencies are summarized in Fig. 6.12.
On the platinum and ruthenium surface, two distinctly different species have
been observed and assigned to terminally bonded and bridging NO, in
accordance with characteristic frequencies of nitrosyls of known structure.
On the platinum surface, the second terminally bonded adsorption state had
a distinctively lower binding energy. The same is probably true also for the
ruthenium surface since there the bridging NO dissociated at a significantly
lower temperature than the terminally bonded species. NO adsorption on
Ni(111) (Fig. 6.13) is also instructive. At small exposures, three losses are

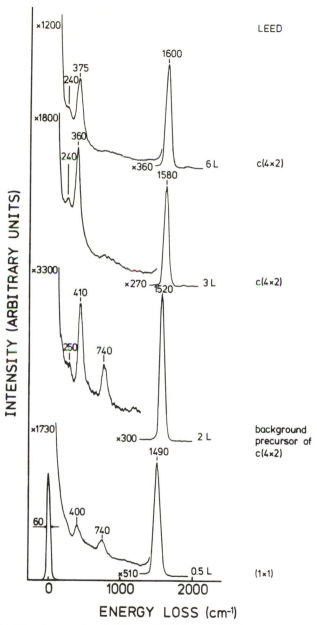

Fig. 6.13   Vibration spectra of NO on Ni(111) as a function of exposure. $T = 150°$K, $E_0 = 2$ eV, $\theta_i = \theta_r = 78°$. The three energy losses (in addition to the phonon loss; cf. Section 5.3.2) in the low coverage spectra are indicative of a bent configuration (cf. Section 4.4). Once the $c(4 \times 2)$ structure begins to form, the NO molecule orients perpendicular to the surface.

observed which are assigned to NO stretch, hindered rotation (740 cm^{-1}), and Ni–N stretch (400 cm^{-1}). The relatively low NO stretching frequency (Fig. 6.12) and the relatively high NO bending frequency (Table B.3) suggests that NO sits in a twofold site. With higher exposure and the formation of the $c(4 \times 2)$ layer, the bending vibration disappears from the spectrum. This disappearance is accompanied by an upward shift of the NO stretching vibration and a downward shift of the Ni–N vibration. Both shifts can be considered as a sign for a reduced bonding to the surface. As discussed in the case of CO on the same crystal face of nickel, the $c(4 \times 2)$ structure concomitant with a single species perpendicularily oriented with respect to the surface is evidence for a twofold bridging site. The stretching frequency is again in accordance with the spectrum of frequencies in nitrosyls (Fig. 6.12).

6.3   FREQUENCY SHIFTS

6.3.1   Origin of Frequency Shifts

In the preceding section we have already noticed that chemisorbed CO and NO exhibit a significant frequency shift of the stretching vibration compared to the gas-phase molecule. The very existence of such a frequency shift is, in fact, the spectroscopic evidence that the intramolecular bond is perturbed by adsorption to the surface atoms. Special cases of such frequency shifts and how they relate to the ligancy of the adsorbed molecule and to lateral interactions have already been discussed in the preceding section and at an earlier stage (Section 3.3.4). Now we wish to address the problem of frequency shifts in a more general manner in order to understand what might be learned regarding the structure, the bond length and the bond strength of adsorbed species. Before such an analysis is undertaken, one must realize, however, that frequency shifts may be caused by a number of different effects. We describe these effects briefly in the following.

Mechanical Renormalization.   By this term we mean frequency shifts introduced by an additional mechanical coupling to the surface without a change in the intramolecular force constants. A simple example again is provided by adsorbed CO. Even if the force constant between the carbon and the oxygen atom is not altered, the coupling of the molecule to the surface via the carbon atom gives rise to a small upward shift of the CO stretching vibration. The magnitude of the effect is easily estimated in a nearest-neighbor free constant model. In Fig. 6.14 we have plotted the frequency shift versus the metal carbon frequency of adsorbed CO, both normalized to the free CO vibration. The shaded area represents the regime typical for adsorbed CO according to Table 6.3. We see that the effect of

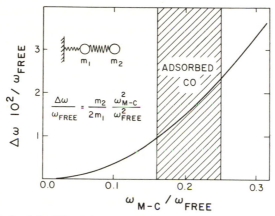

Fig. 6.14   Ratio of the shift of the stretching frequency of adsorbed CO to the frequency of the free molecule versus the ratio of the metal–carbon stretching frequency to the frequency of the free molecule calculated in the nearest-neighbor force constant model assuming the mass of the surface atom to be infinitely large. The shaded area represents the typical range for chemisorbed CO.

renormalization is small as long as the frequencies related to the coupling with the surface are small compared to the innermolecular vibrations (see also Section 4.2.2). In terms of absolute numbers, $40 \, \text{cm}^{-1}$ is the biggest effect of mechanical renormalization to be expected for CO. This shift is small compared to the ligancy effect described in the previous section.

Lateral Interactions.   When adsorbed molecules become more closely packed on the surface, they begin to experience mutual interactions. A special type of interaction, dipole–dipole coupling, was already considered in some detail in Section 3.4.4. In addition, there can be direct chemical interactions as well as indirect interactions through the substrate, as we have seen in Section 5.2.6. Lateral interactions are easily distinguished experimentally from other shifts by observing the spectrum at different coverages. Lateral interactions also give rise to dispersion: the frequency varies when measured as a function of parallel momentum $Q_{\|}$, provided that $Q_{\|}$ is a good quantum number (ordered overlayers). When the overlayer is disordered, lateral coupling will show up as a broadening of the line. Dispersion resulting from dipole–dipole coupling has been studied recently with the $c(2 \times 2)$ CO overlayer on Cu(100) by Andersson and Persson [56]. As expected, the dispersion was relatively small ($\sim 30 \, \text{cm}^{-1}$). Larger effects have been calculated theoretically in lattice dynamical treatments of the surface adlayer system (see Section 5.2.6), but so far no experiments have been reported.

Bond Order.   While mechanical renormalization and lateral interactions have a relatively minor effect on high-frequency intramolecular vibrations, large frequency shifts may occur when the character of the intramolecular bond changes as a result of the bonding to the surface. In Section 4.5.1, we have seen that the characteristic CC stretching frequency of triply bonded —C≡C— group is about a factor of 2 higher than the frequency of a single bonded $\supset$C—C$\subset$ group. We have also already discussed in another context the electron energy loss spectrum of acetylene C_2H_2 (Section 4.5.2) and remarked that the C–C vibration frequency of acetylene adsorbed on nickel is about 1200 cm^{-1} rather than around 2000 cm^{-1}. Consequently, a reduction of the CC bond order to a single bonded CC group may be inferred, which in turn leads to a number of conclusions about the electronic structure of the adsorbed molecule, the internal bond energy, the bond length, and the eventual fate of the molecule when the surface is heated to higher temperature. Though much of this information is only of a qualitative nature, it nevertheless provides valuable information on the chemistry of surfaces and surface reactions, including heterogeneous catalysis. We shall, therefore, be dealing with the bond-order effect more deeply in the upcoming section. We shall also see that the ligancy effect described for CO basically falls into the same category.

Multicentered Bonds—Hydrogen Bonding.   In a number of cases, large frequency shifts have been observed for CH and OH stretching vibrations of adsorbed species. This effect could be correlated with the tendency toward dissociation of the CH and OH, respectively. It is therefore assumed that hydrogen can form a two-centered (or multicentered) bond. By analogy to hydrogen-bonded hydrogen centered between electronegative atoms, one might also call this phenomenon hydrogen bonding, although the frequency shifts associated with it on surfaces are sometimes relatively small, and the hydrogen ligands (carbon and a transition metal) are uncommon in the chemistry of conventional hydrogen-bonded compounds. Since the effect occurs rather frequently with adsorbed hydrogen-containing compounds, it seems to deserve a presentation in a separate section.

6.3.2   Empirical Relations between Force Constants and Other Bond Parameters

For a large number of diatomic molecules, the ground state harmonic frequency ω_e, the equilibrium bond length r_e, and the potential energy D_e or dissociation energy $D_0 \sim D_e - \frac{1}{2}\hbar\omega_e$ are known [Tables B1, B3]. Although these quantities are, in principle, independent parameters of the internuclear

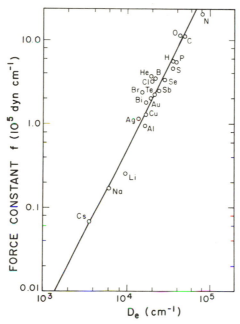

Fig. 6.15 Harmonic force constants versus depth of the potential well D_e for diatomic molecules (data from Ref. [B3] used with permission). Values for the C–C bond in hydrocarbons are also included.

potential, the experimental data show that the parameters are correlated. In Fig. 6.15 we have plotted the force constant as derived from the harmonic frequencies ω_e and the reduced mass versus the depth of the internuclear potential D_e which is approximately equal to the dissociation energy (see Section 5.4). In order to make the plot not too confusing, we have included only the data on homonuclear diatomic molecules [Table B3]; however, the data for the heteronuclear diatomics do not deviate significantly. The solid line is a least-squares fit from which one obtains for the force constant f defined in Table 4.1

$$f \approx 5 \cdot 10^{-4} D_e^2, \tag{6.2}$$

where f is obtained in dynes per centimeter when D_e is inserted in the spectroscopic units cm^{-1} (1 cm^{-1} = 0.1239 eV, 8065.5 cm^{-1} = 1 eV). The equivalent to the dissociation energy of a diatomic molecule in a polyatomic molecule is the total bond energy. In physical chemistry, bond energies are also assigned to individual bonds within a molecule [58]. Although this

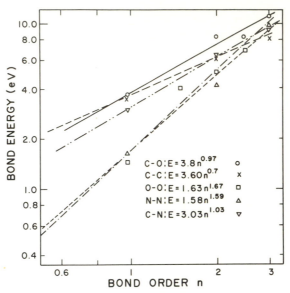

Fig. 6.16 Correlation between bond energies in electron volts and bond order n for five different bonds of particular importance in surface chemistry.

concept is not rigorously defined, it is remarkably successful in calculating total energies of molecules and reaction enthalpies. Tables of standard bond energies are found in Ref. [58] and in many textbooks on physical chemistry. The correlation between bond order and bond energy for five different types of bonds is shown in Fig. 6.16. The bond order (to be defined) can also be correlated with the force constants using data on diatomic molecules and typical force constants related to certain bonds within large molecules. For diatomic molecules, the term "bond order" can be defined in a way which includes positive and negative ions: The valence bond orbital configuration of first-row diatomic elements is $5\sigma_g^2 2\pi_u^4 2\pi_g^{*4}$, where the upper index numbers refer to the maximum occupation. The bond order n is then the number of electron pairs in bonding orbitals minus the number of electron pairs in the antibonding $2\pi_g^*$ orbital. Thus C_2, O_2, CO, and NO achieve bond order 2 and N_2 achieves bond order 3, consistent with the chemical picture. Data for a number of different molecules and bonds are shown in Fig. 6.17. They can be fitted by a common power law,

$$f = 4.2 \times 10^5 n^{1.45} \quad (\text{dyn cm}^{-1}). \tag{6.3}$$

Other well-established correlations exist between the force constant and the bond distance. One of the simplest formulas is Badger's rule [59, 60]

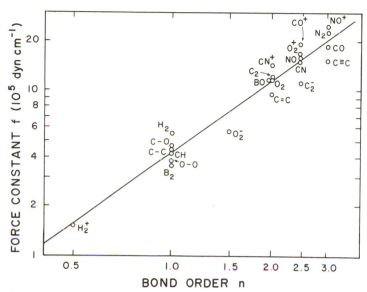

Fig. 6.17 Correlation between bond order as defined in the text and the force constant; again CC— bonds of hydrocarbons and the single bonds of $\geqslant$C—O— and HO—OH are included.

according to which

$$f = 1.86(r_e - d_{ij})^{-3}, \qquad (6.4)$$

where r_e is the equilibrium distance in angstroms, d_{ij} a parameter, and f is in units of 10^5 dyn cm^{-1}. The indices i, j refer to the row of the periodic table of the elements which form the bond. Examples are given in Table 6.4. In Fig. 6.18, Eq. (6.4) is plotted as a full line with $d_{ij} = 0.68$ Å. This parameter provides the best fit of Eq. (6.4) to the data for diatomic molecules. Other fits to the data of diatomic molecules have also been proposed [61] and are claimed to be more accurate. For our purpose the simple form of Eq. (6.4) should suffice. With the exception of CO and C$_2$, the data points in Fig. 6.18, however, do not represent diatomic molecules but CO and CC bonds within polyatomic molecules. We notice that the data points for CO in transition metal carbonyls and for the single bonded CO in alcohols are very close to the line representing Badger's rule, while a slight offset exists for the hydro-carbons. Vibrational spectra and CC bond distances are also known for a number of coordination compounds involving CC groups [63]. Estimates of the force constants for these CC bonds just from the CC stretching vibra-tion also seems to fall into the correlation for uncoordinated CC groups. The

TABLE 6.4

Parameter d_{ij} for Various Combinations
of Elements of the Periodic Table[a,b]

i	j	d_{ij} (Å)	i	j	d_{ij} (Å)
H	H	0.025	H	3T	0.53
H	1	0.36	H	4T	0.61
H	2	0.58	H	5T	0.62
1	1	0.68	1	3T	0.97
1	2	0.92	1	4T	1.08

[a] Ref. [61].
[b] The numbers refer to the row of the periodic
table; T denotes "transition metal."

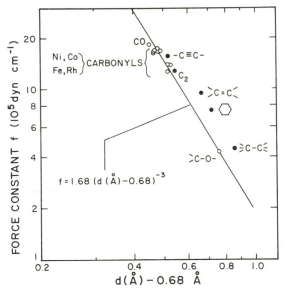

Fig. 6.18 Force constants of CO— and CC— groups versus bond distance. For the
CC— groups data are from Ref. [57] with permission. The carbonyl data are from Ref. [62] with
permission. There the force constants have been calculated disregarding the coupling to the
metal ligand.

data are, however, not included in Fig. 6.18 since the force constants are
more model dependent for the coordinated CC groups.

We now weigh the three correlations discussed in the foregoing against
the background of experimental material available for adsorbates on surfaces

in order to see whether the correlations are likely to be of any predictive value for surface complexes. The first correlation we had established with Eq. (6.2) connects force constants with the dissociation energies. When it comes to "bond energies," the general trend is that higher force constants correlate with higher bond energies. The prefactor and the experiments for a power law fit are, however, different for each type of bond. The concept is therefore not applicable to bonding of atoms in different surface sites since appropriate parameters are not available.

Nevertheless as a qualitative trend that higher vibration frequencies for comparable systems indicate stronger bonding, the correlation, (6.2) still holds. The same seems to be true for intramolecular bonds of adsorbed molecules: It was shown in several examples that when, on a surface, two identical molecular groups with different vibration frequencies exist, the molecular group with the lower vibration frequencies dissociates at a lower temperature. Examples are NO in different sites on Ru(001) [54], CO on a stepped nickel surface [64], and CH groups [65]. We also notice from Table 6.3 that metals which dissociate CO more easily (Fe) lead to lower stretching vibrations of adsorbed CO (in the same sites) than metals which do not easily or at all dissociate CO (Pt, Cu). We shall also return to this point in Section 6.3.4. The application of the force constant–bond order relation has already been discussed by Sexton and Madix [66], in connection with the adsorption of molecular oxygen on silver and platinum. The stretching frequencies of molecular oxygen on these surfaces are at 640 and 870 cm^{-1} for silver and platinum, respectively, while the $Ag-O_2$ and $Pt-O_2$ frequencies are at 240 and 390 cm^{-1} [66, 67]. One may again convert the O–O stretching vibration into a force constant to obtain bond orders n of ~ 0.6 and 0.9 for the O–O bond on Ag and Pt, respectively. According to the definition of the bond order in terms of the occupation of the $2\pi^*$ orbital, these numbers would mean that the $2\pi^*$ orbital should be filled with two electrons with possible further donation into the σ^* orbital in the case of oxygen on silver [66]. In fact it could be shown using photoemission spectroscopy that O_2 on platinum has full π^* orbitals [26], while equivalent experiments for oxygen on silver have not been carried out to date.

The bond-order correlation may probably also be used for an estimate of the amount of charge donation into the π^* orbital of adsorbed CO. In the extreme case of the CO frequency shift on the stepped nickel surface (1530 cm^{-1}), the bond-order relation would predict the π^* orbital to be occupied by more than one electron. As in the case of oxygen on platinum, this occupation of the π^* orbital might give rise to a feature in the photoemission spectrum slightly below the Fermi level.

A dramatic reduction in the bond order is now well documented for several hydrocarbons on transition metals. Several examples where both the CC stretching vibrations and the CH stretching vibration have been determined are listed in Table 6.5. In all cases the CC bond order as estimated from the CC stretching frequency is nearly one. This number is consistent with the observed CH stretching frequencies which, according to Table 4.9 would also indicate the formation of an sp^3 hybrid.

TABLE 6.5

CC Stretching Frequencies, C–C Bond
Order, and CH Stretching Vibrations for
Acetylene and Ethylene on Nickel
(Table 4.13) and Platinum[a]

Adsorption system	ν_{CC}	n	ν_{CH}
C_2H_2–Ni(111)	1200	~ 1.1	2920
C_2H_2–Pt(111)	1310	~ 1.3	3010
C_2H_4–Pt(111)	1230	~ 1.2	2940

[a] Ref. [68], used with permission.

We finally turn to the discussion of surface applications of Badger's rule. Here the possible use could be to provide an estimate of the bond length of molecules or of the bond length between adatoms and surface atoms. From Fig. 6.18 we see that Badger's rule renders a reasonable estimate for the bond length not only for diatomic molecules but also for intramolecular bond lengths of hydrocarbons and CO coordinated to transition metals. In view of the general similarities between coordination compounds and surface complexes, it is unlikely that the correlation of Fig. 6.18 should not

TABLE 6.6

Innermolecular Bond Distance of Adsorbed CO and NiO
Bond Distances as Determined by Electron Diffraction (d_{exp})
Compared to Distances Calculated from the Apparent
Stretching Force Constant using Badger's Rule

Adsorption system	d_{exp} (Å)	d_{th} (Å)	f (10^5 dyn cm^{-1})
CO–Ni(100)	1.10 ± 0.1 [41]	1.14	17.2
CO–Cu(100)	1.15 ± 0.05 [41]	1.14	17.5
O–Ni(100)	1.97 ± 0.05 [70]	1.98 ± 0.03	1.65
O–Ni(111)	1.87 ± 0.05 [71]	1.91 ± 0.03	2.0

apply to adsorbed molecules. The precision in calculating the bond length from Badger's rule is rather high. Though force constants are somewhat model dependent, the scatter is less than 10%. This is equivalent to a precision of ~ 0.02 Å in the bond length. Unfortunately there is little overlap between available vibrational data and bond length determinations for surface complexes. Four cases in which contact can be made are listed in Table 6.6. Of these cases, oxygen on the two nickel surfaces are most significant. Force constants are calculated within the nearest-neighbor model. To a small extent, they depend on an assumed bond length, that is, on the result to be obtained from Badger's rule. It is, nevertheless, easy to achieve a consistent set which satisfies both Badger's rule and the formulas of Table 4.1. The final result compares rather favorably with the bond length determined from electron diffraction.

In summarizing the discussion within this section, it seems that the empirical relations between the force constants and other bond parameters are also applicable to adsorbed species and may, when used with care, allow qualitative predictions about the strength of a bond and a semiquantitative estimate of the bond length.

6.3.3 Two-Centered Bonds of Hydrogen–Hydrogen Bonding

Substantial frequency shifts were observed in several studies for hydrogen bonded (primarily) to an oxygen or carbon atom of an adsorbate, when CH stretching frequencies of adsorbed and gas-phase species are compared. One example is shown in Fig. 6.19. On both the Ni(111) and the Pt(111) surface, broad and intense bands of CH stretching vibrations appear in addition to the normal CH bands of cyclohexane. The frequency shifts are 180 and 310 cm^{-1}, respectively. On both surfaces, cyclohexane seems to be adsorbed intact, since the observed spectrum of the adsorbed species matches the frequency spectrum of gaseous cyclohexane except for the shifted CH stretching vibration. Furthermore, cyclohexane desorbs completely from Ni(111) at about 170°K while it dehydrogenates to benzene on Pt(111) upon heating to roughly 300°K. Broad, shifted hydrogen bands with a high dipole moment are typical for hydrogen bonding [72]. Therefore the spectrum of adsorbed cyclohexane has been considered to be evidence for hydrogen bonding which here, however, appears in the unusual form of a hydrogen bond between carbon and the metal surface atom. Hydrogen bonding of cyclohexane might account for the fact that cyclohexane adsorbs on both surfaces at 140°K, a temperature where physisorbed saturated hydrocarbons would not remain on the surface under ultrahigh vacuum conditions.

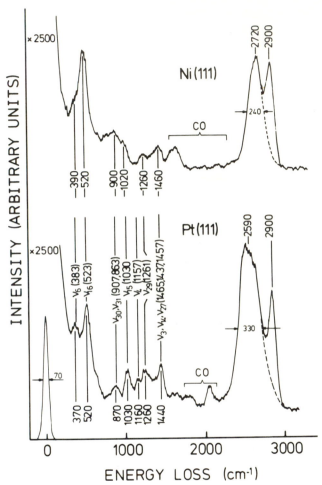

Fig. 6.19 Spectrum of cyclohexane (C_6H_{12}) adsorbed on Ni(111) and Pt(111) at $T = 140°$K ([71] used with permission). All vibrations can be assigned to the A_{1g}, A_{2u}, E_g, and E_u modes of C_6H_{12}. Using the dipole selection rule and the correlation tables (Table 4.6), this indicates a symmetry C_s. It is therefore likely that cyclohexane assumes a canted position with three hydrogens pointing towards the surface. The hydrogen atoms give rise to the strong broad CH stretching band suggesting hydrogen bonding on the molecule to the surface.

Two different forms of hydrogen exist also on ethylene adsorbed on Ni(111) (Fig. 6.20) and Ni(100) (Fig. 6.21). Contrary to cyclohexane, little if any dipole moment is connected with the shifted CH stretching band as the comparison with off-specular spectra shows. The small dipole moment is apparently a genuine feature and not caused by a selection rule effect, since the symmetry of adsorbed ethylene on nickel is sufficiently low in order to

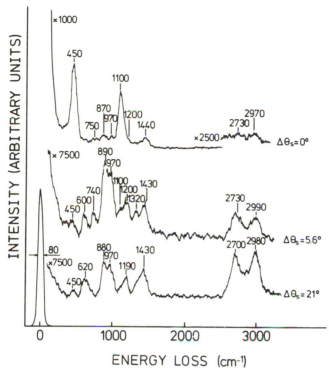

Fig. 6.20 Spectrum of adsorbed ethylene (C_2H_4) on Ni(111) at $T = 150°K$. $E_0 = 2$ eV, $\theta_r = 78°$. A complete mode assignment has not been achieved in this case. Obviously the molecule has two differently bonded forms of hydrogen atoms. The low CH stretching vibration is attributed to hydrogen atoms which have established a partial bonding to the nickel surface atoms which eventually leads to a decomposition of the molecule into C_2H_2 and H_2 at $\sim 230°K$.

allow a perpendicular component of a dipole moment connected with a CH stretching vibration. Another difference with the case of cyclohexane is that the shifted CH stretching band is much less broadened, although the frequency shift is comparable. A possible model to explain the shifted band is that ethylene tilts on the surface to have two hydrogen atoms closely coordinated with the nickel surface so that the hydrogen also bonds weakly with the nickel. This particular type of bonding might be called hydrogen bonding in a more general sense or a two-centered bond. This model is supported by the dissociation of ethylene to acetylene and hydrogen on the nickel surface at about 230°K [73]. The dissociation temperature is significantly lower than that observed with CH bonds on platinum [74] where no frequency shifts are observed. Furthermore, acetylene on Ni(111) shows no shifted CH bands. It tends to disintegrate into CH groups first [74], until ultimately the CH band brakes at about 600°K. It seems therefore that the

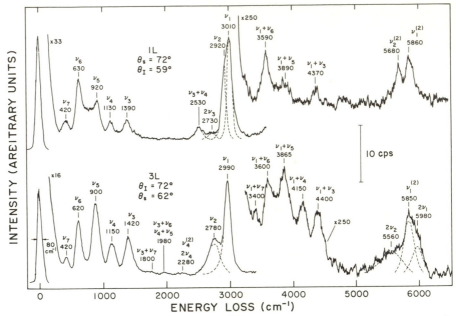

Fig. 6.21 Spectra of fundamentals, overtones, combination bands, and multiple losses of ethylene on Ni(100). For 1 langmuir exposure the intensity of multiple losses seems to be small, the only evidence being possibly the tail at ~6000 cm^{-1}. In the 3-langmuir spectrum, multiple losses appear to be more intense. The hydrogen bonded CH stretch (2760 cm^{-1}) produces a double loss, however, not an overtone in accordance with infrared spectra of hydrogen-bonded molecules.

CH stretching vibration below the normal frequency range is of some predictive value for the further fate of the molecule on the surface at higher temperatures.

Certain forms of lateral hydrogen bonding may also result in frequency shifts. An example is the adsorption of water on Pt(100) [76], Pt(111) [77], Ru(0001) [78], and Ag(110) [79]. A series of spectra of H_2O adsorbed on Pt(100) for submonolayer coverages up to coverages of several layers is shown in Fig. 6.22. A monolayer is reached at approximately 2 langmuirs (1L = 10^{-6} Torr sec). The spectra above one monolayer resemble the infrared spectrum of ice except that the OH stretching vibration of hydrogen-bonded hydrogen is still about 160 cm^{-1} higher than in ice. Obviously some of the hydrogens are also not hydrogen bonded (3680 cm^{-1}). Interestly, the hydrogen-bonded hydrogen groups persist even in submonolayer coverages, which means that the water molecules are sufficiently mobile on the surface to establish hydrogen-bonded H_2O clusters. The energy loss at 2850 cm^{-1} also arises from a OH stretching vibration. This assignment was confirmed

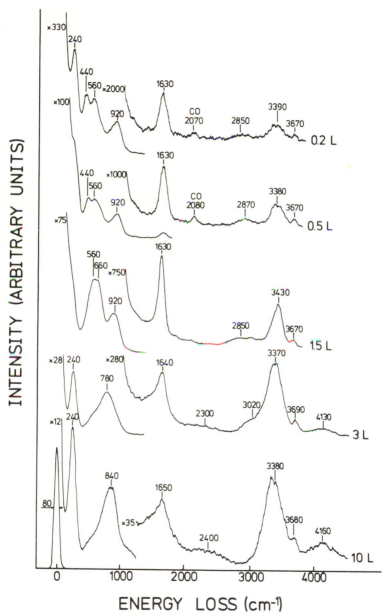

Fig. 6.22 Spectra of adsorbed water on Pt(100) at $T = 150°$K ([76] used with permission). The low coverage spectrum reveals two different types of hydrogen bonding $O—H \cdots O$ (3390 cm^{-1}) and $O—H \cdots Pt$ (2850 cm^{-1}). Some OH groups are not hydrogen bonded (3670 cm^{-1}).

by comparison to spectra of adsorbed D_2O. Since this frequency would be unusually low for hydrogen-bonded water molecules, it is attributed again to a form of hydrogen bonding between oxygen and the platinum surface atoms. A structure model which accounts for this and other details of the spectrum was proposed in Ref. [76].

6.4   OVERTONES AND BOND ENERGY

6.4.1   The Birge–Sponer Extrapolation

With Fig. 6.16, we presented a correlation between the bond energy and the bond order which in turn is correlated with the force constant. With these two correlations one may try to estimate bond energies of individual bonds within an adsorbed molecule from vibrational data. This concept might provide some insight into the relative stability of the intramolecular bonds and thus into the prospects of forming new compounds by breaking bonds in catalytic reactions. On the other hand, we must again appreciate that the magnitude of the force constant in principle is independent of the depth of the potential, and in many situations an uncritical use of the correlations cited above fails to render reasonable estimates. It would therefore be advantageous to have another independent spectroscopic check on the bond energy. In this section, we shall explore the possible use of measuring the overtone frequencies for estimates on the bond energy.

For this purpose we remember the treatment of anharmonic effects for a Morse potential in Section 5.4. The eigenfrequencies of a Morse potential are

$$G(n) = \omega_e(n + \tfrac{1}{2}) - \omega_e x_e(n + \tfrac{1}{2})^2 \tag{6.5}$$

with

$$\omega_e = a\sqrt{D_e h/\pi c M_r} \tag{6.6}$$

and

$$\omega_e x_e = \hbar a^2/c4\pi M_r. \tag{6.7}$$

when $G(n)$, ω_e, and $\omega_e x_e$ are expressed in cm^{-1}. The D_e and a are the parameters characterizing depth and width of the potential. Since the sequence of eigenfrequencies come from an exact solution for the Morse potential, Eq. (6.5) can be used up to the dissociation limit. The dissociation limit is reached when $G(n)$ viewed as a continuous function of n has reached its maximum value. This happens for $n = n_{max}$,

$$dG(n)/dn = 0 = \omega_e - 2\omega_e x_e(n_{max} + \tfrac{1}{2}), \tag{6.8}$$

$$(n_{max} + \tfrac{1}{2}) = \omega_e/2\omega_e x_e. \tag{6.9}$$

The dissociation energy D_0 is the difference between the ground state and $G(n_{max})$,

$$D_0 = (\omega_e^2/4\omega_e x_e) - \tfrac{1}{2}\omega_e + \tfrac{1}{4}\omega_e x_e. \tag{6.10}$$

Thus the dissociation energy for a Morse potential can be expressed in terms of the harmonic frequency ω_e and the anharmonicity parameter $\omega_e x_e$. These two quantities again can be determined from the frequency $v^{(1)}$ of the fundamental and the first overtone $v^{(2)}$:

$$\omega_e = 3v^{(1)} - v^{(2)}, \tag{6.11}$$

$$\omega_e x_e = v^{(1)} - \tfrac{1}{2}v^{(2)}. \tag{6.12}$$

Then D_0 becomes

$$D_0 = \frac{(3v^{(1)} - v^{(2)})^2}{4v^{(1)} - 2v^{(2)}} - \frac{5}{4} v^{(1)} + \frac{3}{8} v^{(2)}. \tag{6.13}$$

This scheme of extrapolating the dissociation limit from the first overtones is known as the Birge–Sponer extrapolation and has found many applications in the physical chemistry of diatomic molecules (for details see Ref. [80]). The accuracy of the procedure improves as one includes more overtones in the analysis, since then the influence of deviations from the linear decrease in the level spacing [Eq. (6.5)] can be considered. The linear extrapolation using Eq. (6.5) tends to overestimate the dissociation energy of diatomic molecules by about 10% on the average but occasionally as much as 50%. This is partly due to deviations of the actual potential from the Morse-type shape partly due to predissociation [80].

For polyatomic molecules, the situation is more complicated. Overtones and combination bands can arise from combinations of all fundamentals. When a certain combination occurs at a similar frequency as a fundamental and belongs to the same representation as the fundamental, the modes couple, introducing additional frequency shifts (Fermi resonance). Any application of the Birge–Sponer extrapolation is therefore confined to cases which more or less resemble the situation of a diatomic molecule and where a vibrational mode can essentially be described with one two-body potential. This is a severe limitation, but there are a number of nontrivial cases in which vibrations of adsorbed molecules are associated with a single bond between two atoms, and a description as a single oscillator without mechanical coupling to the rest of the molecule and the surface is meaningful. These vibrations are the stretching vibrations of adsorbed diatomic molecules, the CH stretching modes of hydrocarbons and other hydrogen stretching modes bonded to, preferable heavy atoms, and possibly also the CC stretching mode of dicarbon species.

6.4.2   Overtone Spectra

We now take a closer look at the available experimental results on overtones of adsorbed molecules. Once again we must distinguish between an overtone and multiple losses. Recall from Section 5.4 that the first overtone appears at a frequency shifted slightly downward from twice the fundamental frequency. It is this shift that allows us to calculate the anharmonicity parameter $\omega_e x_e$, and then to estimate the bond energy. Multiple losses of electrons are sequential inelastic scattering events from different molecules, each being in the ground state prior to the scattering event. This is illustrated in Fig. 5.12b. Therefore multiple losses appear at the exact positions of multiples of the fundamental: $2v_1, 3v_1, 4v_1, \ldots$ (cf. Section 4.4.5). This is true if lateral interactions between the molecules can be ignored. Under certain experimental conditions outlined in Section 4.4.5, multiple losses may prevail entirely to an extent that overtones are masked. This will be the case when more than a single surface layer contributes to a certain loss. For dilute monolayers, multiple losses eventually become rather small compared to single losses. Within a simple isotropic scattering model one may argue that the double loss should scale like θ^2 while, single losses and their overtones are proportional to the coverage θ.

In order to extract the information about anharmonicity from the overtone frequency, the double-loss intensity has to be smaller than the intensity of the overtone loss since otherwise the separation between the two losses would be too small to be resolved in electron energy loss spectroscopy. Unfortunately this restricts overtone observation to small coverages or small scattering cross sections. Since overtones are also weak effects, experiments are not at all easy and it is no surprise that the available data base is still small. Also, because relatively minor frequency shifts need to be evaluated, one cannot sacrifice resolution for intensity.

Despite the restrictions on the adsorption systems and experimental difficulties, several interesting examples can be reported by now. We have already encountered an overtone (Fig. 4.15) of the CH stretching vibration of adsorbed acetylene. An overtone was also reported by Andersson and Davenport [81] for an OH group on an oxidized Ni(100) surface. A more complete study of overtones and multiple losses has been performed recently by Lehwald and Steininger [82] with the system ethylene (C_2H_4) on Ni(100), and O_2 on Pt(111). Excerpts of these spectra are shown in Figs. 6.21 and 6.26. Compared to earlier spectra the signal-to-noise ratio is significantly improved due to a new lens design, and count rates range between 50 and 200 cps even for the off-specular inelastic events, while high resolution was also maintained. As a result, a reasonable signal-to-noise ratio could be carried out into the multiple loss-overtone regime above 3000 cm^{-1}. As a matter of convenience we have enumerated the peaks in Fig. 6.21. This enumeration is not meant to correlate with the frequency enumeration of gaseous ethylene (Appendix B). Also, a mode assignment is not attempted here. It is rather likely, nevertheless, that the 1100–1150 cm^{-1} mode is the CC stretching vibration. We focus our attention mainly on the CH stretching vibrations here. For 1-langmuir exposure, four peaks at 2530, 2730, 2920, and 3010 cm^{-1} are discernable. The low intensity of the first two peaks and the fact that no overtones of these losses are observable, calls for an interpretation as Fermi resonance enhanced $v_4 + v_3$ and $2v_3$ combinations. In the regime above 3000 cm^{-1}, several structures are visible which can be assigned as combinations of v_3, v_5, v_6, with v_1 and overtones of v_1 and v_2. The only evidence for a multiple loss is possibly the tail of the 5860 cm^{-1} band. The 3-langmuir spectrum shows evidence for hydrogen bonding as we have already discussed for the Ni(111) surface. Interestingly, no overtone of this band, which should center around 5400 cm^{-1} or lower, is visible. This seems to be commonplace with hydrogen-bonded systems and may therefore be regarded as additional evidence for the existence of hydrogen bonding on the surface. The $2v$ CH overtone region can be well explained by unfolding the spectrum into (1) double losses of v_2 and v_1, and (2) an overtone of v_1. The overtone frequencies are summarized in Table 6.7 Also included in Table 6.7 is the overtone of the OO stretching vibration of molecular oxygen on Pt(111) (Fig. 6.26). We shall discuss this spectrum in Section 6.5.1 in more detail.

From the overtone frequencies bond energies are estimated with Eq. (6.12). The last column of Table 6.7 lists the standard CH bond energy for a saturated hydrocarbon. For adsorbed acetylene the estimated bond energy is at about the standard bond energy of a CH bond. This consistent with the fact that acetylene is fairly stable on this surface. Ethylene, however, tends

TABLE 6.7

Fundamental Frequencies and First Overtones
for Several CH Groups and Molecular Oxygen
on Silver[a]

Adsorption system	$v^{(1)}$	$v^{(2)}$	D_0 (eV)	D (eV)
C_2H_2–Ni(111)	2920	5730	5.0 ± 0.5	4.27 [58]
1 L C_2H_4–Ni(100)	2920	5680	3.5 ± 0.5	
	3010	5860	3.7 ± 0.5	
3 L C_2H_4–Ni(100)	2780			
	2990	5850	4.5 ± 0.5	
O_2–Pt(111)	875	1645	0.5 ± 0.08	~0.35 [26]

[a] Bond energies D_0 are estimated with Eq. (6.12) and compared with the standard CH bond energy and the activation energy for dissociation of molecular oxygen on platinum [26], used with permission.

to dehydrogenerate on nickel to form acetylene. The medium coverage data for 1-langmuir exposure seem to indicate a weakening of the CH bonds. Also, the molecule must have at least two inequivalent CH groups at medium coverage which is somewhat surprising in view of the high symmetry of the surface. With higher coverage the CH groups become even more inequivalent since one (pair?) of CH groups forms the hydrogen bond to the surface. The estimated bond energy for the normal CH groups is then back in the range of standard bond energies. These features are probably best interpreted by assuming that the molecule tilts and thereby accommodates a pair of hydrogen atoms in a fourfold site while the other pair tilts further away from the surface. More data on the system must, however, be collected before any more detailed model should be put forward.

The estimated bond energy of molecular oxygen on platinum comes out to be rather low: The bond energy of doubly bonded oxygen is 4.25 eV and of singly bonded oxygen 1.44 eV. On the other hand, the bond energy estimated from the Birge–Sponer extrapolation is at least not inconsistent with an analysis of desorption data by Gland et al. [26]. They obtained an activation barrier for the dissociation on the surface of 0.35 eV.

6.5   CHEMICAL ANALYSIS OF ADSORBED SPECIES

6.5.1   Simple Dissociation Reactions

In the preceding sections we have, for the most part, tacitly assumed that a molecule retains its basic chemical nature when adsorbed on the surface and does not dissociate, partially decompose, or form new bonds to any other preadsorbed or gas-phase molecules. In practical research, it is always the first step of analysis to prove or disprove whether such proposition would be consistent with the spectrum. Unlike almost all other surface analytical tools, vibration spectroscopy is a specific probe of the chemical bonds which link surface atoms. Therefore the most important contribution to surface science that vibration spectroscopy has made and will make in the future is to establish the chemical identity of species on the surface. In the following, we shall present a number of examples to demonstrate how vibration spectroscopy can be used for surface chemical analysis. Again, vibration spectroscopy should not be the only technique to be used as the source of information. For the purpose of structure analysis, we have found the combination with electron diffraction particularly useful. For the chemical analysis the *in situ* comparison to Auger electron spectroscopy and flash desorption spectroscopy is very fruitful.

We begin our presentation with examples for simple dissociation reactions of the diatomic molecules CO, NO, and O_2. The presence or absence of a vibrational loss of the stretching vibration is the most significant information obtained from vibration spectroscopy. If the molecule is dissociated on the surface, vibrations of the adsorbed C, N, or O atoms may also be observed. However, as we have seen in Section 6.2, these vibrations are less characteristic of the type of atom which actually causes the vibrational loss. Vibration spectroscopy therefore does not necessarily discriminate between different possible reaction mechanisms: For example, in order to establish which of the two reactions

$$CO_{ads} \rightarrow C_{ads} + O_{ads}, \qquad 2CO_{ads} \rightarrow CO_2 + C_{ads}$$

has taken place, the combination with other techniques is necessary. By using Auger spectroscopy, the two cases are discriminated by either finding oxygen and carbon in about equal amounts or merely carbon on the surface. Desorption spectroscopy on the other hand would be able to detect the recombinative desorption of carbon and oxygen if both species are present.

Examples of spectroscopic evidence for dissociative adsorption were already studied in the very first experimental work on electron energy loss spectroscopy on surfaces (Propst and Piper [83]). There, however, energy resolution was but 400 cm^{-1}. The same W(100) surface was later reinvestigated by Froitzheim *et al.* [38]. The result for CO adsorption is shown in Fig. 6.23. Up to exposures of about 1 langmuir no CO stretching vibration, yet two relatively intense losses at ~ 500 and ~ 630 cm^{-1} are observed. The 550-cm^{-1} loss could also be produced by contaminating the surface with carbon via acetylene decomposition. In initial stages of oxygen adsorption a loss at ~ 610 cm^{-1} appeared. Furthermore, CO desorbs from the surface when heated to $\sim 1200°$K (Fig. 6.24) leaving a clean surface behind. These data are clear evidence for dissociative adsorption of CO with the oxygen and carbon atoms occupying a fourfold hollow site. At the time when these spectra were first published, the evidence for dissociative adsorption was a rather significant result. In the literature up to 1972, researchers apparently were misled by the fact that CO desorbs as molecular CO and believed CO to adsorb nondissociatively [84, 85]. The various flash desorption peaks were assumed to represent different binding states of molecular CO. Counterevidence to the binding site model for the desorption peaks was also produced by Froitzheim *et al.*: after the CO-covered surface was heated to 1000°K, i.e. after the β_1 desorption peak was flashed off, the spectrum after cooling down to 300°K, was left unchanged. This demonstrates that different β desorption peaks do not refer to different binding states of the adsorbate, but rather to various stages of the recombination-desorption process itself.

Once the surface is saturated with about a monolayer of dissociated CO so that the dissociative fourfold hollow sites are blocked, the surface becomes less vigorously active as a dissociation catalyst and allows for additional

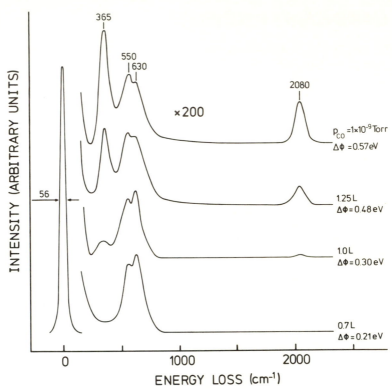

Fig. 6.23 Spectra of W(100) after CO exposure at ~ 300°K. At low coverage no molecular CO is observed. The two energy losses at 550 and 630 cm⁻¹ are identified as due to vibrations of carbon and oxygen by decomposing C_2H_2 on the surface and exposure to oxygen, respectively [38] (used with permission). With higher exposure molecular CO is also adsorbed. Saturation coverage is not reached at room temperature without an ambient pressure of CO.

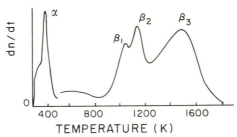

Fig. 6.24 Flash desorption trace of CO on W(100) after Clavenna and Schmidt ([85] used with permission). All peaks refer to molecular desorption of CO and the surface is clean after the CO desorption is completed. This has mislead researchers to assume that the "β states" refer to different forms of *molecular* CO on the surface. Vibration spectroscopy has shown however that only the α peak represents desorption of molecular CO which was in a *molecular state prior* to the desorption process. The different structures in the high temperature regime all result from recombinative desorption of oxygen and carbon. The structures reflect the kinetics of this process, not different types of binding states at room temperature.

weakly bound molecular CO. This weak bonding can be seen from the low temperature of the α desorption peak ("α-CO") but also from the relatively unperturbed CO stretching vibration and the low metal-carbon frequency (cf. Table 6.3). An even more weakly bound CO state was found at low temperatures using infrared spectroscopy ($\nu_{CO} = 2130$ cm^{-1} [86]).

Our second example for illustrating dissociation of diatomic molecules demonstrates a situation where dissociation does not occur immediately in the adsorption process, but instead, after thermal processing. The system NO on Ni(111) has already been discussed in connection with the dipole selection rule (Section 4.4.4) and structure analysis (Section 6.2.2). Once again, a spectrum for Ni(111) dosed with a larger amount of NO so as to produce the $c(4 \times 2)$ diffraction pattern with the NO in upright position in bridging sites is shown in Fig. 6.25. This structure is stable only for temperatures below $\sim 300°$K. When the surface is annealed to temperatures above $300°$K, the NO stretching loss disappears and the spectrum turns into the spectrum shown in Fig. 6.25 as the second and third trace. This spectrum remains essentially unchanged up to annealing temperatures of $\sim 600°$K. The spectrum displays three losses whose nature is not immediately obvious from vibration spectroscopy alone. Yet Auger spectroscopy indicates that N and O atoms in about equal amounts are on the surface. After a short anneal to $645°$K, nitrogen is partially desorbed and the typical oxygen vibration peak at $570-580$ cm^{-1} begins to appear. The final spectrum represents a state where no Auger-detectable amounts of nitrogen are on the surface. The spectrum is then identical to spectra obtained after oxygen adsorption and the diffraction pattern has turned into a $p(2 \times 2)$ [53]. The extra energy loss at 950 cm^{-1} is likely subsurface oxygen produced in the reaction, as discussed before. The loss at 1500 cm^{-1} is some molecular NO which readsorbs from the residual gas on the cold surface during the time needed for recording the spectra.

Apart from the obvious evidence for dissociation of NO, the experiment reveals several other interesting features. During the dissociation and subsequent desorption process, no oxygen atoms leave the surface. Except for the small amount of incorporated oxygen the fractional coverage of the surface with oxygen atoms before and after the process must be the same. This connects the coverages of the O-$p(2 \times 2)$ layer with the NO-$c(4 \times 2)$ layer. As a function of exposure prior to the O-$p(2 \times 2)$ structure, a $\sqrt{3} \times \sqrt{3}$ structure is observed. This structure can only be produced for a coverage of $\theta = \frac{1}{3}$. The $p(2 \times 2)$ layer must therefore correspond to $\theta = \frac{1}{2}$ which means that the actual surface unit mesh is $p(2 \times 1)$ rather than $p(2 \times 2)$. (The $p(2 \times 1)$ and the $p(2 \times 2)$ structures are indistinguishable by electron diffraction because the $p(2 \times 1)$ structure has three equivalent domains on a (111) fcc surface.) The $c(4 \times 2)$ structure of NO therefore also corresponds

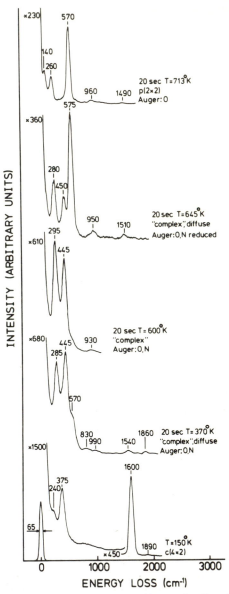

Fig. 6.25 Decomposition of NO on Ni(111) at 6 langmuirs. $E_0 = 2$ eV, $\theta_i = \theta_r = 78°$. NO dissociates at about room temperature leaving a densely packed coadsorption layer of oxygen and nitrogen behind. After heating to 700°K nitrogen is desorbed from the surface. The vibration spectrum and the diffraction pattern are then characteristic of a (nearly saturated) oxygen layer.

to half a monolayer coverage. The intermediate coadsorption system of nitrogen and oxygen is thus a rather densely packed structure with a total coverage of one. Such a coverage is not achieved by exposing the surface to oxygen alone. The unusual frequencies of 295 and 445 cm^{-1} for the intermediate system, compared to 410 and 580 cm^{-1} for nitrogen and oxygen alone up to coverages of about $\theta = \frac{1}{2}$ seem to reflect a substantial weakening of the surface bond of both nitrogen and oxygen which in turn could facilitate further reactions of these species with other surface species on gas-phase molecules.

One final example in this section is the adsorption of oxygen on silver and platinum, which we had touched upon briefly in Section 6.3.3. Platinum and silver are both oxidation catalysts. Platinum catalyses the H_2, O_2 reaction to H_2O even at room temperature and is currently used by the automotive industry for complete oxidation of the automobile exhaust gas (CO and NO). Silver seems to be the most effective material for catalyzing the epoxidation of ethylene to ethylene oxide. Both materials can adsorb oxygen dissociately, like almost all metals. But as was shown by vibration spectroscopy recently [66, 67, 26], these metals can also adsorb oxygen in a molecular form in which the bond order of the OO bond is about one or even less. The spectrum of molecular oxygen for the Pt(111) surface [82] is shown in Fig. 6.26. When the surface is annealed, the molecular form is partly dissociated to give rise to the loss at 490 cm^{-1}. This oxygen is relatively strongly bonded since it desorbs (as O_2) at about 700°K. The desorption temperature is larger than the oxygen desorption temperature on silver, 550°K [26]). On the other hand, the molecular phase of oxygen seems to be more strongly bonded on the silver surface since the desorption temperature of molecular oxygen is 185°K on silver and 150°K on platinum. This is consistent with the lower bond order of the OO bond on silver which we had estimated in Section 6.3.2.

These new spectroscopic findings can be correlated with the old question of why silver is a good catalyst for ethylene epoxidation, while platinum is not. Both materials have in common the capability of dissociating oxygen without binding the oxygen too rigidly. The two materials also have the molecular phase in common. However, we do not know whether it plays a direct role in the process of ethylene oxidation. Silver and platinum differ, however significantly in their attitude toward ethylene (Fig. 4.9). Platinum bonds ethylene strongly and reduces the internal CC bond order to about one (Table 6.5), while silver adsorbs ethylene rather weakly without an appreciable distortion of the molecule (Section 4.4.3). Complete oxidation to CO or CO_2—which is the competing process with epoxidation—involves CC bond breaking. This process should proceed with a smaller rate on the silver surface (where the ethylene double bond is retained) than on the

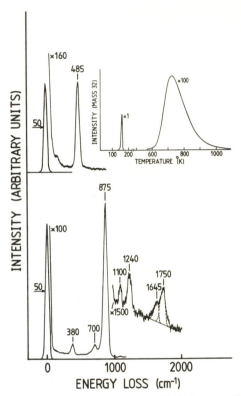

Fig. 6.26 Spectra of a saturated oxygen layer on Pt(111) at $T = 92°K$ and 6 langmuirs after Lehwald *et al.* ([82] used with permission). At low temperatures oxygen is completely in a molecular state with an unusually low OO stretching vibration at 875 cm^{-1}. The feature at ~700 cm^{-1} decreases in intensity with increasing coverage. It is assigned to the O_2 stretching vibration of oxygen adsorbed in a different site [82]. The spectrum shows also several multiple losses and the overtone of the 875 cm^{-1} loss at 1645 cm^{-1} (see Section 6.4.2). When the surface is flashed to higher temperature oxygen partially desorbs, partially dissociates. The upper spectrum was obtained after heating to 300°K.

platinum surface. On the other hand, atomic oxygen on silver can easily slip into the CC bond of ethylene to produce the epoxide. Apparently, silver makes such a unique catalyst for the epoxidation process because the surface dissociates oxygen but binds ethylene weakly (see also Ref. [67]).

6.5.2 New Surface Species

Upon adsorption, diatomic molecules may either dissociate or adsorb nondissociatively. This represents a relatively simple situation and it is easy to distinguish between the two cases. The number of possible reactions and reaction intermediates, however, increases dramatically with the number of atoms of the molecule adsorbed initially. At the same time the chemical analysis of a surface species becomes more and more difficult, the more atoms that are involved. The analysis becomes almost hopeless (with vibration spectroscopy) if one has reason to believe that more than a single species is present on the surface. Therefore the process of preparation of unique surface phases, with a single species only, is a step of equal importance to the analysis itself. The method by which a certain phase is prepared frequently offers as many clues for the analysis as the spectrum itself. A good indication as to whether the surface phase consists of a single species only is also the existence of a well-ordered diffraction pattern. Furthermore, well-ordered layers, preferable of small unit cells, typically lead to high intensity in the reflected beam and thus to a high intensity of the dipole lobe of the inelastic spectrum. It is also helpful to start the preparation of a new phase with an associative adsorption of the gas-phase molecule. As we have seen in a number of examples, the vibration spectrum of the adsorbed molecule often provides valuable hints to the relative strength of inner-molecular bonds which in turn can serve as clues for the further fate of the molecule. Preparation of identifiable surface species therefore often begins with adsorption of a molecule at low temperatures. The new species is then produced by either simple thermal processing or exposing the surface to another gas at low or elevated temperatures.

Both examples we wish to discuss here involve preparation by thermal processing and deal with the adsorption and stepwise decomposition of methanol (CH_3OH) and acetylene (C_2H_2) on nickel. Here it is useful to remember that the ultimate fate of a hydrocarbon on nickel is the state of lowest free energy, reached when the molecule is decomposed to produce (graphitic) carbon on the surface and H_2 in the gas phase or CO and H_2 in the gas phase for CH_3OH. When the pressure is not essentially zero as in ultrahigh vacuum, the ultimate fate of the molecule is dictated by the minimum of the free enthalpy which, depending on the partial pressures of the constituents of the gas phase, may involve complex hydrocarbons on the surface. As thermal processing in ultrahigh vacuum leads to a very well-known final end, the interesting aspect is not the ultimate product, but intermediate metastable species through which the product is obtained. Frequently a certain metastable intermediate can be isolated to be the prevailing species by annealing the low-temperature surface phase to a certain higher temperature for some time and allowing the surface to cool down again to freeze-in the metastable state thereby obtained.

Spectra of three intermediate steps in the decomposition of CH_3OH on nickel are shown in Fig. 6.27. The upper left spectrum identifies adsorbed

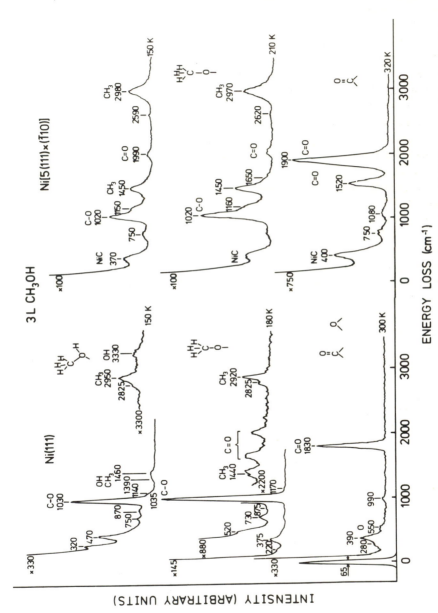

Fig. 6.27 The decomposition of methanol on Ni(111) and a stepped nickel surface with (111) terraces via an intermediate methoxy group. Earlier spectra of methanol on Ni(111) ([87] used with permission) were not completely free of water which is quite noticeable even when the fractional coverage is only 1%.

CH_3OH [87]. The OH stretching frequency is shifted downward from its gas phase value (3681 cm^{-1}) and is at the same position where it appears for liquid CH_3OH (3328 cm^{-1}). This might indicate lateral hydrogen bonding to neighboring CH_3OH molecules. On the other hand, the same frequency shift is also observed for low coverages of methanol, and the OH bond brakes after thermal processing to 180°K. Therefore the frequency shift might also indicate hydrogen bonding to the surface. The spectrum after processing to 180°K still contains the characteristic frequencies of a CH_3 group and also the strong CO stretching mode while the OH band has gone. The surface species is thus identified as CH_3O—. The small amount of CO in the surface is likely caused by some CO pickup from the gas phase during the annealing process rather than produced by further decomposition of the adsorbed molecule. Further decomposition is, however, achieved after processing to 300°K. It is interesting to note that some atomic oxygen is also present on the surface after this step, while molecular CO alone would not dissociate at 300°K. It seems therefore that the lower CO bond energy of the H_3CO group allows CO dissociation to take place as a side reaction. The surface may also still have adsorbed hydrogen which is, however, invisible on the Ni(111) surface probably because of the small dipole moment [7]. Further annealing of the surface would desorb the hydrogen, the molecular CO and ultimately also carbon and oxygen by recombinative desorption of CO [88]. Part of the oxygen may also recombine to produce O_2 and leave carbon behind to be dissolved in the bulk.

It is interesting to compare this sequence of events with the stepped nickel surface (right-hand side of Fig. 6.27). Here no OH stretching vibration is noticeable at 150°K, indicating the greater activity of step atoms for dehydrogenation which has also been observed with other hydrocarbons on the same stepped surface [89]. The intensity of the CH vibrations is considerably larger than on the flat surface (consider magnification scale). This is, however, somewhat misleading since the reflectivity of the stepped surface is lower. This tends to increase the relative intensity of losses which are predominantly caused by impact scattering, and the CH vibrations are likely candidates for impact scattering (Section 3.6.1). Further analysis of the angular profiles is not yet available for this system. In any case a small amount of molecular CO seems to be formed in the adsorption process at 150°K, itself also (1990 cm^{-1}). Annealing the surface to 210°K does not change the adsorbed layer much. Further annealing to 320°K, however, again dehydrogenates the methoxy groups to leave molecular CO in step and terrace sites (1520 cm^{-1}, 1990 cm^{-1} [64]). The rest of the energy losses have also been found with CO adsorption only on this particular stepped surface and they may all be caused by hindered rotations of CO and the metal carbon vibration of CO. An alternative interpretation for the

1080 cm^{-1} loss being due to adsorbed hydrogen (and to a hydrogen contamination in "pure" CO adsorption experiments) cannot be excluded. In summary, the following reaction paths have been established:

1. Ni(111)

    $$CH_3OH_{gas} \xrightarrow[150°K]{} CH_3OH$$

    $$CH_3OH \xrightarrow[180°K]{} CH_3O + H$$

    $$CH_3O \xrightarrow[300°K]{} CO + H + C + O$$

    $$2H \xrightarrow[380°K]{} H_{2gas}$$

    $$CO \xrightarrow[460°K]{} CO_{gas} \quad [88]$$

    $$C + O \xrightarrow[850°K]{} CO_{gas} \quad [88],$$

2. Ni[5(111) × (110)]

    $$CH_3OH_{gas} \xrightarrow[<150°K]{} CH_3O + H + CO$$

    $$CH_3O \xrightarrow[320°K]{} CO + H + \text{desorption processes.}$$

The temperatures given for these processes depend of course on the time scale of annealing and should be considered as a guideline only.

An interesting feature of the catalytic properties of the stepped nickel surface seems to be that the OH group is dissociated for the entire adsorbed methanol, and not only for that fraction of the molecules which adsorbs near the steps. The same is found with the other example of an intermediate which we want to present now. We have already discussed the adsorption of acetylene C_2H_2 on Ni(111) in connection with the angular profiles (Section 4.5.3) and the frequency shifts (Section 6.4.2) and have also mentioned that upon thermal processing to $\sim 400°K$ the CC bond breaks and CH groups are formed. Hydrogen is also released in that process. No decomposition of acetylene is found up to $300–350°K$ on the flat Ni(111) surface. On the stepped Ni[5(111) × ($\bar{1}$10)] surface, however, dehydrogenation occurs in the adsorption process at $150°K$ [89]. The spectrum in Fig. 6.28 shows—among some amount of normal acetylene—a species characterized by a pair of vibrations at 350 and 2220 cm^{-1}. Occasionally this species is formed even without normal C_2H_2 at all or any other CH vibration. Comparison to the adsorption of C_2D_2 [89] shows that the new species must be a C_2 group with the 350 cm^{-1} and the 2220 cm^{-1} frequency being the metal carbon and the carbon stretching modes, respectively. The high frequency of the CC stretching vibration indicates triple bonding between the carbon atoms. Nevertheless, the bond is rather fragile with respect to further decomposition into carbon atoms which takes place at about $180°K$ (Fig. 6.28). At the same time a loss at 520 cm^{-1} develops, representing the isolated carbon atom with no hydrogen attached to it.

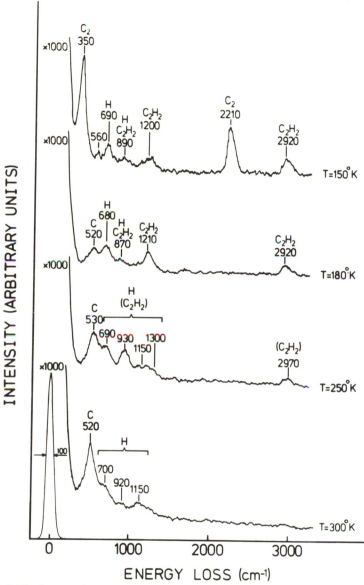

Fig. 6.28 Spectra of a stepped nickel surface after exposure to acetylene at 150°K at 0.6 langmuir; Ni[5(111) × (110)]. A new C_2 surface group is formed in the adsorption process by dehydrogenating acetylene (2210 and 350 cm^{-1}). The C_2 group dissociates into two surface carbon atoms (520 cm^{-1}) at about 180°K. The small amount of undehydrogenated acetylene present at low temperatures decomposes easily on the stepped surface when the temperature is raised, which is contrary to acetylene on the flat Ni(111) surface. After annealing to 300°K only carbon and hydrogen but no CH groups are on the surface.

6.5.3   Surface Species on Pt(111)—A Case Study

This final section on the applications of electron energy loss spectroscopy is devoted to the analysis of a certain surface species on Pt(111). The species is produced by adsorbing ethylene or acetylene near room temperature and has been investigated by six independent research groups experimentally and theoretically [68, 90–102]. The matter has been rather controversial in the past five years and no less than five chemically different structures have been proposed. The issue has gained further importance since the same species apparently can also be produced on Pt(100) [96], on Rh(111) [97], and possibly also on palladium [98]. While the question is still not settled completely, the different research groups recently seem to be acquiring a converging set of viewpoints.

In this section we want to give a brief account of this work, partly because one of the authors of this volume was engaged in generating the controversy, partly because the analysis of the surface species on platinum is a very nice demonstration of how the contributions of different experimental techniques and theoretical studies may eventually be tied together into a consistent pattern.

The first serious attempt to solve the structure of the surface species was undertaken by the group of Somorjai [90] in 1976 using low-energy electron diffraction. The surface species then was prepared by adsorbing acetylene on Pt(111) at room temperature and heating the sample to about 400°K for 1 hour. During this procedure a sharp (2×2) diffraction pattern developed. This surface phase was named "stable acetylene" by this group. Identical diffraction intensities were obtained later by the same group [91] by exposing the surface to ethylene. The diffraction intensities as a function of electron beam energy ("I–V profiles") were compared to calculated I–V profiles for acetylene in various positions on the surface until the relatively best agreement was found. Thus the structure was analyzed with the *a priori* assumption that the surface species was C_2H_2, which at the time of this analysis seems to have been the general belief.

This assumption was challenged by Ibach *et al.* [92] on the basis of the vibration spectrum of the surface species. In a subsequent paper [68] it was furthermore shown that while the surface species could readily be formed from ethylene—either by exposing at room temperature or by annealing the surface after ethylene adsorption at lower temperature—the surface species could not be produced from acetylene unless additional hydrogen was provided (Fig. 6.29). This extra hydrogen could either be offered as atomic hydrogen or hydrogen preadsorbed on the surface. It was therefore argued that in the previous work, acetylene had also been converted by unintentionally dosing the surface with the (partly atomic) hydrogen from

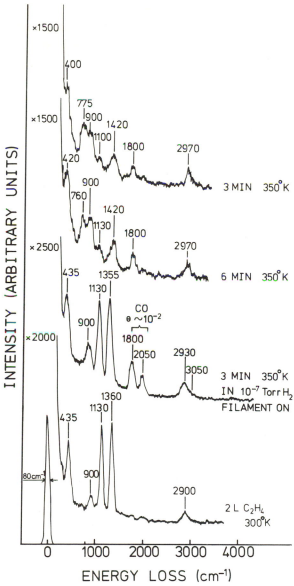

Fig. 6.29 Spectra of acetylene on Pt(111) at $T = 140°K$ and 2 langmuirs after temperature processing in vacuum and in a hydrogen atmosphere. Acetylene plus hydrogen and ethylene produce essentially the same spectrum at room temperature ([68], used with permission).

TABLE 6.8

Characteristic Energy Losses of the Room Temperature
Phase Formed from Ethylene on Pt (III) and Several
Attempts to Assign the Observed Modes to Normal
Modes of Various Surface Species[a]

$\hbar\omega$ (cm^{-1})	Comment	[68] CH–CH$_3$	[95] C–CH$_3$	[94] CH–CH$_2$	[99] C–CH$_3$
3050	Observed occasionally	ν-CH	ν_a-CH$_3$	ν_a-CH$_2$	—
2890	Specular	ν-CH$_3$	ν-CH$_3$	ν_s-CH ν_s-CH$_2$	ν_s-CH$_3$
2950	Off specular				ν_a-CH$_3$
1420	Off specular	δ_d-CH$_3$	δ_d-CH$_3$	δ-CH$_2$	δ_d-CH$_3$
1350	Specular	δ_s-CH$_3$	δ_s-CH$_3$	δ-CH	δ_s-CH$_3$
1130	Specular	δ-CH	ρ-CH$_3$	ρ_w, ρ_t-CH$_2$	ν-CC
980	Off specular				ρ-CH$_3$
900	Specular, sometimes not present	ν-CC	ν-CC	ν-CC	—

[a] We have obtained in- and off-specular spectra of the surface species with much improved resolution and sensitivity. Data in Tables 6.8 and 6.9 are taken from these results.

the residual gas. The surface species formed is this process was proposed to be ethylidene (CH$_3$–CH$\langle$) and a mode assignment based on the comparison to CH$_3$–CHCl$_2$ was also attempted (Table 6.8). We notice that a tail in the energy loss spectrum around 3050 cm^{-1} plays a rather crucial role in this assignment. We will come back to this point shortly.

In the meantime, Kesmodel *et al.* [95] had renewed their efforts on the structure analysis by electron diffraction testing a number of other geometries. Reasonable agreement between experiment and theory was found for a CC species oriented perpendicularly above a threefold site with a CC distance of 1.50 ± 0.05 Å. As low-energy electron diffraction is not sensitive to hydrogen atoms, nothing could be said about the chemical composition of this species. Nevertheless, the (vertical) CC distance of 1.50 Å made an ethylidyne group (CH$_3$–C$\leqq$) more likely than CH$_3$–CH since the latter ought to be somewhat canted with respect to the surface normal, which would let the vertical CC distance appear smaller than 1.50 Å. Based on this model Kesmodel *et al.* also proposed another mode assignment (Table 6.8).

A still different species was suggested by Demuth [93, 94]. His most significant contribution, in retrospect, was probably the observation that hydrogen was released from the surface at about 300°K, when an ethylene-exposed surface was annealed to produce the species in question. The amount of hydrogen released was approximately one-fourth of the total amount of hydrogen on the surface. This suggested a species of the composition C_2H_3. Demuth, however, did not endorse the ethylidyne (CH_3–C≣) model, but rather proposed —CH_2–CH for which he obtained better agreement between photoemission spectra and a theoretical model for the electronic orbitals. He also proposed a mode assignment (Table 6.8).

Flash desorption studies were also attempted by Baro and Ibach [100] who found the hydrogen release at room temperature rather variable with the ethylene coverage and sometimes as high as 50% of the total amount. A significantly more careful flash desorption study by Steininger [101], however, did not reproduce this result, but rather reestablished the hydrogen loss to be $\sim 25\%$ for initial ethylene exposures between 0.5 and 3 langmuirs (see Fig. 6.30). In this latter study the mass spectrometer viewed the surface through a small-diameter glass tube and was shown to be selective to desorption from the flat portions of the crystal only (rather than the rim as well, which involves other crystal faces). Steininger also detected a release of ethylene prior to the formation of the surface species and also a small ethylene release at about 450°K.

Baro and Ibach [100] also reported an off-specular spectrum of the surface species (Fig. 6.31) which confirmed that the two most prominent losses in the specular spectrum (1130 and 1360 cm^{-1}) were dipole enhanced, while

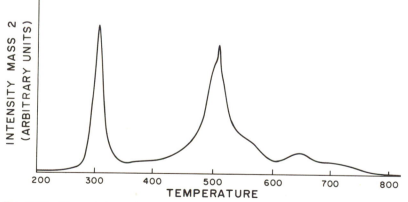

Fig. 6.30 Thermal desorption spectrum of hydrogen after dosing the surface with 1.3 langmuirs of ethylene at 100°K ([101] used with permission). Heating rate was about 10°K/sec. This spectrum shows that the surface species produced from ethylene by adsorption at room temperature, or after heating to room temperature for a few minutes, must have lost approximately 25% of its hydrogen content.

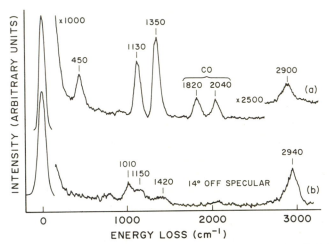

Fig. 6.31   (a) Spectrum obtained by dosing a Pt(111) with 2 langmuirs of ethylene (C_2H_4) at 150°K and annealing the surface to 350°K for 4 min in 10^{-7} Torr hydrogen. Under these conditions the 900 cm^{-1} loss has almost, if not completely, disappeared. There is also hardly any tail of the CH stretching loss at 3050 cm^{-1}. This suggests that both features are possibly associated with the beginning decomposition of the surface species into CH groups or incomplete conversion of ethylene into ethylidine. (b) Off-specular spectrum after adsorption of ethylene at room temperature ($\Delta\theta = 14°$) ([100] used with permission). The CH stretching loss now appears at 2940 cm^{-1} indicating a second nondipole active CH stretch vibration. The spectrum further shows that the additional losses at 1010 and 1420 cm^{-1} are most likely nondipole active as well, while the losses at 1130 and 1360 cm^{-1} are definitely dipole active (cf. Tables 6.8 and 6.9).

a new loss at 1010 cm^{-1} was also observed. In addition, the center of the CH stretching loss shifted from 2900 to 2940 cm^{-1}. Since dispersion effects of this magnitude are not expected for a CH stretching vibration this spectrum indicates another probably nondipole active mode in the CH stretching regime. Baro and Ibach also studied the decomposition of the surface phase at higher temperatures. Decomposition began at about 450°K. It was accompanied by the appearance of relatively strong peaks at 800 and 3000 cm^{-1}, which with increasing temperature, shifted to 850 and $\sim$3100 cm^{-1}. These peaks were assigned to a CH group in accordance with the results on Ni(111) [75] and a recent study of the vibrational spectrum of $CH_3CO_3(CO)_9$ [102]. Since further dehydrogeneration of the surface species produces loss intensity in the regime above 3000 cm^{-1}, the tail sometimes observed earlier (Fig. 6.29) might not be a feature of the surface phase in question. Finally, Lehwald [103] prepared the species by exposing the surface to ethylene at low temperatures followed by an annealing to 450°K in 10^{-7} Torr of hydrogen. The in-specular spectrum after this preparation is also shown in Fig. 6.31. While the characteristic losses at 1130, 1360 (plus its tail at 1420 cm^{-1}), and 2900 cm^{-1} are present, and at about the same intensity level the 900 cm^{-1} loss has disappeared. Also, the 2900 cm^{-1} definitely has no tail in the 3050 cm^{-1} region. This indicates that the 900 cm^{-1} loss and the 3050 cm^{-1} tail might not be essential features of the surface phase, but rather due to an admixture or due to partial decomposition.

This reasoning would still be rather speculative, indeed, were it not for a very careful and complete study by the group of Sheppard [Skinner et al., 99] on the organometallic compound $CH_3CCo_3(CO)_9$ which involved infrared absorption measurements and a normal–coordinate analysis of the vibration spectrum. Frequency and symmetry assignments for the fundamentals are summarized in Table 6.9 and compared to the frequencies and apparent symmetries of the surface species. We see that the assignment not only accounts for the frequencies of the surface species, including the deuterated one, but also for the intensities. Modes designated as being of A_1 symmetry in $CH_3CCo_3(CO)_9$ are observed as relatively strong dipole active modes for the surface species. All modes designated as E are either not observed in specular direction at all or very weakly. The assignment also accounts for the fact that in the deuterated surface species only one loss is intense, while the other A_1 mode is weak. This agreement is indeed a very satisfying one. The assignment does not account for the 900 cm^{-1} loss nor for the occasionally observed 3050 cm^{-1} tail. We have seen, however, that both losses are not necessarily genuine features of the surface phase since it can be prepared without these features (Fig. 6.31). So the end conclusion of this major effort of many groups seems to be that the surface species is ethylidyne

TABLE 6.9

Vibration Spectrum of Ethylidyne–Tricobalt–Nonacarbonyl and Mode Assignment According to Skinner et al.[a] Compared to Energy Loss Spectrum of Surface Species on Platinum.[b]

Assignment	$CH_3CCo_3(CO)_9$	Surface species	$CD_3CCo_3(CO)_9$	Surface species
ν_d-CH_3	2930(m) E	2950 E	2192(w) E	2220 E
ν_s-CH_3	2888(m) A_1	2890(m) A_1		2080(w) A_1
δ_d-CH_3	1420(m) E	1420 E	1031(w) E	1030(w)
δ_s-CH_3	1356(m) A_1	1350(ms) A_1	1002(vw) A_1	990(w)
ν-CC	1163(m) A_1	1130(ms) A_1	1182(ms) A_1	1160(ms) A_1
ρ-CH_3	1004(s) E	980 E	828(s) E	790(w)

[a] Ref. [99], used with permission.

[b] The symmetry assignment of the surface species is based on the disappearance of the modes designated as A_1 in the off-specular spectrum. Data are, however, not as complete as e.g. for C_2H_2 on Ni(111) (Fig. 4.16). The assignment proposed by Skinner et al. not only accounts for the frequencies but also for the intensities and the apparent symmetry of the modes.

(CH$_3$–C$\lessgtr$). It is furthermore interesting to note that quite independent of these more recent experimental results, that in two theoretical studies employing the extended Hückel model the conclusion was made that in a threefold site, CH$_3$–C$\leqq$ is the most likely species to form [104, 105].

A case study such as this one would be incomplete without a few considerations about what lessons may be learned for future chemical analysis of surface compounds using vibration spectroscopy. The first lesson is that more attention probably ought to be paid to the conditions of occurrence of a surface species itself. Ultrahigh vacuum in the 10^{-11} Torr range and a molecular-beam-type dosing of the surfaces (in order to keep accidental exposure to background gases low) at certain temperatures or before a temperature treatment might be important. The effect of varying each of these parameters on the vibration spectrum ought to be studied before a certain set of frequencies is established to belong to one phase. While this requirement is probably self-evident to a chemist, it is somewhat new in surface physics since so far relatively simple systems have been studied. We further notice that access to data of an electron diffraction analysis, and to flash desorption data can be important for the interpretation. Finally a comparison with analogous organometallic compounds and a normal coordinate analysis can be very helpful.

In retrospect, one might therefore be somewhat critical about the original vibrational analysis of the surface species on platinum by Ibach *et al.* [92]. On the other hand, their study was the first attempt to apply electron energy loss spectroscopy to a complex hydrocarbon phase of unknown composition. In fact, in the meantime electron spectrometers have significantly improved. As already demonstrated in Fig. 6.21, off-specular spectra are now obtained with a better signal-to-noise ratio than was available in earlier in-specular data on the platinum surface phase. Renewed studies employing this more advanced technology are likely to resolve remaining uncertainties about simple surface compounds formed from hydrogenation, dehydrogenation, and hydrogenolysis of C$_2$ hydrocarbons.

REFERENCES

1. H. Froitzheim, Electron Spectroscopy for Surface Analysis *in* "Topics in Current Physics" (H. Ibach, ed.), Vol. 4, p. 205. Springer-Verlag, Berlin and New York, 1977.
2. H. Ibach, H. Hopster, and B. Sexton, *Appl. Surface Sci.* **1**, 1 (1977).
3. W. H. Weinberg, *in* "Experimental Methods of Surface Physics" (R. L. Park and M. G. La Gally, eds.). Academic Press, to be published.
4. J. Pritchard, *in* "Modern Methods of Surface Analysis" (Dechema Monographs **78**, 231), Dechema, Frankfurt, 1975.
5. H. Froitzheim, H. Ibach, and S. Lehwald *Phys. Rev. B* **14**, 1362 (1976).

6. S. Andersson, *Solid State Commun.* **20**, 229 (1976).
7. H. Ibach and D. Bruchmann, *Phys. Rev. Lett.* **44**, 36 (1980).
8. S. Lehwald and H. Ibach, *in* "Vibrations at Surfaces" (R. Caudano, R. Gilles, A. A. Lucas, eds.), p. 137. Plenum, New York, 1982.
9. G. B. Fisher, B. A. Sexton, and J. L. Gland, *J. Vac. Sci. Technol.* **17**, 144 (1980).
10. W. Erley, H. Ibach, *Solid State Commun.* **37**, 937 (1981).
11. G. E. Thomas and W. H. Weinberg, *J. Chem. Phys.* **69**, 3611 (1978).
12. B. A. Sexton, *J. Vac. Sci. Technol.* **16**, 1033 (1979).
13. S. Andersson, *Proc. Int. Vac. Congr., 7th Int. Conf. Solid Surfaces, 3rd, Vienna*, p. 1019, *1977*.
14. A. M. Baro, H. Ibach, and H. D. Bruchmann, *Surface Sci.* **88**, 384 (1979).
15. S. Lehwald, personal communication.
16. S. Andersson, *Surface Sci.* **79**, 385 (1979).
17. See, e.g., M. van Hove, *in* "The Nature of the Surface Chemical Bond" (T. N. Rhodin and G. Ertl, eds.), p. 275 ff. North-Holland Publ., Amsterdam, 1979.
18. D. J. M. Fassaert and A. van der Avoird, *Surface Sci.* **55**, 313 (1976).
19. T. H. Upton and W. A. Goddard, *Phys. Rev. Lett.* **42**, 472 (1979).
20. M. A. Jayasooriya, M. A. Chesters, M. W. Howard, S. F. A. Kettle, D. B. Powell, and N. Sheppard, *Surface Sci.* **93**, 526 (1980).
21. M. W. Howard, U. A. Jayasooriya, S. F. A. Kettle, D. B. Powell, and N. Sheppard, *J. C. S. Chem Comm.* **18**, 1979.
22. R. F. Willis, *Surface Sci.* **89**, 457 (1979).
23. See, e.g., C. E. Smith, J. P. Biberian, and G. A. Somorjai, *J. Catalysis* **57**, 426 (1979).
24. S. Lehwald, personal communication.
25. G. Dalmai-Imelik, J. C. Bertolini, and J. Rousseau, *Surface Sci.* **63**, 67 (1977).
26. J. L. Gland, B. A. Sexton, and G. B. Fisher, *Surface Sci.* **95**, 587 (1980).
27. S. Onari, T. Arai, and K. Kudo, *Phys. Rev. B* **16**, 1717 (1977).
28. J. Pritchard, *J. Vac. Sci. Technol.* **9**, 895 (1972).
29. R. P. Eischens, W. A. Pliskin, and S. A. Francis, *J. Chem. Phys.* **22**, 194 (1954).
30. T. T. Nguyen and N. Sheppard, *in* "Advances in Infrared and Raman Spectroscopy" (R. E. Hester and R. H. J. Clark, eds.), Vol. 5. Heyden, London, 1978.
31. W. Erley, *J. Vac. Sci. Technol.* **18**, 472 (1981).
32. S. Andersson, *Solid State Commun.* **21**, 75 (1977).
33. W. Erley, H. Wagner, and H. Ibach, *Surface Sci.* **80**, 612 (1979).
34. S. Andersson, *Surface Sci.* **89**, 477 (1979).
35. G. E. Thomas and W. H. Weinberg, *J. Chem. Phys.* **70**, 954 (1979).
36. L. H. Dubois and G. A. Somorjai, *Surface Sci.* **91**, 514 (1980).
37. A. M. Baro and H. Ibach, *J. Chem. Phys.* **71**, 4812 (1979).
38. H. Froitzheim, H. Ibach, and S. Lehwald, *Surface Sci.* **63**, 56 (1977).
39. See, e.g., S. Y. Tong, *Progr. Surface Sci.* **7**, 1 (1975).
40. M. Passler, A. Ignatiev, F. Jona, D. W. Jepsen, and P. M. Marcus, *Phys. Rev. Lett.* **43**, 360 (1979).
41. S. Andersson and J. B. Pendry, *Phys. Rev. Lett.* **43**, 363 (1979).
42. F. C. Campuzano and R. G. Greenler, *Surface Sci.* **83**, 301 (1979).
43. A. M. Bradshaw and F. M. Hoffman, *Surface Sci.* **72**, 513 (1978).
44. K. Horn and J. Pritchard, *Surface Sci.* **55**, 701 (1976).
45. J. C. Tracy, *J. Chem. Phys.* **56**, 2748 (1972).
46. J. Pritchard, *Surface Sci.* **79**, 231 (1979).
47. J. P. Biberian and M. A. van Hove, *in* "Vibrations at Surfaces" (R. Caudano, R. Gilles, A. A. Lucas, eds.). Plenum, New York, 1982.

48. K. Nakamoto, "Infrared and Raman Spectra of Inorganic and Coordination Compounds," 3rd ed. Wiley (Interscience), New York, 1978.

49. D. M. Adams, "Metal–Ligand and Related Vibrations." Arnold, London, 1967.

50. F. A. Cotton and G. Wilkinson, "Advanced Inorganic Chemistry," 3rd ed. Wiley, New York, 1972.

51. G. Pirug, H. P. Bonzel, H. Hopster, and H. Ibach, *J. Chem. Phys.* **71**, 593 (1979).

52. H. Ibach and S. Lehwald, *Surface Sci.* **76**, 1 (1978).

53. S. Lehwald, J. T. Yates, and H. Ibach, *Proc. of ECOSS3, Suppl. Rev. Le Vide, les Couches Minces* **201**, 211 (1980).

54. G. E. Thomas and W. H. Weinberg, *Phys. Rev. Lett.* **41**, 1181 (1978).

55. T. L. Einstein and J. R. Schrieffer, *Phys. Rev. B* **7**, 3629 (1973).

56. S. Andersson and B. N. J. Persson, *Phys. Rev. Lett.* **45**, 1421 (1980).

57. G. Herzberg, Infrared and Raman Spectra of Polyatomic Molecules, *in* "Molecular Spectra and Molecular Structure," Vol. II, p. 193. Van Nostrand–Reinhold, Princeton, New Jersey, 1945.

58. R. T. Sanderson, "Chemical Bonds and Bond Energy." Academic Press, New York, 1971.

59. R. M. Badger, *J. Chem. Phys.* **2**, 128 (1934); **3**, 710 (1935); *Phys. Rev.* **48**, 284 (1935).

60. J. Waser and L. Pauling, *J. Chem. Phys.* **18**, 618 (1950).

61. D. R. Herschbach and V. W. Laurie, *J. Chem. Phys.* **35**, 458 (1961).

62. G. Brodèn, G. Pirug, and H. Bonzel, *Chem. Phys. Lett.* **51**, 250 (1977).

63. G. Davidson, *Organometal. Chem. Rev. A* **8**, 303 (1972).

64. W. Erley, H. Ibach, S. Lehwald, and H. Wagner, *Surface Sci.* **83**, 585 (1979).

65. J. E. Demuth, H. Ibach, and S. Lehwald, *Phys. Rev. Lett.* **40**, 1044 (1978).

66. B. A. Sexton and R. F. Madix, *Chem. Phys. Lett.* **76**, 294 (1980).

67. C. Backx, C. P. M. de Groot, and P. Biloen, *Appl. Surface Sci.* **6**, 256 (1980).

68. H. Ibach and S. Lehwald, *J. Vac. Sci. Technol.* **15**, 407 (1978).

69. J. E. Demuth, D. W. Jepsen, and P. M. Marcus, *Phys. Rev. Lett.* **31**, 540 (1973).

70. P. M. Marcus, J. E. Demuth, and D. W. Jepsen, *Surface Sci.* **53**, 501 (1975).

71. J. E. Demuth, H. Ibach, and S. Lehwald, *Phys. Rev. Lett.* **40**, 1044 (1978).

72. See, e.g., S. N. Vinogoradov and R. H. Linnell, "Hydrogen Bonding." Van Nostrand–Reinhold, Princeton, New Jersey, 1971; M. D. Joesten and L. J. Schaad, "Hydrogen Bonding." Dekker, New York, 1974.

73. J. E. Demuth and D. E. Eastman, *Phys. Rev. Lett.* **32**, 1132 (1974).

74. T. E. Fischer, S. R. Kelemen, and H. P. Bonzel, *Surface Sci.* **64**, 157 (1977).

75. J. E. Demuth and H. Ibach, *Surface Sci.* **78**, L238 (1978).

76. H. Ibach and S. Lehwald, *Surface Sci.* **91**, 187 (1980).

77. B. A. Sexton, *Surface Sci.* **94**, 435 (1980).

78. P. A. Thiel, F. M. Hoffmann, and W. H. Weinberg, "Vibrations at Surfaces" (R. Caudano, R. Gilles, A. A. Lucas, eds.). Plenum, New York, 1982.

79. E. M. Stuve, R. J. Madix, and B. A. Sexton, to be published.

80. A. G. Gaydon, "Dissociation Energies." Chapman and Hall, London 1968.

81. S. Andersson and J. W. Davenport, *Solid State Commun.* **28**, 677 (1978).

82. S. Lehwald, H. Steininger, and H. Ibach, *Surface Sci.*, to be published.

83. F. M. Propst and Th. C. Piper, *J. Vac. Sci. Technol.* **4**, 53 (1967).

84. J. Andersson and P. J. Estrup, *J. Chem. Phys.* **46**, 563 (1973).

85. L. R. Clavenna and L. D. Schmidt, *Surface Sci.* **33**, 11 (1972).

86. J. T. Yates, R. G. Greenler, J. Ratajczykowa, and D. A. King *Surface Sci.* **36**, 739 (1973).

87. J. E. Demuth and H. Ibach, *Chem. Phys. Lett.* **60**, 395 (1979).

88. W. Erley and H. Wagner, *Surface Sci.* **74**, 333 (1978).

89. S. Lehwald and H. Ibach, *Surface Sci.* **89**, 425 (1979).
90. L. L. Kesmodel, P. C. Stair, R. C. Baetzhold, and G. A. Somorjai, *Phys. Rev. Lett.* **36**, 1316 (1976).
91. L. L. Kesmodel, R. C. Baetzhold, and G. A. Somorjai, *Surface Sci.* **66**, 299 (1977).
92. H. Ibach, H. Hopster, and B. Sexton, *Appl. Surface Sci.* **1**, 1 (1977).
93. J. E. Demuth, *Surface Sci.* **80**, 387 (1979).
94. J. E. Demuth, *Surface Sci.* **93**, 282 (1980).
95. L. L. Kesmodel, L. H. Dubois, and G. A. Somorjai, *J. Chem. Phys.* **70**, 2180 (1979).
96. H. Ibach, *Proc. Int. Conf. Vibrations Adsorbed Layers, Jülich, 1978* p. 64.
97. L. H. Dubois, D. G. Castner, and G. H. Somorjai, *J. Chem. Phys.* **72**, 5234 (1980).
98. P. Hansma, private communication.
99. P. Skinner, M. W. Howard, I. A. Oxton, S. F. A. Kettle, D. B. Powell, and N. Sheppard, *J. Chem. Soc. Faraday Trans. II*, **77**, 1203 (1981).
100. A. M. Baro and H. Ibach, *Surface Sci.*, to be published.
101. H. Steininger, "Thesis, Technische Hochschule Aachen", 1981, to be published.
102. M. W. Howard, S. F. A. Kettle, I. A. Oxton, D. B. Powell, N. Sheppard, and P. Skinner. *J. Chem. Soc. Faraday Trans. II*, **77**, 397 (1981).
103. S. Lehwald, private communication.
104. A. Gavezotti and M. Simonetta, *Surface Sci.* **99**, 453 (1980).
105. A. B. Anderson and A. T. Hubbard, *Surface Sci.* **99**, 384 (1980).

CHAPTER 7

OUTLOOK

At the time of writing this volume, the field of electron energy loss spectroscopy is in rapid expansion and development. As the number of researchers involved in the field is growing, areas of possible applications are likely to spread. Also, instrumental technology is going to improve further. Any attempt to provide an outlook on the future development of the field might therefore be daring.

We shall nevertheless try to convey our personal view of where the field is likely to move. We also use the opportunity to point out a few areas of research which have not been covered in this volume, not because they are not important but because there has simply been no experimental material available until now.

Further instrumental development is likely to follow several routes. One is of an electron optical nature. Combination of different energy analyzers and improved lens design are currently being pursued. Presently, when a spectral range of 3000 cm^{-1} is scanned with a resolution of 60 cm^{-1}, only 2% of all inelastically scattered electrons are analyzed at any time. Multiple detection schemes might improve this fraction, if not up to 100%, then possibly by an order of magnitude. Higher count rates might then be used for a fast instrumental response toward real-time spectroscopy or for improved resolution below the current "record" of 30 cm^{-1}. A second line of development is probably that of attempting to reduce adjustment time and computer-controlled operation including digital data processing. There is also no principal obstacle for constructing a simple electron energy loss spectrometer as an "add on" system in order to complete a set of other ultrahigh vacuum analysis instruments. An example of this type of instrument was described recently by Sexton [1]. Contrary to spectrometers discussed in this volume, the Sexton spectrometer does not allow one to vary the angle between the incident beam and the analyzer position. While this is a definite disadvantage in single-crystal work, where angular profiles have proved to

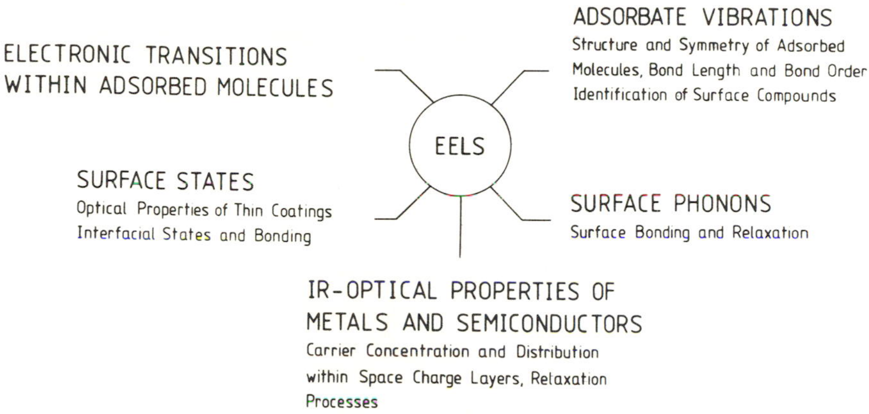

ADSORBATE VIBRATIONS
Structure and Symmetry of Adsorbed
Molecules, Bond Length and Bond Order
Identification of Surface Compounds

ELECTRONIC TRANSITIONS
WITHIN ADSORBED MOLECULES

EELS

SURFACE STATES
Optical Properties of Thin Coatings
Interfacial States and Bonding

SURFACE PHONONS
Surface Bonding and Relaxation

IR-OPTICAL PROPERTIES OF
METALS AND SEMICONDUCTORS
Carrier Concentration and Distribution
within Space Charge Layers, Relaxation
Processes

Fig. 7.1   Surface excitations which can be probed by electron energy loss spectroscopy.

be an important asset, fixed-angle spectrometers are perfectly adequate for studies on disordered ("real") surfaces. Somewhat different spectrometers may also be employed when elementary excitations other than surface vibrations are studied. As we have already mentioned in Chapters 3 and 5, electron energy loss spectroscopy, in principle, probes all elementary excitations. Presently, studies of electronic transitions of adsorbed molecules, surfaces, and interfaces (Fig. 7.1) are confined to a few papers, but these applications may gain importance in the future.

Since electron energy loss spectroscopy was developed as an instrument of basic research, almost the entire research focused on single-crystal surfaces. Yet single-crystal substrates are not a requirement of the technique as such. In a recent study Dubois *et al.* [2] prepared a special substrate by evaporating an aluminum film on a flat surface. This film was subsequently oxidized to produce Al_2O_3 which is a typical support material of a catalyst. Upon this surface Dubois *et al.* evaporated a small amount of rhodium in an atmosphere of 10^{-5} Torr CO. The rhodium clustered into small particles on the Al_2O_3 surface. Thus, the substrate modeled a supported metal catalyst with rhodium as the catalytically active material. An electron energy loss spectrum of this CO-covered substrate is shown in Fig. 7.2 together with an inelastic tunneling spectrum and an infrared spectrum of a supported catalyst. While the electron energy loss spectrum is a little wanting in resolution, it shows that the application of electron energy loss spectroscopy to "real" surfaces is not meeting significant obstacles. The degradation of resolution sometimes encountered when electrons are scattered from rough surfaces is caused by electrons which traverse the analyzer at large input angles (Chapter 2). This effect can be avoided by

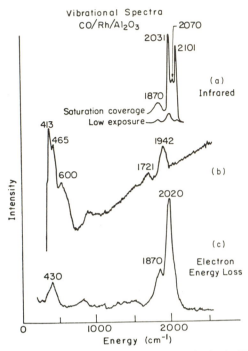

Fig. 7.2 Spectrum of CO adsorbed on a model catalyst obtained with (a) infrared absorption spectroscopy, (b) inelastic tunnel spectroscopy, and (c) electron energy loss spectroscopy.

proper lens design. In the work of Dubois *et al.* the surface was simply covered with CO. It would certainly be even more interesting to probe the surface after it has participated in a catalytic reaction. Experimental equipment which allows for this type of investigation is currently about to be completed in several laboratories. A three-story design with the sample on a push-rod is shown in Fig. 7.3 as an example. It features a reaction chamber in the top story with gas inlets and gas-chromatographic analysis of the reaction products, an ultra high vacuum preparation stage with electron diffraction optics, Auger spectrometer, and a quadrapole mass spectrometer, and finally electron energy loss spectroscopy in the basement.

Throughout this volume, we have totally neglected inelastic scattering of electrons from molecules via a "resonance" process. Such processes are well known in electron scattering from gas-phase molecules [4]. There, for certain electron energies the electron and the molecule form a compound state with a lifetime of the order of 10^{-10} to 10^{-15} sec. For a chemisorbed molecule, the resonance is quenched by coupling of the molecule to the surface. This is illustrated in Fig. 7.4a, where we show the energy variation for off-specular scattering from the CO stretching mode, with the CO

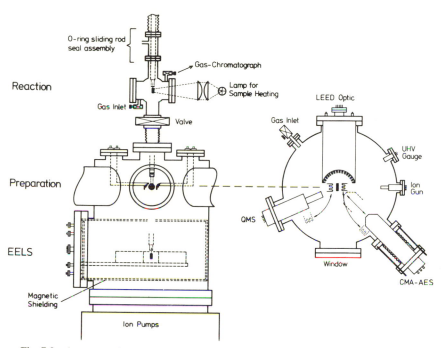

Fig. 7.3 Apparatus featuring a reaction chamber, a preparation stage, and electron energy loss spectrometer (after Erley [3], used with permission).

chemisorbed on the Pt(111) surface [5]. One sees a cross section substantially larger than estimated from calculations such as those displayed in Fig. 3.17, but at the same time substantially smaller than that appropriate to the gas phase shape resonance. In essence, the shape resonance characteristic of the gas phase is dramatically broadened upon chemisorption. On the other hand, negative ion resonances have been observed recently [6, 7] in studies of physisorbed species. We illustrate an example in Fig. 7.4b, where we show the energy variation of the excitation cross section for N_2 physisorbed on a silver surface [7].

From the standpoint of surface analysis, the vibrational cross sections for resonance scattering are rather high; from Fig. 7.4a at 2 eV the cross section exceeds 10^{-17} cm^2 sr^{-1}, which is at least two orders of magnitude higher than typical impact scattering cross sections at this energy (see Fig. 3.17). In Section 3.6.1, we have concluded that the sensitivity for excitation via impact scattering is such that vibrations of 1/10 of a monolayer of adsorbate could be detected. With recent improvements (Fig. 6.22) the limit may actually be close to 1/100 of a monolayer. For physisorbed molecules excited via resonance scattering therefore, the detection limit would be

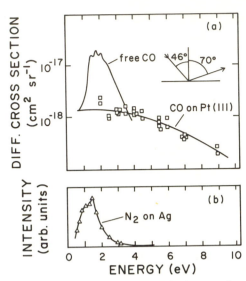

Fig. 7.4 The energy variation of the cross sections for large-angle scattering from (a) CO chemisorbed on Pt (111) ([5] used with permission), and (b) N_2 physisorbed on Ag (111) ([7] used with permission). In (a), the energy variation of the gas phase cross section for CO is also given. Note the strong negative ion resonance near 2 eV.

between 10^{-3} and 10^{-4} of a monolayer even with present-generation spectrometers. At the same time, the method can be still viewed as being nondestructive, since even if one allows for one molecule being dissociated or desorbed per incident electron, only a fraction of 10^{-3} of the adsorbed molecules would be affected during a typical time of measurement. The combination of high sensitivity and nondestructiveness could make electron energy loss spectroscopy a unique research instrument for physisorbed layers as well.

It is interesting to consider which modes of a molecule are excited in a resonance scattering process. In a long-lived negative ion resonance, the wave function of the captured electron must be a basis function for an irreducible representation of the symmetry group of the molecule. With the molecule in this state, the nuclei are not in equilibrium; the extra force field on the nuclei reflects the new charge configuration of the molecule, which is given by the square of the wave function of the captive electron. Thus, we have a selection rule which operates in such a resonance: the vibrational normal modes strongly excited on resonance are those which appear in the decomposition of the direct product of the irreducible representation of the wave function of the negative ion with itself [8]. If the symmetry group of the molecule (on the surface) admits only nondegenerate representations, it follows that just as in dipole scattering, only totally symmetric vibrational modes are strongly excited on resonance. For the surface point groups we have considered C_{3v}, C_{4v}, and C_{6v} admit (twofold) degenerate representations. For C_{3v}, in a negative ion resonance of E symmetry, all modes are strongly excited, while for C_{4v} the vibrational modes of E symmetry are "silent" for a resonance of E symmetry. For C_{6v}, a negative ion resonance of either E_1 or E_2 symmetry will excite modes of A_1, A_2, and E_2 character.

At the time of this writing, very nearly all studies of electron energy loss from clean and absorbate-covered surfaces explore electrons scattered very near the specular direction by the dipole mechanism and chemisorbed molecules. Recent studies explore large-angle deflections where the impact mechanism operates. The number of such studies reported so far on simple and easily characterized adsorbate–substrate combinations is quite limited at this time. It is clear from the data available so far, and the discussions in Chapters 3–6, that data on inelastic losses at large deflection angles, with the selection rules that operate there, in combination with that on the small-angle dipole-dominated scatterings provides one with access to a rich storehouse of information on surface structure, lateral interactions between adsorbates, and other aspects of the surface environment. We can expect major advances in our knowledge of surface dynamics as more data of this kind appear. So far, the development of electron energy loss spectroscopy at large deflection angles has been hampered by weak signals, which were hard to detect. It is possible that theoretical studies of the angle and energy variation of the cross section, based on a formalism similar to that described in the last half of Chapter 3, may suggest scattering geometries that lead to stronger signals, or they may suggest new ways of extracting information from the data in hand. During the next decade, we may expect that the study of electron energy losses at large deflection angles will be an exciting and rapidly developing area of surface spectroscopy. We may expect that by the end of the next decade, our knowledge of excitations at the crystal surface will have been advanced in a most fundamental fashion by such studies.

REFERENCES

1. B. A. Sexton, *J. Vac. Sci. Technol.* **16**, 1033 (1979).
2. L. H. Dubois, P. K. Hansma, and G. A. Somorjai, *Appl. Surface Sci,* to be published.
3. W. Erley, private communication.
4. G. J. Schulz, *Rev. Mod. Phys.* **45**, 423 (1973).
5. H. Ibach, *Proc. EUCMOS XV, Norwich 1981.* Elsevier, Amsterdam, to be published.
6. L. Sanche and M. Michaud, *Phys. Rev. Lett.* **47**, 1008 (1981).
7. J. E. Demuth, D. S. Schmeisser, and Ph. Avouris, *Phys. Rev. Lett.* **47**, 1166 (1981).
8. See, e.g., I. C. Walker, A. Stamatovic, and S. F. Wong, *J. Chem. Phys.* **69**, 5532 (1978).

EVALUATION OF THE FUNCTION
$P(\mathbf{Q}_{||}, \omega)$ FOR THE TWO-LAYER
MODEL

As remarked just before Eq. (2.10), the function $P(\mathbf{Q}_{||}, \omega)$ may be expressed in terms of correlation functions $\langle E_\alpha(\mathbf{x}t), E_\beta(\mathbf{x}'t') \rangle$ of the electric field functions in the medium. Thus, our primary task is the evaluation of these correlation functions.

Through use of a procedure outlined by Abrikosov *et al.* [1], this may be done as follows. Suppose we have a medium, possibly nonuniform, described by a dielectric constant $\varepsilon(\mathbf{x}, \omega)$, where $\varepsilon(\mathbf{x}, \omega)$ may be complex. Then we construct a set of Green's functions $D_{ij}(\mathbf{xx}'; \omega)$ by solving the differential equations

$$\sum_k \left\{ \frac{\partial^2}{\partial x_i \, \partial x_k} - \delta_{ik} \nabla^2 + \frac{\omega^2}{c^2} \varepsilon(\mathbf{x}, \omega) \delta_{ik} \right\} D_{kj}(\mathbf{xx}'; \omega) = 4\pi \, \delta_{ij} \, \delta(\mathbf{x} - \mathbf{x}'). \quad \text{(A.1)}$$

For ω complex, these equations are to be solved subject to the boundary conditions that $D_{ij}(\mathbf{x}, \mathbf{x}'; \omega)$ vanish as either $|\mathbf{x}|$ or $|\mathbf{x}'|$ becomes infinite. Once the functions $D_{ij}(\mathbf{xx}'; \omega)$ are known, for ω at a general point in the complex ω plane, the next step is to form a set of spectral density functions $A_{ij}(\mathbf{xx}'; \omega)$ defined by

$$A_{ij}(\mathbf{xx}'; \omega) = (\omega^2/ic^2)[D_{ij}(\mathbf{xx}'; \omega + i\eta) - D_{ij}(\mathbf{xx}'; \omega - i\eta)], \quad \text{(A.2)}$$

where η is a positive infinitesimal. Now for any model which retains translational invariance in the two directions parallel to the surface, the functions $A_{ij}(\mathbf{xx}'; \omega)$ are necessarily functions of only the difference $\mathbf{x}_{||} - \mathbf{x}'_{||}$, where $\mathbf{x}_{||}$ and $\mathbf{x}'_{||}$ are the projections of $\mathbf{x}$ and $\mathbf{x}'$ onto a plane parallel to the surface. The Green's functions themselves share this property, so we write

$$A_{ij}(\mathbf{xx}'; \omega) = \int \frac{d^2 Q_{||}}{(2\pi)^2} \exp[i\mathbf{Q}_{||} \cdot (\mathbf{x}_{||} - \mathbf{x}'_{||})] A_{ij}(\mathbf{Q}_{||}\omega; zz'). \quad \text{(A.3)}$$

We then have the identify [1]

$$[1 + n(\omega)]A_{ij}(\mathbf{Q}_{||}\omega; zz')$$

$$= \frac{1}{\hbar} \int d^2x_{||}\, dt\, \exp[i\mathbf{Q}_{||} \cdot (\mathbf{x}_{||} - \mathbf{x}'_{||})]e^{-i\omega t}\langle E_i(\mathbf{x}t)E_j(\mathbf{x}', 0)\rangle. \quad \text{(A.4)}$$

Thus, once the spectral density functions just introduced are constructed, we may find the electric field correlation functions that enter $P(\mathbf{Q}_{||}, \omega)$. The prescription outlined above allows the electric field correlation functions to be found, with the effect of retardation included fully. For the purposes of the electron scattering theory, we have seen that retardation effects are unimportant. Thus, we take the limit as the velocity of light $c \to \infty$ in the discussion that follows.

Through use of the identities already given, we may write the function $P(Q_{||}, \omega)$ in the form

$$P(\mathbf{Q}_{||}, \omega) = i^{-1}[1 + n(\omega)][\mathscr{P}(\mathbf{Q}_{||}, \omega + i\eta) - \mathscr{P}(\mathbf{Q}_{||}, \omega - i\eta)], \quad \text{(A.5)}$$

where $n(\omega) = [\exp(\hbar\omega/k_B T) - 1]^{-1}$ is the Bose–Einstein function. After some algebra, and the use of Eq. (3.10), one finds

$$\mathscr{P}(\mathbf{Q}_{||}, \omega) = \frac{\hbar}{16\pi^2} \frac{\omega^2}{c^2} \int_{-\infty}^{0+} dz' \int_{-\infty}^{0-} dz''\, e^{Q_{||}(z+z')}$$

$$\times \left\{ Q_{||}^2 D_{xx}(\mathbf{Q}_{||}\omega; z'z'') + iQ_{||}\frac{\partial}{\partial z'} D_{xz}(\mathbf{Q}_{||}\omega; z'z'') \right.$$

$$\left. - iQ_{||}\frac{\partial}{\partial z} D_{zx}(\mathbf{Q}_{||}\omega; z'z'') + \frac{\partial^2}{\partial z\, \partial z'} D_{zz}(\mathbf{Q}_{||}\omega; z'z'') \right\}. \quad \text{(A.6)}$$

So far, our discussion is quite general. We now turn to the two-layer model illustrated in Fig. 3.2, for which the functions $D_{ij}(\mathbf{Q}_{||}\omega; z'z'')$ have been obtained elsewhere [2]. This has been done with retardation fully included, and here we quote the form valid when retardation is ignored. It is then useful to introduce a new set of response functions $g_{ij}(\mathbf{Q}_{||}\omega; zz')$ set up so the velocity of light disappears from all expressions that appear as intermediate steps in the calculation:

$$g_{ij}(\mathbf{Q}_{||}\omega; zz') = (\omega^2/c^2)D_{ij}(\mathbf{Q}_{||}\omega; zz'). \quad \text{(A.7)}$$

The required functions may be expressed in terms of $W(Q_{||}, \omega)$, and certain electric fields $\mathbf{E}^>(Q_{||}, z)$ and $\mathbf{E}^<(Q_{||}, z)$. We define

$$W(\mathbf{Q}_{||}\omega) = \frac{\varepsilon_s}{2Q_{||}} [(\varepsilon_s + \varepsilon_b)(\varepsilon_s + 1) + (\varepsilon_b - \varepsilon_s)(\varepsilon_b - 1)e^{-2Q_{||}d}], \quad \text{(A.8)}$$

and let

$$E_z^>(Q_{||}, z) = \begin{cases} e^{-Q_{||}z}, & z > 0, \\ A_+ e^{+Q_{||}z} + A_- e^{-Q_{||}z}, & -d < z < 0, \\ B_+ e^{+Q_{||}z} + B_- e^{-Q_{||}z}, & z < -d, \end{cases} \tag{A.9}$$

and

$$E_x^>(Q_{||}, z) = \begin{cases} -ie^{-Q_{||}z}, & z > 0 \\ iA_+ e^{Q_{||}z} - iA_- e^{-Q_{||}z}, & -d < z < 0 \\ iB_+ e^{Q_{||}z} - iB_- e^{-Q_{||}z}, & z < -d, \end{cases} \tag{A.10a}$$

with

$$B_\pm = \frac{e^{\pm Q_{||}d}}{2}\left[\left(\frac{1}{\varepsilon_b} \pm 1 \right) \cosh(Q_{||}d) + \left(\frac{\varepsilon_s}{\varepsilon_b} \mp \frac{1}{\varepsilon_s} \right) \sinh(Q_{||}d) \right]. \tag{A.10b}$$

The field $E^<(Q_{||}, z)$ has components given by

$$E_z^<(Q_{||}, z) = \begin{cases} D_+ e^{+Q_{||}z} + D_- e^{-Q_{||}z}, & z > 0, \\ C_+ e^{Q_{||}z} + C_- e^{-Q_{||}z}, & -d < z < 0, \\ e^{Q_{||}z}, & z < -d, \end{cases} \tag{A.11a}$$

and

$$E_x^<(Q_{||}, z) = \begin{cases} i(D_+ e^{Q_{||}z} - D_- e^{-Q_{||}z}), & z > d, \\ i(C_+ E^{Q_{||}z} - C_- e^{-Q_{||}z}), & -d < z < 0, \\ ie^{Q_{||}z}, & z < -d, \end{cases} \tag{A.11b}$$

where

$$D_\pm = \tfrac{1}{2} e^{-Q_{||}d}\left[(\varepsilon_b \pm 1) \cosh(Q_{||}d) + \left(\varepsilon_s \pm \frac{\varepsilon_b}{\varepsilon_s} \right) \sinh(Q_{||}) \right] \tag{A.12a}$$

and finally

$$C_+ = \tfrac{1}{2}(\varepsilon_b/\varepsilon_s + 1), \tag{A.12b}$$

$$C_- = \tfrac{1}{2}(\varepsilon_b/\varepsilon_s - 1)e^{-2Q_{||}d}. \tag{A.12c}$$

In terms of the quantities introduced above, we have

$$g_{xx}(Q_{||}\omega; zz') = \frac{4\pi}{W(Q_{||}, \omega)} \left[E_x^>(Q_{||}, z)E_x^<(Q_{||}, z')\theta(z - z') \right.$$
$$\left. + E_x^<(Q_{||}, z)E_x^>(Q_{||}, z')\theta(z' - z) \right], \tag{A.13}$$

with similar expressions for $g_{xz}(\mathbf{Q}_{||}\omega; zz')$ and $g_{zx}(\mathbf{Q}_{||}\omega; z'z)$. Then $g_{zz}(\mathbf{Q}_{||}\omega; z'z)$ is given by

$$g_{zz}(\mathbf{Q}_{||}\omega; z'z) = \frac{4\pi}{W(Q_{||},\omega)} \left[E_z^>(Q_{||}, z) E_z^<(Q_{||}, z) E_z^<(Q_{||}, z')\theta(z - z') \right.$$

$$\left. + E_z^<(Q_{||}, z) E_z^>(Q_{||}, z'(Q_{||}, z')\theta(z' - z) \right] + \frac{4\pi}{\varepsilon(z)} \delta(z - z'),$$

(A.14)

where $\varepsilon(z)$ is unity for $z > 0$, equals $\varepsilon_s(\omega)$ for $-d \leq z < 0$, and equals $\varepsilon_b(\omega)$ for $z < -d$.

It is now a straightforward, but tedious matter to insert the forms above into the expression for $\mathscr{P}(\mathbf{Q}_{||}, \omega)$ given in Eq. (A.5). One finds the result

$$\mathscr{P}(\mathbf{Q}_{||},\omega) = -\frac{Q_{||}\hbar}{2\pi} \left\{ \frac{(\varepsilon_s - 1)(\varepsilon_s + \varepsilon_b) + (\varepsilon_s + 1)(\varepsilon_b - \varepsilon_s)e^{-2Q_{||}d}}{(\varepsilon_s + \varepsilon_b)(\varepsilon_s + 1) + (\varepsilon_s - 1)(\varepsilon_b - \varepsilon_s)e^{-2Q_{||}d}} \right\}. \quad (A.15)$$

After some rearrangement, $P(\mathbf{Q}_{||}, \omega)$ itself may be cast into the form

$$P(Q_{||}, \omega) = (2\hbar Q_{||}/\pi)[1 + n(\omega)] \, \text{Im} \left\{ \frac{-1}{\tilde{\varepsilon}(Q_{||}, \omega) + 1} \right\}, \quad (A.16)$$

where $\tilde{\varepsilon}(Q_{||}\omega)$ is an effective, wave-vector-dependent dielectric constant of the structure "seen" by the incoming electron. We have

$$\tilde{\varepsilon}(Q_{||}, \omega) = \varepsilon_s(\omega) \left[\frac{1 + \Delta(\omega) \exp(-2Q_{||}d)}{1 - \Delta(\omega) \exp(-2Q_{||}d)} \right], \quad (A.17)$$

where

$$\Delta(\omega) = \frac{\varepsilon_b(\omega) - \varepsilon_s(\omega)}{\varepsilon_b(\omega) + \varepsilon_s(\omega)}. \quad (A.18)$$

Notice that

$$\lim_{d \to 0} \tilde{\varepsilon}(Q_{||}, \omega) = \varepsilon_b(\omega), \quad (A.19)$$

and also

$$\lim_{d \to \infty} \tilde{\varepsilon}(Q_{||}, \omega) = \varepsilon_s(\omega). \quad (A.20)$$

REFERENCES

1. A. A. Abrikosov, L. P. Gor'kov, and I. Dzyaloshinski, "Methods of Quantum Field Theory in Statistical Physics," Chapter 6. Prentice-Hall, Englewood Cliffs, New Jersey, 1963.
2. D. L. Mills and A. A. Maradudin, *Phys. Rev. B* **12**, 2943 (1975).

VIBRATIONAL FREQUENCIES OF SELECTED MOLECULES

This appendix consists of tables and references, and is presented as Tables B1 through B12 on the following pages.

TABLE B.1

Fundamental Frequencies $\nu = \omega_e - 2x_e\omega_e$ of Heteronuclear Diatomic Molecules

(hydrides and oxides)

—— ν(X–H) ——
—— ν(X–O) ——

Each cell lists the element symbol with ν(X–H) (upper value) and ν(X–O) (lower value).

1	2	3	4	5	6	7	8	9	10	11	12	13	14	15	16	17
H 4159 / 3571																
Li 1359	Be 1988 / 1464											B 2282 / 1862	C 2733 / 2143	N 3120 / 1576	O 3571 / 1556	F 3959
Na 1133	Mg 1432 / 775											Al 1624 / 965	Si 1972 / 1230	P 2270 / 1220	S 2582 / 1136	Cl 2886 / (780)
K 955	Ca 1260 / 722	Sc 957	Ti 999	V 1003	Cr 1538 / 886	Mn 1490 / 831	Fe 870	Co (1890) / (838)	Ni (1926) / (615)	Cu 1866 / 622	Zn 1497	Ga 1547 / 755	Ge 1834 / 977	As 2116 / 957	Se 906	Br 2559 / 757
Rb 908	Sr 1172 / 645	V 858	Zr 930	Nb 982	Mo	Tc	Ru 865	Rh	Pd	Ag 1692 / 484	Cd 1338	M 1425 / 696	Sn 1655 / 815	Sb 807	Te 789	I 2230 / 673
Cs 916	Ba 1140 / 666	La	Hl	Ta 1021	W 1047	Re	Os	Ir	Pt (2250) / 841	Au 2219	Hg 1221	Te 1345	Pb 1505 / 714	Bi 1636 / 684	Po	Ac

TABLE B.2

Vibrational Frequencies of XY$_2$ Molecules (Point Group C_{2v})

Molecule	$\nu_1(A_1)$ s stretch	$\nu_2(A_1)$ bend	$\nu_3(B_1)$ a stretch
H$_2$O	3657	1595	3756
Liq.	3219	1627	3445
D$_2$O	2671	1178	2788
H$_2$S	2615	1183	(2627)
D$_2$S	1892	934	2000
N^{14}O$_2$	1318	750	1618
N^{15}O$_2$	1306	740	1580
SO$_2$	1151	518	1362
SO^{16}O^{18}	1122	507	1341

TABLE B.3

Vibrational Frequencies of Bent XYZ Molecules (Point Group C_s)

Molecule	ν_1 stretch	ν_2 stretch	ν_3 bend
FNO	1844	520[a]	766[a]
ClNO	1800	332[a]	596[a]
BrNO	1799	266[a]	542[a]
FOH	3537	1393	886
ClOH	3609	1242	725

[a] Modes are assigned according to L. H. Jones, L. B. Asprey, and R. R. Ryan, *J. Chem. Phys.* **49**, 581 (1968).

TABLE B.4

Vibrational Frequencies of XY$_3$ Molecules (Point Group C_{3v}).

Molecule	$\nu_1(A_1)$ s stretch	$\nu_2(A_1)$ s deform	$\nu_3(E)$ d stretch	$\nu_4(E)$ d deform
NH$_3$	3337[a]	950[a]	3414	1628
ND$_3$	2420	747[a]	2556	1191
PH$_3$	2327	991	2421	1121
PD$_3$	1694	730	(1698)	806
AsH$_3$	2122	906	2185	1005

[a] Frequencies split by inversion doubling.

TABLE B.5

Acetylene: CHCH (Point Group $D_{\infty h}$)

Repr.	No.	Type of mode	C$_2$H$_2$	C$_2$D$_2$	IR	R
Σ_g^+	ν_1	CH s stretch	3374	2701		S, p
	ν_2	CC stretch	1974	1762		VS, p
Σ_u^+	ν_3	CH a stretch	3289	2439	S, z	
Π_g	ν_4	CH bend	612	505		VW, dp
Π_u	ν_5	CH bend	730	537	VS, x, y	

Formaldehyde: H_2CO (Point Group C_{2v})

Repr.	No.	Type of mode	H_2CO	D_2CO	IR	R
A_1	ν_1	CH_2 s stretch	2783	2056	S, z	S, p
	ν_2	CO stretch	1746	1700	VS, z	W, p
	ν_3	CH_2 sciss	1500	1106	S, z	M, p
B_1	ν_4	CH_2 a stretch	2843	2160	VS, x	W, dp
	ν_5	CH_2 rock	1249	990	S, x	
B_2	ν_6	CH_2 wagg	1167	988	S, y	

TABLE B.7

Formic Acid: HCOOH (Point Group C_s)

Repr.	No.	Type of mode	HCOOH	DCOOD	IR	R
A'	ν_1	OH stretch	3570	2632	M, x, z	
	ν_2	CH stretch	2943	2232	M, x, z	
	ν_3	C=O stretch	1770	1742	VS, x, z	
	ν_4	CH bend	1387	945	VW, x, z	
	ν_5	OH bend	1229	1040	W, x, z	
	ν_6	C—O stretch	1105	1171	S, x, z	
	ν_7	OCO bend	625	558	M, x, z	
A''	ν_8	CH bend	1033	873	W, y	
	ν_9	Torsion	638	491	S, y	

TABLE B.8

Methanol: CH_3OH (Point Group C_s)

Repr.	No.	Type of mode	CH_3OH	CD_3OD	IR	R
A'	ν_1	OH stretch	3681	2724	M, x, z	
	ν_2	CH_3 d stretch	3000	2260	M, x, z	
	ν_3	CH_3 s stretch	2844	2080	S, x, z	
	ν_4	CH_3 d deform	1477	1024	M, x, z	
	ν_5	CH_3 s deform	1455	1135	M, x, z	
	ν_6	OH bend	1345	1060	S, x, z	
	ν_7	CH_3 rock	1060	776	W, x, z	
	ν_8	CO stretch	1033	983	VS, x, z	
A''	ν_9	CH_3 d stretch	2960	2228	S, y	
	ν_{10}	CH_3 d deform	1477	1080	M, y	
	ν_{11}	CH_3 rock	1165	892		
	ν_{12}	Torsion	—	196		

TABLE B.9

Ethylene: CH_2CH_2 (Point Group D_{2h})

Repr.	No.	Type of mode	C_2H_4	C_2D_4	IR	R
A_g	ν_1	CH_2 s stretch	3026	2251		VS, p
	ν_2	CC stretch	1623	1515		VS, p
	ν_3	CH_2 sciss	1342	981		M, p
A_u	ν_4	CH_2 twist	1023	728		
B_{1g}	ν_5	CH_2 a stretch	3103	2304		W, dp
	ν_6	CH_2 rock	1236	1009		dp
B_{1u}	ν_7	CH_2 wagg	949	720	M, z	
B_{2g}	ν_8	CH_2 wagg	943	780		W, dp
B_{2u}	ν_9	CH_2 a stretch	3106	2345	S, y	
	ν_{10}	CH_2 rock	826	586	W, y	
B_{3u}	ν_{11}	CH_2 s stretch	2989	2200	S, x	
	ν_{12}	CH_2 sciss	1444	1078	S, x	

TABLE B.10

Propylene: CH_3—CH=CH_2 (Point Group: C_s)

Repr.	No.	Type of mode	C_3H_6	IR, x, z	R
A'	ν_1	CH_2 a stretch	3081	M, x, z	W, p
	ν_2	CH stretch	3012	M, x, z	p
	ν_3	CH_2 s stretch	2979	S, x, z	W, p
	ν_4	CH_3 d stretch	2916	S, x, z	VS, p
	ν_5	CH_3 s stretch	2852	M, x, z	VS, p
	ν_6	C=C stretch	1647	S, x, z	VS, p
	ν_7	CH_3 d deform	1448	S, x, z	W, p
	ν_8	CH_2 sciss	1416	W, x, z	M, p
	ν_9	CH_3 s deform	1399	W, x, z	
	ν_{10}	CH bend	1297	W, d, z	VS, p
	ν_{11}	(CH_2 rock)	1224	M, x, z	
	ν_{12}	(CH_3 rock)	1043	VW, x, z	
	ν_{13}	C—C stretch	919	S, x, z	S, p
	ν_{14}	CCC bend	417	M, x, z	M, p
A''	ν_{15}	CH_3 d stretch	2960	M, y	VW, dp
	ν_{16}	CH_3 d deform	1472	M, y	
	ν_{17}	(CH_2 rock)	1166	W, y	
	ν_{18}	(CH_3 rock)	996	W, y	
	ν_{19}	CH bend	936	VW, y	
	ν_{20}	CCH_2 twist	578	S, y	
	ν_{21}	(CCH_3 twist)	177		dp

TABLE B.11

Benzene: C_6H_6 (Point Group: D_{6h})

Repr.	No.	Type of mode	C_6H_6	C_6D_6	IR	R
A_{1g}	ν_1	CH stretch	3062	2293		VS, p
	ν_2	Ring stretch	992	943		VS, p
A_{2g}	ν_3	CH Bend	1326	1037		
A_{2u}	ν_4	CH bend	673	497	S, z	
B_{1u}	ν_5	CH stretch	3068	2292		
	ν_6	Ring deform	1010	969		
B_{2g}	ν_7	CH bend	995	827		
	ν_8	Ring deform	703	601		
B_{2u}	ν_9	Ring stretch	1310	1286		
	ν_{10}	Ring stretch	1150	824		
E_{1g}	ν_{11}	CH bend	849	662		M, dp
E_{1u}	ν_{12}	CH stretch	3063	2287	S, x, y	
	ν_{13}	Ring stretch and deform	1486	1335	S, x, y	
	ν_{14}	CH bend	1038	814	S, x, y	
E_{2g}	ν_{15}	CH stretch	3047	2265		S, dp
	ν_{16}	Ring stretch	1596	1552		S, dp
	ν_{17}	CH bend	1178	867		S, dp
	ν_{18}	Ring deform	606	577		S, dp
E_{2u}	ν_{19}	CH bend	975	795		
	ν_{20}	Ring deform	410	352		

TABLE B.12

Cyclohexane: C_6H_{12} (Point Group: D_{3d})

Repr.	No.	Type of mode	C_6H_{12}	C_6D_{12}	IR	R
A_{1g}	ν_1	CH$_2$ a stretch	2930	2152		VS, p
	ν_2	CH$_2$ s stretch	2852	2082		VS, p
	ν_3	CH$_2$ sciss	1465	1117		M, p
	ν_4	CH$_2$ rock	1157	1012		S, p
	ν_5	CC stretch	802	723		VS, p
	ν_6	CC deform, torsion	383	298		M, p
A_{1u}	ν_7	CH$_2$ twist	1383	864		
	ν_8	CH$_2$ wagg	1157	842		
	ν_9	CC stretch, torsion	1057	1187		
A_{2g}	ν_{10}	CH$_2$ wagg	1437	1126		
	ν_{11}	CH$_2$ twist	1090	778		
A_{2u}	ν_{12}	CH$_2$ a stretch	2915	2206	M, z	
	ν_{13}	CH$_2$ s stretch	2860	2108	M, z	
	ν_{14}	CH$_2$ sciss	1437	1091	M, z	
	ν_{15}	CH$_2$ rock	1030	917	M, z	
	ν_{16}	CCC deform	523	395	W, z	
E_g	ν_{17}	CH$_2$ a stretch	2930	2199		M, dp
	ν_{18}	CH$_2$ s stretch	2897	2104		S, dp
	ν_{19}	CH$_2$ sciss	1443	1071		S, dp
	ν_{20}	CH$_2$ wagg	1347	1212		VS, dp
	ν_{21}	CH$_2$ twist	1266	937		VS, dp
	ν_{22}	CC stretch	1027	795		VW, dp
	ν_{23}	CH$_2$ rock	785	637		S, dp
	ν_{24}	CC deform, torsion	426	373		
E_u	ν_{25}	CH$_2$ a stretch	2933	2206		VS, x, y
	ν_{26}	CH$_2$ s stretch	2863	2108		VS, x, y
	ν_{27}	CH$_2$ sciss	1457	1069		VS, x, y
	ν_{28}	CH$_2$ wagg	1355	1165		W, x, y
	ν_{29}	CH$_2$ twist	1261	991		S, x, y
	ν_{30}	CH$_2$ rock	907	687		S, x, y
	ν_{31}	CC stretch	863	720		S, x, y
	ν_{32}	CC deform, torsion	248	203		VW, x, y

REFERENCES

1. G. Herzberg, "Molecular Spectra and Molecular Structure," Vols. I and II. Van Nostrand–Reinhold, Princeton, New Jersey, 1950 and 1945.
2. T. Shimanouchi, "Tables of Molecular Vibrational Frequencies," Vol. I. NSRDS-NBS39, 1972.
3. B. Rosen, "International Tables of Selected Constants," Vol. 17. Pergamon, Oxford, 1970.
4. K. Nakamoto, "Infrared Spectra of Inorganic and Coordination Compounds." Wiley, New York, 1963.
5. T. Shimanouchi, Tables of Molecular Vibrational Frequencies, II. *J. Phys. Chem. Ref. Data* **6** (1977).

CONVERSION OF FREQUENCY UNITS

λ^{-1} [cm^{-1}]	λ [μm]	ν[THz]	hν[meV]	hν[10^{-3}H]
10	1000.000	0.300	1.240	0.046
20	500.000	0.600	2.480	0.091
30	333.333	0.899	3.720	0.137
40	250.000	1.199	4.959	0.182
50	200.000	1.499	6.199	0.228
60	166.667	1.799	7.439	0.273
70	142.857	2.099	8.679	0.319
80	125.000	2.398	9.919	0.365
90	111.111	2.698	11.159	0.410
100	100.000	2.998	12.399	0.456
110	90.909	3.298	13.638	0.501
120	83.333	3.597	14.878	0.547
130	76.923	3.897	16.118	0.592
140	71.429	4.197	17.358	0.638
150	66.667	4.497	18.598	0.683
160	62.500	4.797	19.838	0.729
170	58.824	5.096	21.077	0.775
180	55.556	5.396	22.317	0.820
190	52.632	5.696	23.557	0.866
200	50.000	5.996	24.797	0.911
210	47.619	6.296	26.037	0.957
220	45.455	6.595	27.277	1.002
230	43.478	6.895	28.517	1.048
240	41.667	7.195	29.756	1.094
250	40.000	7.495	30.996	1.139
260	38.462	7.795	32.236	1.185
270	37.037	8.094	33.476	1.230
280	35.714	8.394	34.716	1.276
290	34.483	8.694	35.956	1.321
300	33.333	8.994	37.195	1.367
310	32.258	9.293	38.435	1.412
320	31.250	9.593	39.675	1.458
330	30.303	9.893	40.915	1.504
340	29.412	10.193	42.155	1.549
350	28.571	10.493	43.395	1.595
360	27.778	10.792	44.635	1.640
370	27.027	11.092	45.874	1.686
380	26.316	11.392	47.114	1.731
390	25.641	11.692	48.354	1.777
400	25.000	11.992	49.594	1.823
410	24.390	12.291	50.834	1.868
420	23.810	12.591	52.074	1.914
430	23.256	12.891	53.314	1.959

λ^{-1} [cm^{-1}]	λ [μm]	ν [THz]	$h\nu$ [meV]	$h\nu$ [10^{-3}H]
440	22.727	13.191	54.553	2.005
450	22.222	13.491	55.793	2.050
460	21.739	13.790	57.033	2.096
470	21.277	14.090	58.273	2.141
480	20.833	14.390	59.513	2.187
490	20.408	14.690	60.753	2.233
500	20.000	14.990	61.993	2.278
510	19.608	15.289	63.232	2.324
520	19.231	15.589	64.472	2.369
530	18.868	15.889	65.712	2.415
540	18.519	16.189	66.952	2.460
550	18.182	16.488	68.192	2.506
560	17.857	16.788	69.432	2.552
570	17.544	17.088	70.671	2.597
580	17.241	17.388	71.911	2.643
590	16.949	17.688	73.151	2.688
600	16.667	17.987	74.391	2.734
610	16.393	18.287	75.631	2.779
620	16.129	18.587	76.871	2.825
630	15.873	18.887	78.111	2.870
640	15.625	19.187	79.350	2.916
650	15.385	19.486	80.590	2.962
660	15.152	19.786	81.830	3.007
670	14.925	20.086	83.070	3.053
680	14.706	20.386	84.310	3.098
690	14.493	20.686	85.550	3.144
700	14.286	20.985	86.789	3.189
710	14.085	21.285	88.029	3.235
720	13.889	21.585	89.269	3.281
730	13.699	21.885	90.509	3.326
740	13.514	22.184	91.749	3.372
750	13.333	22.484	92.989	3.417
760	13.158	22.784	94.229	3.463
770	12.987	23.084	95.468	3.508
780	12.821	23.384	96.708	3.554
790	12.658	23.683	97.948	3.599
800	12.500	23.983	99.188	3.645
810	12.346	24.283	100.428	3.691
820	12.195	24.583	101.668	3.736
830	12.048	24.883	102.908	3.782
840	11.905	25.182	104.147	3.827
850	11.765	25.482	105.387	3.873
860	11.628	25.782	106.627	3.918
870	11.494	26.082	107.867	3.964
880	11.364	26.382	109.107	4.010
890	11.236	26.681	110.347	4.055
900	11.111	26.981	111.587	4.101
910	10.989	27.281	112.826	4.146
920	10.870	27.581	114.066	4.192
930	10.753	27.880	115.306	4.237
940	10.638	28.180	116.546	4.283
950	10.526	28.480	117.786	4.328
960	10.417	28.780	119.026	4.374
970	10.309	29.080	120.265	4.420
980	10.204	29.379	121.505	4.465
990	10.101	29.679	122.745	4.511
1000	10.000	29.979	123.985	4.556
1010	9.901	30.279	125.225	4.602
1020	9.804	30.579	126.465	4.647
1030	9.709	30.878	127.705	4.693
1040	9.615	31.178	128.944	4.739
1050	9.524	31.478	130.184	4.784
1060	9.434	31.778	131.424	4.830

λ^{-1} [cm^{-1}]	λ [μm]	ν [THz]	$h\nu$ [meV]	$h\nu$ [10^{-3}H]
1070	9.346	32.078	132.664	4.875
1080	9.259	32.377	133.904	4.921
1090	9.174	32.677	135.144	4.966
1100	9.091	32.977	136.383	5.012
1110	9.009	33.277	137.623	5.057
1120	8.929	33.576	138.863	5.103
1130	8.850	33.876	140.103	5.149
1140	8.772	34.176	141.343	5.194
1150	8.696	34.476	142.583	5.240
1160	8.621	34.776	143.823	5.285
1170	8.547	35.075	145.062	5.331
1180	8.475	35.375	146.302	5.376
1190	8.403	35.675	147.542	5.422
1200	8.333	35.975	148.782	5.468
1210	8.264	36.275	150.022	5.513
1220	8.197	36.574	151.262	5.559
1230	8.130	36.874	152.502	5.604
1240	8.065	37.174	153.741	5.650
1250	8.000	37.474	154.981	5.695
1260	7.937	37.774	156.221	5.741
1270	7.874	38.073	157.461	5.786
1280	7.812	38.373	158.701	5.832
1290	7.752	38.673	159.941	5.878
1300	7.692	38.973	161.180	5.923
1310	7.634	39.272	162.420	5.969
1320	7.576	39.572	163.660	6.014
1330	7.519	39.872	164.900	6.060
1340	7.463	40.172	166.140	6.105
1350	7.407	40.472	167.380	6.151
1360	7.353	40.771	168.620	6.197
1370	7.299	41.071	169.859	6.242
1380	7.246	41.371	171.099	6.288
1390	7.194	41.671	172.339	6.333
1400	7.143	41.971	173.579	6.379
1410	7.092	42.270	174.819	6.424
1420	7.042	42.570	176.059	6.470
1430	6.993	42.870	177.299	6.515
1440	6.944	43.170	178.538	6.561
1450	6.897	43.470	179.778	6.607
1460	6.849	43.769	181.018	6.652
1470	6.803	44.069	182.258	6.698
1480	6.757	44.369	183.498	6.743
1490	6.711	44.669	184.738	6.789
1500	6.667	44.968	185.977	6.834
1510	6.623	45.268	187.217	6.880
1520	6.579	45.568	188.457	6.926
1530	6.536	45.868	189.697	6.971
1540	6.494	46.168	190.937	7.017
1550	6.452	46.467	192.177	7.062
1560	6.410	46.767	193.417	7.108
1570	6.369	47.067	194.656	7.153
1580	6.329	47.367	195.896	7.199
1590	6.289	47.667	197.136	7.244
1600	6.250	47.966	198.376	7.290
1610	6.211	48.266	199.616	7.336
1620	6.173	48.566	200.856	7.381
1630	6.135	48.866	202.096	7.427
1640	6.098	49.166	203.335	7.472
1650	6.061	49.465	204.575	7.518
1660	6.024	49.765	205.815	7.563
1670	5.988	50.065	207.055	7.609
1680	5.952	50.365	208.295	7.655

λ^{-1} [cm^{-1}]	λ[µm]	ν[THz]	$h\nu$[meV]	$h\nu$[10^{-3}H]
1690	5.917	50.665	209.535	7.700
1700	5.882	50.964	210.775	7.746
1710	5.848	51.264	212.014	7.791
1720	5.814	51.564	213.254	7.837
1730	5.780	51.864	214.494	7.882
1740	5.747	52.163	215.734	7.928
1750	5.714	52.463	216.974	7.973
1760	5.682	52.763	218.214	8.019
1770	5.650	53.063	219.453	8.065
1780	5.618	53.363	220.693	8.110
1790	5.587	53.662	221.933	8.156
1800	5.556	53.962	223.173	8.201
1810	5.525	54.262	224.413	8.247
1820	5.495	54.562	225.653	8.292
1830	5.464	54.862	226.893	8.338
1840	5.435	55.161	228.132	8.384
1850	5.405	55.461	229.372	8.429
1860	5.376	55.761	230.612	8.475
1870	5.348	56.061	231.852	8.520
1880	5.319	56.361	233.092	8.566
1890	5.291	56.660	234.332	8.611
1900	5.263	56.960	235.572	8.657
1910	5.236	57.260	236.811	8.702
1920	5.208	57.560	238.051	8.748
1930	5.181	57.859	239.291	8.794
1940	5.155	58.159	240.531	8.839
1950	5.128	58.459	241.771	8.885
1960	5.102	58.759	243.011	8.930
1970	5.076	59.059	244.250	8.976
1980	5.051	59.358	245.490	9.021
1990	5.025	59.658	246.730	9.067
2000	5.000	59.958	247.970	9.113
2010	4.975	60.258	249.210	9.158
2020	4.950	60.558	250.450	9.204
2030	4.926	60.857	251.690	9.249
2040	4.902	61.157	252.929	9.295
2050	4.878	61.457	254.169	9.340
2060	4.854	61.757	255.409	9.386
2070	4.831	62.057	256.649	9.431
2080	4.808	62.356	257.889	9.477
2090	4.785	62.656	259.129	9.523
2100	4.762	62.956	260.368	9.568
2110	4.739	63.256	261.608	9.614
2120	4.717	63.555	262.848	9.659
2130	4.695	63.855	264.088	9.705
2140	4.673	64.155	265.328	9.750
2150	4.651	64.455	266.568	9.796
2160	4.630	64.755	267.808	9.842
2170	4.608	65.054	269.047	9.887
2180	4.587	65.354	270.287	9.933
2190	4.566	65.654	271.527	9.978
2200	4.545	65.954	272.767	10.024
2210	4.525	66.254	274.007	10.069
2220	4.505	66.553	275.247	10.115
2230	4.484	66.853	276.487	10.160
2240	4.464	67.153	277.726	10.206
2250	4.444	67.453	278.966	10.252
2260	4.425	67.753	280.206	10.297
2270	4.405	68.052	281.446	10.343
2280	4.386	68.352	282.686	10.388
2290	4.367	68.652	283.926	10.434
2300	4.348	68.952	285.165	10.479
2310	4.329	69.251	286.405	10.525

λ^{-1} [cm^{-1}]	λ [µm]	ν [THz]	$h\nu$ [meV]	$h\nu$ [10^{-3}H]
2320	4.310	69.551	287.645	10.571
2330	4.292	69.851	288.885	10.616
2340	4.274	70.151	290.125	10.662
2350	4.255	70.451	291.365	10.707
2360	4.237	70.750	292.605	10.753
2370	4.219	71.050	293.844	10.798
2380	4.202	71.350	295.084	10.844
2390	4.184	71.650	296.324	10.889
2400	4.167	71.950	297.564	10.935
2410	4.149	72.249	298.804	10.981
2420	4.132	72.549	300.044	11.026
2430	4.115	72.849	301.284	11.072
2440	4.098	73.149	302.523	11.117
2450	4.082	73.449	303.763	11.163
2460	4.065	73.748	305.003	11.208
2470	4.049	74.048	306.243	11.254
2480	4.032	74.348	307.483	11.300
2490	4.016	74.648	308.723	11.345
2500	4.000	74.948	309.962	11.391
2510	3.984	75.247	311.202	11.436
2520	3.968	75.547	312.442	11.482
2530	3.953	75.847	313.682	11.527
2540	3.937	76.147	314.922	11.573
2550	3.922	76.446	316.162	11.618
2560	3.906	76.746	317.402	11.664
2570	3.891	77.046	318.641	11.710
2580	3.876	77.346	319.881	11.755
2590	3.861	77.646	321.121	11.801
2600	3.846	77.945	322.361	11.846
2610	3.831	78.245	323.601	11.892
2620	3.817	78.545	324.841	11.937
2630	3.802	78.845	326.081	11.983
2640	3.788	79.145	327.320	12.029
2650	3.774	79.444	328.560	12.074
2660	3.759	79.744	329.800	12.120
2670	3.745	80.044	331.040	12.165
2680	3.731	80.344	332.280	12.211
2690	3.717	80.644	333.520	12.256
2700	3.704	80.943	334.759	12.302
2710	3.690	81.243	335.999	12.347
2720	3.676	81.543	337.239	12.393
2730	3.663	81.843	338.479	12.439
2740	3.650	82.142	339.719	12.484
2750	3.636	82.442	340.959	12.530
2760	3.623	82.742	342.199	12.575
2770	3.610	83.042	343.438	12.621
2780	3.597	83.342	344.678	12.666
2790	3.584	83.641	345.918	12.712
2800	3.571	83.941	347.158	12.758
2810	3.559	84.241	348.398	12.803
2820	3.546	84.541	349.638	12.849
2830	3.534	84.841	350.878	12.894
2840	3.521	85.140	352.117	12.940
2850	3.509	85.440	353.357	12.985
2860	3.497	85.740	354.597	13.031
2870	3.484	86.040	355.837	13.076
2880	3.472	86.340	357.077	13.122
2890	3.460	86.639	358.317	13.168
2900	3.448	86.939	359.556	13.213
2910	3.436	87.239	360.796	13.259
2920	3.425	87.539	362.036	13.304
2930	3.413	87.838	363.276	13.350
2940	3.401	88.138	364.516	13.395

λ^{-1} [cm^{-1}]	λ [µm]	ν [THz]	$h\nu$ [meV]	$h\nu$ [10^{-3}H]
2950	3.390	88.438	365.756	13.441
2960	3.378	88.738	366.996	13.487
2970	3.367	89.038	368.235	13.532
2980	3.356	89.337	369.475	13.578
2990	3.344	89.637	370.715	13.623
3000	3.333	89.937	371.955	13.669
3010	3.322	90.237	373.195	13.714
3020	3.311	90.537	374.435	13.760
3030	3.300	90.836	375.675	13.805
3040	3.289	91.136	376.914	13.851
3050	3.279	91.436	378.154	13.897
3060	3.268	91.736	379.394	13.942
3070	3.257	92.036	380.634	13.988
3080	3.247	92.335	381.874	14.033
3090	3.236	92.635	383.114	14.079
3100	3.226	92.935	384.353	14.124
3110	3.215	93.235	385.593	14.170
3120	3.205	93.534	386.833	14.216
3130	3.195	93.834	388.073	14.261
3140	3.185	94.134	389.313	14.307
3150	3.175	94.434	390.553	14.352
3160	3.165	94.734	391.793	14.398
3170	3.155	95.033	393.032	14.443
3180	3.145	95.333	394.272	14.489
3190	3.135	95.633	395.512	14.534
3200	3.125	95.933	396.752	14.580
3210	3.115	96.233	397.992	14.626
3220	3.106	96.532	399.232	14.671
3230	3.096	96.832	400.472	14.717
3240	3.086	97.132	401.711	14.762
3250	3.077	97.432	402.951	14.808
3260	3.067	97.732	404.191	14.853
3270	3.058	98.031	405.431	14.899
3280	3.049	98.331	406.671	14.945
3290	3.040	98.631	407.911	14.990
3300	3.030	98.931	409.151	15.036
3310	3.021	99.230	410.390	15.081
3320	3.012	99.530	411.630	15.127
3330	3.003	99.830	412.870	15.172
3340	2.994	100.130	414.110	15.218
3350	2.985	100.430	415.350	15.263
3360	2.976	100.729	416.590	15.309
3370	2.967	101.029	417.829	15.355
3380	2.959	101.329	419.069	15.400
3390	2.950	101.629	420.309	15.446
3400	2.941	101.929	421.549	15.491
3410	2.933	102.228	422.789	15.537
3420	2.924	102.528	424.029	15.582
3430	2.915	102.828	425.269	15.628
3440	2.907	103.128	426.508	15.674
3450	2.899	103.428	427.748	15.719
3460	2.890	103.727	428.988	15.765
3470	2.882	104.027	430.228	15.810
3480	2.874	104.327	431.468	15.856
3490	2.865	104.627	432.708	15.901
3500	2.857	104.926	433.948	15.947
3510	2.849	105.226	435.187	15.992
3520	2.841	105.526	436.427	16.038
3530	2.833	105.826	437.667	16.084
3540	2.825	106.126	438.907	16.129
3550	2.817	106.425	440.147	16.175
3560	2.809	106.725	441.387	16.220
3570	2.801	107.025	442.626	16.266

λ^{-1} [cm^{-1}]	λ [μm]	v [THz]	hv [meV]	hv [10^{-3}H]
3580	2.793	107.325	443.866	16.311
3590	2.786	107.625	445.106	16.357
3600	2.778	107.924	446.346	16.403
3610	2.770	108.224	447.586	16.448
3620	2.762	108.524	448.826	16.494
3630	2.755	108.824	450.066	16.539
3640	2.747	109.124	451.305	16.585
3650	2.740	109.423	452.545	16.630
3660	2.732	109.723	453.785	16.676
3670	2.725	110.023	455.025	16.721
3680	2.717	110.323	456.265	16.767
3690	2.710	110.623	457.505	16.813
3700	2.703	110.922	458.745	16.858
3710	2.695	111.222	459.984	16.904
3720	2.688	111.522	461.224	16.949
3730	2.681	111.822	462.464	16.995
3740	2.674	112.121	463.704	17.040
3750	2.667	112.421	464.944	17.086
3760	2.660	112.721	466.184	17.132
3770	2.653	113.021	467.423	17.177
3780	2.646	113.321	468.663	17.223
3790	2.639	113.620	469.903	17.268
3800	2.632	113.920	471.143	17.314
3810	2.625	114.220	472.383	17.359
3820	2.618	114.520	473.623	17.405
3830	2.611	114.820	474.863	17.450
3840	2.604	115.119	476.102	17.496
3850	2.597	115.419	477.342	17.542
3860	2.591	115.719	478.582	17.587
3870	2.584	116.019	479.822	17.633
3880	2.577	116.319	481.062	17.678
3890	2.571	116.618	482.302	17.724
3900	2.564	116.918	483.542	17.769
3910	2.558	117.218	484.781	17.815
3920	2.551	117.518	486.021	17.861
3930	2.545	117.817	487.261	17.906
3940	2.538	118.117	488.501	17.952
3950	2.532	118.417	489.741	17.997
3960	2.525	118.717	490.981	18.043
3970	2.519	119.017	492.220	18.088
3980	2.513	119.316	493.460	18.134
3990	2.506	119.616	494.700	18.179
4000	2.500	119.916	495.940	18.225
4010	2.494	120.216	497.180	18.271
4020	2.488	120.516	498.420	18.316
4030	2.481	120.815	499.660	18.362
4040	2.475	121.115	500.899	18.407
4050	2.469	121.415	502.139	18.453
4060	2.463	121.715	503.379	18.498
4070	2.457	122.015	504.619	18.544
4080	2.451	122.314	505.859	18.590
4090	2.445	122.614	507.099	18.635
4100	2.439	122.914	508.339	18.681
4110	2.433	123.214	509.578	18.726
4120	2.427	123.513	510.818	18.772
4130	2.421	123.813	512.058	18.817
4140	2.415	124.113	513.298	18.863
4150	2.410	124.413	514.538	18.908
4160	2.404	124.713	515.778	18.954
4170	2.398	125.012	517.017	19.000
4180	2.392	125.312	518.257	19.045
4190	2.387	125.612	519.497	19.091
4200	2.381	125.912	520.737	19.136

λ^{-1} [cm^{-1}]	λ [μm]	ν [THz]	$h\nu$ [meV]	$h\nu$ [10^{-3}H]
4210	2.375	126.212	521.977	19.182
4220	2.370	126.511	523.217	19.227
4230	2.364	126.811	524.457	19.273
4240	2.358	127.111	525.696	19.319
4250	2.353	127.411	526.936	19.364
4260	2.347	127.711	528.176	19.410
4270	2.342	128.010	529.416	19.455
4280	2.336	128.310	530.656	19.501
4290	2.331	128.610	531.896	19.546
4300	2.326	128.910	533.135	19.592
4310	2.320	129.209	534.375	19.637
4320	2.315	129.509	535.615	19.683
4330	2.309	129.809	536.855	19.729
4340	2.304	130.109	538.095	19.774
4350	2.299	130.409	539.335	19.820
4360	2.294	130.708	540.575	19.865
4370	2.288	131.008	541.814	19.911
4380	2.283	131.308	543.054	19.956
4390	2.278	131.608	544.294	20.002
4400	2.273	131.908	545.534	20.048
4410	2.268	132.207	546.774	20.093
4420	2.262	132.507	548.014	20.139
4430	2.257	132.807	549.254	20.184
4440	2.252	133.107	550.493	20.230
4450	2.247	133.407	551.733	20.275
4460	2.242	133.706	552.973	20.321
4470	2.237	134.006	554.213	20.366
4480	2.232	134.306	555.453	20.412
4490	2.227	134.606	556.693	20.458
4500	2.222	134.906	557.932	20.503
4510	2.217	135.205	559.172	20.549
4520	2.212	135.505	560.412	20.594
4530	2.208	135.805	561.652	20.640
4540	2.203	136.105	562.892	20.685
4550	2.198	136.404	564.132	20.731
4560	2.193	136.704	565.372	20.777
4570	2.188	137.004	566.611	20.822
4580	2.183	137.304	567.851	20.868
4590	2.179	137.604	569.091	20.913
4600	2.174	137.903	570.331	20.959
4610	2.169	138.203	571.571	21.004
4620	2.165	138.503	572.811	21.050
4630	2.160	138.803	574.051	21.095
4640	2.155	139.103	575.290	21.141
4650	2.151	139.402	576.530	21.187
4660	2.146	139.702	577.770	21.232
4670	2.141	140.002	579.010	21.278
4680	2.137	140.302	580.250	21.323
4690	2.132	140.602	581.490	21.369
4700	2.128	140.901	582.729	21.414
4710	2.123	141.201	583.969	21.460
4720	2.119	141.501	585.209	21.506
4730	2.114	141.801	586.449	21.551
4740	2.110	142.100	587.689	21.597
4750	2.105	142.400	588.929	21.642
4760	2.101	142.700	590.169	21.688
4770	2.096	143.000	591.408	21.733
4780	2.092	143.300	592.648	21.779
4790	2.088	143.599	593.888	21.824
4800	2.083	143.899	595.128	21.870
4810	2.079	144.199	596.368	21.916
4820	2.075	144.499	597.608	21.961
4830	2.070	144.799	598.848	22.007

λ^{-1} [cm^{-1}]	λ [µm]	ν [THz]	$h\nu$ [meV]	$h\nu$ [10^{-3}H]
4840	2.066	145.098	600.087	22.052
4850	2.062	145.398	601.327	22.098
4860	2.058	145.698	602.567	22.143
4870	2.053	145.998	603.807	22.189
4880	2.049	146.298	605.047	22.235
4890	2.045	146.597	606.287	22.280
4900	2.041	146.897	607.526	22.326
4910	2.037	147.197	608.766	22.371
4920	2.033	147.497	610.006	22.417
4930	2.028	147.796	611.246	22.462
4940	2.024	148.096	612.486	22.508
4950	2.020	148.396	613.726	22.553
4960	2.016	148.696	614.966	22.599
4970	2.012	148.996	616.205	22.645
4980	2.008	149.295	617.445	22.690
4990	2.004	149.595	618.685	22.736
5000	2.000	149.895	619.925	22.781

INDEX

363